Advances in Agriculture Environment and Health

Advances in Agriculture Environment & Health

Fruits, Vegetables, Animals and Biomedical Sciences

Editors:

SHASHI BALA SINGH
O.P. CHAURASIA
ASHISH YADAV
AGNES M. RIMANDO
THOMAS H. TERRILL

2009

SATISH SERIAL PUBLISHING HOUSE
403, Express Tower, Commercial Complex, Azadpur, Delhi-110033 (India)
Phone : 011-27672469, 27672852, Fax : 91-11-27672046
E-mail :info@satishserial.com, hkjain1975@yahoo.com
Website : www.satishserial.com

Published by :

SATISH SERIAL PUBLISHING HOUSE
403, Express Tower, Commercial Complex, Azadoyr, Delhi - 110033 (INDIA)
Phone : 011-27672852 Fax : 91-11-27672046
E-mail : info@satishserial.com, hkjain1975@yahoo.com

ISBN 81-89304-59-3

Typeset at: Laxmi Art Creations, Delhi
Printed at: Salasar Imaging Systems, Delhi-35

डॉ० डब्ल्यू. सेल्वामूर्ति

विशिष्ट वैज्ञानिक एवं मुख्य नियंत्रक अनुसंधान तथा विकास

Dr. W. Selvamurthy

Distinguished Scientist & Chief Controller Research & Development

भारत सरकार
रक्षा मंत्रालय
रक्षा अनुसंधान तथा विकास संगठन
201, डी.आर.डी.ओ. भवन
नई दिल्ली–110011

Government of India
Ministry of Defence
Defence Research & Development Organisation
201, DRDO Bhawan
New Delhi-110011

FOREWORD

The Defence Research and Development Organization (DRDO) occupies a premier position as the country's topmost research organization and is engaged in developing major weapon systems for our Defence Forces. The motto of the organization is to achieve self-reliance in critical areas of defence technology for the benefit of the nation. DRDO, ever since its inception, has been aiming to provide the much needed life support systems to our soldiers. To meet the cherished goals, DRDO in addition to its systems labs, has nine life sciences laboratories that are actively engaged in meeting the food and agriculture requirements of our troops working in inhospitable terrains, besides focusing on biomedical technology and military psychology.

The high altitude region of the Himalayas is characterized by the most difficult terrain in the world and severe cold arid climatic conditions prevail in the region. The region is characterized by low atmospheric pressure, low partial pressure of oxygen, low relative humidity, low annual precipitation, high wind velocity, long frozen winters, high ultra violet radiations with high light intensity. Making food and quality medical care available in such regions is a greatest concern.

Food and health security is a global issue today and targeting the most food insecure areas is the need of the hour. Food security is influenced by poverty, health, education and agriculture, among other socio-economic factors. Non-food factors like access to sanitation, primary healthcare and education also play a role. To focus the public and political attention on food and health security related issues in high altitudes, awareness needs to be generated on issues of biodiversity conservation and sustainable management of resources in high altitudes to meet the future food, health and livelihood security.

Developing novel approaches for food and health security in the high altitude region would help provide food and good health for the local population, where problems are of an entirely different nature, and this in turn will reduce dependence on conventional agricultural and health practices resorted to in the plains. This edited volume touches upon areas envisaged to meet tomorrow's requirements and offers guidelines for further research. The book comprises chapters contributed by experts on the subject on issues that are vital to meeting the current and futuristic challenges of the high altitude region, mainly for achieving sustainable agriculture, animal and health security. I am sure the compilation will add new dimensions to the high altitude research and researchers, academicians, industrialists, and policy makers will find the book very valuable, practical and functional.

23rd August, 2008

(W. SELVAMURTHY)

PREFACE

High altitude agro-animal and biomedical research is gaining momentum in India. High altitude cold desert regions are found in all the continents and circumpolar regions of the world. This treeless world possesses extremely hostile conditions and men, animals and plants of the region face many biotic and abiotic stresses. Not only atmospheric pressure decreases with rise in altitude, but there is also fall in the partial pressure of oxygen, carbon dioxide, which leads to the poor performance efficiency of man, animals and plants. Area-specific multidisciplinary research has been conducted around the globe to provide sustainable solutions to the high altitude problems.

The present book has been deliberated and designed to convene the long-felt need for a book covering major aspects of high altitude related issues. A multi-disciplinary group of eminent scientists from India and abroad have contributed chapters to bring out this vital book. Chapters deal with the various subject's viz. exploitation of heterosis breeding to increase productivity, horticulture development in high altitudes, production of quality fruits by using phyto-hormones, sustainable fruits and vegetables production and it's postharvest management. Some chapters also describe the researches on seabuckthorn and its sustainable resource management. Importance of medicinal plant biodiversity of trans Himalaya, carbon sequestration of orchards, stress induced changes in burley tobacco seed development, agriculture extension through local institutions and storage studies of biodiesel in cold desert has also been discussed. The importance of biotechnological approaches for enhancing crop productivity and sustainable agriculture has been emphasized. The unraveling of rice-wheat synteny using single copy genes, molecular biology of rutin biosynthesis in buck wheat, development of short duration quality protein maize hybrids, and molecular characterization of indigenous crops of Ladakh has been elaborated. The sustainable fodder production for livestock, strategies for poultry production, broiler rabbit as a source of meat & fur, and fisheries potential in high altitudes has been presented in detail. The book also deals with biomedical aspects viz. bio-medical systems for high altitude applications, endothelin-1 and high altitude pulmonary hypertension, hypoxic trainings and essential hypertension, and genetics of high altitude pulmonary hypertension.

It is a well-prepared and comprehensive book with updated scientific and general information's. The book will make the readers acquainted with the current researches and developments in this field. We are grateful to all the contributors for their dedicated efforts and submission of chapters within time frame. The editors have made utmost efforts to correct errors in chapters; still it is possible that the readers will find them in few places, we request the readers to kindly pardon us.

We extend our sincere gratitude and thanks to Shri M. Natarajan, SA to RM, Secretary Defence R&D & DG R&D, Dr. W. Selvamurthy, CC R&D (LS & HR), Shri G. Elangovan CC R&D (R&M), Dr. Narendra Kumar, DOP and Dr. R.C. Sawhney, DLS, of DRDO HQ for technical guidance and encouragement. We also extend our sincere thanks to all the scientists and staff of DIHAR who have timely and valuably supported us in editing this book.

Leh-Ladakh **Editors**
September 2008

CONTENTS

1

Commercial and Strategic Exploitation of Heterosis Breeding in Major Vegetable Crops

HARI HAR RAM
Krishidhan Vegetable Seeds, Shivaji Nagar, Pune, Maharashtra, India

INTRODUCTION

The phenomenon of heterosis also referred to as hybrid vigour has been the basis of much of the yield improvements in crop plants including vegetables in the last century. Commercial and strategic exploitation of heterosis is of vital importance to increase productivity and thereby to meet the challenge of providing food and nutritional security for an ever-increasing human population particularly in the Indian sub-continent. Hybrid technology is one of the potential and pragmatic options to meet the huge food, fodder and fibre requirements needed in the decades ahead. The effective use of hybrid technology has accelerated the development and expansion of seed industry worldwide. Rapid advances in plant breeding and associated seed production technologies have served to enhance the competitiveness of hybrids in terms of increased yield and improving seed quality. The aim of this paper is to understand the advantages of hybrids and the powerful impact of the heterosis and to find out how advances in hybrid seed production technologies affect the productivity and quality of selected vegetable crops, namely, cucurbits, tomato, hot-pepper, and okra with emphasis on research and development (R&D) programs in private sector seed industries in India. It is further hoped that this article will act as a catalyst for further R&D in this field and contribute towards better public-private-partnership mode in order to leverage advantages to both sectors.

General Advantages of Hybrids

Breeding of hybrid varieties in vegetables were reviewed and has following general advantages:

i) Hybrids offer new possibilities for genetic improvement in productivity.

ii) Hybrids provide earliness.

iii) Hybrids are uniform.

iv) Hybrids are superior in quality.

v) Hybrids provide opportunities for the rapid deployment of dominant genes conferring resistance to diseases and pests.

vi) Hybrids in certain cases are adapted to mechanical harvesting.

However, it must be well understood and kept in mind that the advantages of F1 hybrids in terms of yield superiority may be difficult to establish in certain cases. Yield *per se* is often of little significance, since many vegetable crops are harvested before maturity and may be graded before marketing. Thus, marketable yield is variable proportion of the total yield at harvest which may not in any case represent the biological potential of the crop.

SCOPE OF HYBRIDS

Cucurbits

Richard W. Robinson, an authority on cucurbits has presented an extensive review on hybrid cucurbits including importance of hybrid cucurbits, inbreeding depression, heterosis and hybrid seed production technology. Hybrid cucurbits started gaining importance after World War II. Slowly summer squash, pickling cucumbers, melons and watermelons occupied increasing areas at the world level. Lately, winter squash, pumpkin and other cucurbits have been added to the list. In India, hybrid breeding in vegetables got a fillip when the government of India, Ministry of Agriculture and Cooperation announced a new policy on import of seed/planting material on Sep. 16, 1988 and the New Seed Policy came into effect from 1 October,

1988 which prompted a large number of private sector seed companies to enter into vegetable seed business on a large scale. This trend got further boost when Protection of Plant Varieties and Farmers' Rights (PPVFR) Act-2001 came on the statute book in India. Accordingly, several private sector seed companies in India have strong R&D, seed production, quality assurance, processing and marketing programmes on hybrids in bottle gourd, bitter gourd, smooth gourd, ridge gourd, cucumber, watermelon, muskmelon, pumpkin, tomato, hot-pepper, okra and eggplant.

It has been reported hybrids comprised 43 percent of the cucumber seed, 56 percent of the summer squash seed, 10 percent of the winter squash seed, but only about 1 percent of the melon seed produced in the USA in 1980. Hybrid melon seed was mostly imported from countries with less labour costs, and only about 1000 pounds of hybrid melon seed was then produced in the US. The use of hybrid cucurbits has increased considerably in Japan and Taiwan over the years. Japanese grow almost 100 percent hybrids in cucumber, watermelon, melons and summer squash and almost all of the melons, watermelons, and cucumbers grown in Taiwan are hybrids.

In India presently a large number of companies are engaged in hybrid seed business of vegetable crops including cucurbits. In cucumber, the area under hybrid seeds is about 24,000 ha with hybrid seed market of 18 tons. The corresponding values for other cucurbits are 45,000 ha and 70 tons of bitter gourd, 40,000 ha and 65 tons of bottle gourd, 11,000 ha and 17 tons of smooth gourd, 40,000 ha and 48 tons of watermelon and 6000 ha and 8 tons of muskmelon. These show the tremendous scope of hybrid seeds in cucurbits in India where the current seed replacement by the hybrids in general is estimated to be approximately 40 percent. The entire hybrid seed production in India is through hand pollination using trained farmers who are contract growers for seed companies. The area where the hybrid seed production of vegetables is concentrated is the Ranibennur region of Karnataka.

There are a large number of research papers, specially from the Indian sub-continent, on level of inbreeding depression and heterosis in most of the cucurbits and the results have tremendous variability. Research papers on these aspects are also available from the US, European, Japanese and Chinese scientists. *C. pepo* has been found to be more prone to inbreeding depression than other cucurbits. However, in general, cucurbits have been considered to not lose vigor when inbred despite being cross-pollinated species. More recent investigations, however, have shown that inbreeding depression can occur in the Cucurbitaceae. Inbreds show variability for this and some may be very productive. Less or no inbreeding depression in cucurbits may be due to the fact that only small populations of cucurbits are often grown due to large amount of space required for cucurbit vines and this results in some inbreeding despite floral mechanism favouring out-crossing.

No inbreeding depression has been found in inbred lines developed from a pickling cucumber population, perhaps because cucumbers are often grown in small groups of plants where there is little out-crossing. There have been reports of heterosis in specific hybrid combinations in cucumber, specially when the parental lines are diverse. Where breeders have worked to develop elite hybrids, heterosis may average only a 5 percent advantage in cucumbers. In cucumber, the hybrid seed is produced on commercial scale mostly by using a gynoecious parent as a female parent and the monoecious parent as the male parent and honey bees as pollen vectors. Gynoecious inbreds can be developed by self-pollination of plants that have been treated with silver nitrate or other ethylene inhibitors. The advantage of producing gynoecious x monoecious crosses is that the resultant hybrid will be gynoecious having pistillate flowers at every node. A monoecious pollinizer will have to be mixed in around 10 percent with the hybrid seed if it is not parthenocarpic. Gynoecious cultivars have earliness and sometimes higher yield than monoecious cultivars as they have pistillate flowers at every node.

Melon normally does not show any inbreeding depression. However, some crosses have been reported to exhibit heterosis of 3 percent for earliness and 8 percent for yield. The lack of inbreeding depression and consequently heterosis in melons is primarily due to 65 to 80 percent self-pollination in natural populations. Hybrids are produced by hand pollination of emasculated perfect flowers on the female parent using staminate flowers from the male parent. Gynoecious inbreds have been reported but fruits from pistillate flowers are oval, rather than round like those produced from perfect flowers.

Summer squash has been shown to have heterosis of 11-84 percent for yield in zucchini types and 0-82 percent yield advantage over open-pollinated types in yellow types. In hybrid seed production of summer squash, one of the inbreds is rendered as female for first two to three weeks of flowering by spraying the plants with ethephon at the two and four- leaf stage. Ethephon does not work properly to suppress induction of male flowers in *C. maxima* and *C. moschata.*

In watermelon, there is very little inbreeding depression. Yield heterosis is expressed in certain parental combinations averaging around only 10 percent. However, hybrids are common due to improvement in quality, earliness and offering the advantage in protection of the parental lines and also for production of seedless watermelon, where tetraploid female is crossed with a diploid male to have triploid watermelons which are triploid and seedless when planted for commercial production.

Tomato

Yardanov has given a comprehensive account of heterosis and its exploitation in tomato. It is now recognized that heterosis manifestation in tomato is in the form of the greater vigor, faster growth, earliness, productivity and higher resistance to diseases and unfavourable conditions. Historically, Bulgaria took lead in large scale extension of hybrid tomato cultivars and organization of proper hybrid seed production of tomato. After the Second World War, hybrid breeding in tomato was adopted on a large scale in the Netherlands, England, France, USA, Japan, etc. In the USA, effectively 100 percent of fresh market and 80 percent of processing tomato are F1 hybrids. Most hybrids are produced by manual emasculation but a few are assisted with the use of genetic male sterility. Most of the fresh market tomato hybrids imported from Mexico are in *rin* heterozygotes to increase shelf life and because superior genetic background have outperformed the FlavrSavr tomato. F1 hybrids offer the advantages of complementation for disease resistance and other traits, heterosis for yield, and specially yield stability. Major seed companies include Seminis (Asgrow-Petoseed-Royal Sluis) and Hazera (Israel).

In India, Indo American Hybrid Seed Company took lead by introducing tomato hybrid cultivar on commercial scale in 1973. Now hybrid tomatoes are expanding at a much faster rate in India where several private sector seed companies are engaged in hybrid tomato research and seed production and marketing. At present, In India approximately 3,50,000 ha area is under hybrid tomato with market size of 52 tons of hybrid seed. The predominant tomato hybrids in India are Shaktiman, Pratham, Abhinav, Suraksha, Mahalaxmi,

Kundal, Rassam, Padmavati, JK Desi, Navin 2000, Krishna, Surya, etc. from different companies.

For successful hybrid seed production, there should be day temperature of 21-25° C, night temperature of 15-20° C and RH of about 60 per cent at fruit maturity and harvesting.

Hand-pollination is practiced effectively for commercial hybrid seed production in most countries. The planting ratio of female and male parent is usually 3:1. Female seedlings are planted at a distance of 60 x 50 cm. Male parent is planted in the adjoining and a separate plot at a spacing of 150 x 40 cm. Usually male parent seedlings are planted 10-15 days earlier than the female parent to obtain adequate pollen from the very beginning of the pollination programme. The female parent is staked after the opening of flowers in the first cluster. Staking facilitates plant handling during emasculation and pollination and field management operations. Both the plots are inspected regularly and the off types are rogued out. Sharp pointed forceps are used to split open the anther cone. After this, the anther cone is pulled out leaving calyx, corolla and pistil intact. Sometimes, corolla could also be taken out in the same operation. To help identify the hybrid fruits from the selfed fruits at the time of harvest, all the sepals or 2 sepals of the calyx of the emasculated buds are cut. Emasculation is done in the afternoon followed by pollination the next morning. The best time of pollen collection is during the early morning hours when the anthers have dehisced. The anther cones are removed from freshly opened flowers and put are put in suitable containers such as cellophane or paper bags or plastic pans. These are dried under a 100 watt bulb about 30 cm above the bag during the night. Ideal temperature for drying is 30° C. The dried anther cones are put in a plastic cup. This cup is covered with a fine mesh screen and then sealed with a similar tight fitting cup serving as lid. Both containers are shaken so that the pollen is collected in the lid-cup. Pollen so collected are transferred into a small convenient to handle container for pollination.

As a rule-of-thumb, 20 fruits/plant are maintained for female parent with large fruit, 30 fruits/plant for medium size fruit female and more than 30 fruits/plant for parent with small fruit.

Hot-Pepper / Chilli

Chilli hybrids are very popular in India and other parts of the world, for example, China, Korea, Japan and other countries. In India the open pollinated (OP) cultivars account for approximately 1,60,000 ha with seed requirement of 238 tons whereas the area under hybrids is about 1,84,000 ha accounting for 46 tons of hybrid seed market. The hybrid chilli market is dominated by the following segments of the products:

A. Red Dry/Dual Purpose Products

i) *Indam-5 Segment:* The fruits of this segment are characterized by the fruits with 12-14 cm length, 1.2-1.3 cm diameter, smooth skin, bright dark red colour after drying, medium pungency and 95-100 seeds/fruit. The plants have deep root system and are tolerant to powdery mildew, leaf curl virus, die-back, anthracnose and sucking pests.

ii) *Wonder-Hot Segment:* The fruits are 14-16 cm long with 1.3 cm diameter, smooth skin, bright dark red colour after drying, low to medium pungency and 100-105 seeds/fruit. The plants have deep root system and are tolerant to powdery mildew, leaf curl virus, die-back, anthracnose and sucking pests.

iii) *HPH-12, HPH-117 and HPH-4884 Segment:* The fruits of the hybrids in this segment are 8-10 cm long with 1.1 cm diameter, pointed tips, smooth skin, green colour at fresh stage and bright dark red colour after drying. The pungency of the green pods is medium to high. Average number of seeds/fruit is 95-100. The plants have deep root system and are tolerant to powdery mildew, leaf curl virus, die-back, anthracnose and sucking pests.

B. Fresh Green Purpose Products

i) *Jwala Segment:* This segment is typically named based on the features of the popular variety Pusa Jwala. The fruits are 12-15 cm long with 1.0-1.1 cm diameter, highly wrinkled parrot green colour skin, medium to high pungency, and 80-90 seeds/fruit. The plants have deep root system and tolerance to powdery mildew, leaf curl virus, die-back, anthracnose and sucking pests.

ii) *Sitara Segment:* The hybrids under this segment have fruits with 12-14 cm length, 1.2 cm diameter, smooth skin, light green colour, medium pungency, and 90-100 seeds/pod. The roots are deeply placed. These products are tolerant to common diseases like powdery mildew, leaf curl virus, die-back, anthracnose and sucking pests.

C. Badyagi Segment

i) *Devnor Delux, Bejo-304 and 305 Segment:* The cultivars in this segment have fruits having 14-16 cm length, 1.2-1.3 cm diameter, very heavy wrinkles on fruit skin with elastic property. The fruits become blood dark red after drying. The pungency level is low. The average number of seeds/fruit is 95-100. As in other segments, the plants have deep root system, and tolerance to powdery mildew, leaf curl virus, die-back, anthracnose and sucking pests.

It may be mentioned that the nomenclature of the segments is purely empirical based on the characteristic features of the products and this system may vary from individual to individual depending upon the grouping of the products.

There are three systems of hybrid seed production in chilli. These include hand emasculation and pollination, use of GMS (spell out) and that of CMS (spell out). In hand emasculation and pollination system, also known as line x line system, the flower buds of the female parent are emasculated and covered with small paper bags in the afternoon and the following morning, these are pollinated with the pollen grains collected from freshly opened flowers of the male parent. Prior to emasculation, previously set fruits and open flowers are removed.

It should be kept in mind that chilli requires slightly humid warm climate for growth and flowering. The crop can withstand heat but below 25 degree centigrade temperature, flowering may be hampered. Taking seed production in consideration, the Kharif season planting, i.e. nursery sowing in the month of May – June, is ideal for crop growth, flowering and ultimately higher seed yield in the Maharastra and Karnataka regions where bulk of the commercial hybrid seed of hot-pepper is produced. It can be successfully grown on medium to heavy soil with good drainage. Twenty tons of FYM (spell out) and 200:100:100 kg of NPK / ha will help in achieving good crop. In the main field, the crop should be protected from powdery mildew by spraying Sulfex 2 g/ l of water. During crop growth it is very necessary to protect the crop from sucking pests and Spodoptera, which is the mainnuisance in seed production under net house. For sucking pests, intermittent sprayings with Trizophos and Confidor (0.5 ml/ l of water) maybe used. For controlling Spodoptera, the crop should be initially sprayed with Thiodon, Monocrotophos, and finally by any synthetic pyrethroids, specially for seed production in net houses. The surface of the covering net, from the outside up to the crop height level, should also be sprayed as the adults lay the eggs on the outside surface of the net. Seed rate is 250-300 g / ha. The seed need to be planted on 25 to 30 cm raised beds of 5 x 0.5 m size, which will provide seedlings sufficient for planting a one-hectare land. The required isolation distance is 1000 m. In case of seed production under the net, there is no need for isolation distance. The required spacing is 75 x 45 cm. The crop should be regularly inspected at various stages. At least 2 pre-flowering inspections are required, and the volunteers and morphological variants are removed.

While using GMS, at least two inspections are required for complete flower checking (with the consideration of non-availability of sterility marker) for sterile/ fertile plants. At this time, male fertile plants from 75 percent of planted area are removed and sterile plants from the remaining 25 percent of the planted area from one particular side of the plot are also removed. Fertile plants from this area will act as a pollinator (B line) counterpart (male) for seed production of the female line which shall be used for commercial hybrid seed production the following year by planting and crossing with a well recognized male parent after removal of 50 percent of fertile plants.

At least one inspection is required at fruit development stage of the plants to be used as pollinator for production of female parent seed. Pollination work should not be started until at least one fruit with complete skin colour and fruit shape developed is seen on the each pollinator plant. After careful examination, all the variants are removed. In this process, there may be a delay of 10-15 days for pollination work. In order to avoid this, male seedlings can be nursery sown and transplanted 15 days earlier than the female seedlings. Two inspections are required during pollination. The first is to explain the procedure for pollen extraction from male flowers and actual pollination work and the second inspection is to confirm that the work is being doneaccording to prescribed procedures. Prior to harvest, one inspection is required to instruct on the harvesting of only pollinated and completely red ripe fruits.

The pollination work is carried out using the following steps. One day before starting the actual pollination work in the morning, pluck all the open-pollinated set of fruits and open flowers from the female and male plants. Then in the evening generally at 4 pm pluck all the flower buds which are about to open in next day's morning in a plastic bowl. Then separate the anthers from the flowers and keep them in a plastic bowl in a room or at a protected area. The following morning take the bowl with the anthers to the field and expose to sunlight for 15 to 30 minutes or until the anthers start to burst. Alternatively, take two plastic or steel containers like *Bowls* (of about 300-500 ml) with equal mouth diameter. Put some of the anthers in one container and cover its mouth with muslin cloth. Put the other container over the mouth of other container containing the anthers in an inverted position. Hold these two containers in an inverted position and shake them vigorously so that the pollen grains will accumulate in the lower empty container. Collect the pollen grain powder and fill in the small plastic cups specially designed for filling the pollen grains and bearing them like rings for pollination purpose. For a 10 R (in Maharashtra, 1 R = 100 square m) area, 10 to 20 workers are needed for pollination work, depending upon the stage of the plant and flush of the flowers.

After filling the pipes (50 percent plugged with cotton lint) by pollen grains insert the ring in a finger. Form the pairs of workers. One worker of a pair will tie a thread to the pedicel of the flower and the other will pollinate the stigma by inserting it in a pipe attached to ring and filled with pollen grains. Pollinate flowers that are completely open, and with receptive stigma. Threads need to be tied prior to pollinating the flowers to avoid further handling of the pollinated flower. To achieve the desired fruit load, in general, a 30 day-pollination work is sufficient.

Pluck the fully red ripe fruits (pedicel tied with the thread) and dry them in sunlight. For emergency, power driven dryers can also be used.

In a small scale production, after complete drying of the fruits the seeds are separated by beating the fruits followed by winnowing. Here power driven machines are also used for seed extraction. After extraction, the seeds should be well cleaned, then graded, dried and stored after proper labeling. Prior to use, the seed should be tested for compliance with seed standards. Under normal conditions one can get average yield of 10 kg in a 10 R area. In case of production under net house, depending upon the variety, the seed yield can be more.

For hybrid seed production any proven and desirable male fertile line other than the maintainer can be used. Female to male ratio of 3 : 1 can be used.

Cytoplasmic genetic male sterility is also used in hybrid seed production in chilli where the offsprings of male sterile plants are not necessarily male sterile but can be male fertile when certain stocks are used as pollinators. Here both nuclear and cytoplasmic factors interact. This system has been historically discovered in onion. The mode of operation of this nuclear-cytoplasmic system can be understood by considering the evidence on inheritance of male sterility in onions. The recessive gene ms converts the normal hermaphrodite onion into a potentially male-sterile form but the homozygous recessive genotype, msms can only express itself in one particular type of cytoplasm designated as S type. In normal cytoplasm designated as N, the msms genotypes produce normal male fertile plants. Plants of the genetic constitution MsMs or Msms are male fertile irrespective of whether the cytoplasm is S or N type. The genotypes and phenotypes based on this system are as follows:

Male Fertile

Cytoplasm	Nuclear genotype
N	MsMs, Msms, msms
S	MsMs, Msms

Male Sterile

Cytoplasm	Nuclear genotype
S	msms

Cytoplasmic genetic male often abbreviated as CMS has been reported in a large number of vegetable crops.

In principle, the system of hybrid seed production using CMS is typically called as three line hybrid breeding system where the three lines used are as follows:

Line A, Sterile cytoplasm as S with nuclear gene as msms; Line B, Normal fertile cytoplasm as N with nuclear gene as msms; Line C or R, Normal fertile cytoplasm as N with nuclear gene as MsMs. Line A is normally referred to as the female for hybrid seed production, line B is the maintainer line to maintain the A line, and the C or R line is the restorer line used as male parent for commercial hybrid seed production with line A as the female parent.

In CMS method, the sterility of the female is 100 percent and there is no need to rouge out fertile plants (except volunteers) from the female population. There is a separate maintainer line called ' B ' line, which maintains the sterility of the female. An advantage of this system over the GMS system is that the field does not remain gappy due to roguing of male fertile plants from the female area under seed production. Female to male ratio is 3:1, and the rest of the activities are same as in the parental line seed increase in the GMS system.

For hybrid seed production using CMS system, any proven and desirable male fertile counterpart line having the gene for fertility restoration can be used. Female to male ratio of 3:1 can be used to have sufficient male flowers.

Okra

The hybrids have become very popular in okra particularly in India where, per reliable estimates, the area under hybrids is about 1,10,000 ha with market potential of 700 tons of hybrid seed. The hybrids are in demand due to higher yield potential, plants with short internodes and shiny green pods, along with tolerance to yellow mosaic virus.

For hybrid seed production, manual method of emasculation and pollination is used. Emasculation is done in the after-noon/evening where two longitudinal and opposite cuts are applied on the calyx and both segments are pulled downward. The corolla thus exposed are removed using a pair of forceps followed by removal of anthers with forceps and/or tapping of the buds. The emasculated buds are covered with paper bag. Next morning the pollination is carried out. For pollination, the unopened buds from the male parent are covered a day prior to pollination and the next day the opened flowers are taken out and the anther column is rubbed against the stigma of the emasculated flowers which are again covered with paper bag.

Brinjal

Brinjal hybrids are very popular especially in Asia where Takii has been the major developer of the hybrid seeds. Hybrids have become very much in demand in India where the area under hybrids is estimated to be 1,10,000 ha with a hybrid seed market potential of around 23 tons. The hybrid seed is produced through manual method using hand emasculation and pollination. Emasculation of long styled mature flower buds of the female parent is done in the afternoon on a large scale. For pollination, the anthers from open flowers are collected in good amount in large petri-dishes. These anthers are kept under light during the night to dehisce the pollen grains. These are sieved through muslin cloth using same method described for tomato and chilli and collected into a container and used for hand-pollination. All the non-crossed flowers and the fruits already set are removed from the female parent. Fruits are ready for harvest 70-80 days after pollination. Usually on natural ripening, brinjal fruit skin color attains light yellow or deep yellow/orange color. For seed extraction, fruits are cut, crushed and macerated with water. The seeds are removed during washing of crushed fruits. Normally 150-200 g of hybrid seed is enough to plant 1 ha of land. On the average, each crossed fruit yields 3 g seed meaning thereby that 70 crossed fruits shall produce about 210 g seed which is sufficient to plant 1 ha land.

STRATEGIC USE OF DOMINANT GENES IN HYBRID BREEDING

Cucurbits

In the earlier part of this article, it was pointed out that the hybrids offer advantages in utilizing dominant genes conferring resistance to diseases and pests as the F1 hybrids manifest the resistance immediately. Fortunately, resistance to many diseases in cucurbits is dominant. This permits development of hybrid cultivars resistant to a disease when only one of the parents is resistant. It also permits development of hybrids with multiple disease resistance when a line with dominant resistance to one disease is crossed with another line dominant for resistance to other disease.

Single dominant gene has been reported to confer resistance in cucumber for diseases like anthracnose, bacterial wilt, scab, target leaf spot, Fusarium wilt, watermelon mosaic virus, and papaya ring spot virus. Likewise, there are reports of single dominant genes for resistance to anthracnose, Fusarium wilt, and zucchini yellow mosaic virus in watermelon. In melon, dominant genes include those conditioning resistance to powdery mildew, Alternaria leaf blight, Fusarium wilt, gummy stem blight, papaya ring spot virus, and zucchini yellow mosaic virus.

Dominant genes have also been reported for resistance/tolerance to insects, for example, for aphid resistance in melons, and squash vine borer and squash bug resistance in squash. Dominant gene for resistance to red pumpkin beetle has been reported in squash, melon, and bottle gourd and for resistance to fruit fly in squash, melon, and watermelon.

Other dominant genes having potential advantages in hybrid development in cucurbits are *F* for gynoecious sex expression and *Pc* for parthenocarpic fruit development in cucumber. The dominant *B* gene of squash has been reported to be of significance in hybrid cultivars of squash as the fruits become yellow early in development, have increased vitamin A content and do not display disfiguring symptoms of watermelon mosaic virus when infected.

Reciprocal crosses in cucurbits often do not play any significant role except in some cases where it seems better to produce hybrid seeds with one line as the female parent. Usually, the parent with larger seed is used as female if there are differences with respect to seed size in the parents. In this context, one can quote the example of hybrid seed production of watermelon hybrid NS 295 (Namdhari Seeds) and Tambola (Krishidhan Vegetable Seeds), where the female parent is large seeded and the male parent is small seeded.

Tomato

In tomato there are reports of several dominant genes conferring resistance to major disease, for example, the genes *Cf*-1, *Cf*-2, *Cf*-3 and *Cf*-4 for resistance to leaf mold, *I*-1 and *I*-2 for Fusarium immunity race 1 and race 2 respectively, *Ve* for resistance to Verticillium, *Se* for resistance to Septoria, *Ph*-1 for resistance to late blight, *Ad* for resistance to Alternaria, *Tm* and *Tm*-2 for resistance to tobacco mosaic virus and *Mi* for resistance to nematodes. These genes can easily be deployed for disease resistance breeding through development of tomato hybrids.

Hot-Pepper

The dominant genes with potential to be exploited in developing hybrids of hot-pepper include the disease resistance genes *Bs* 1 (resistance to bacterial spot-*Xanthomonas campestris* pv *vesticatoria* race 1 in *C. chacoense* PI 260435), *Bs* 2 (resistance to bacterial spot race 2) and *N* (resistance to nematode- *M. incognita*). Pendent fruiting habit is dominant over the upright fruiting habit and provides a greater opportunity to use upright fruiting habit cultivars with high level of tolerance to leaf curl virus as one of the parents without affecting the pendent fruiting habit commonly accepted by the consumers.

Brinjal

In brinjal, there are reports of a few dominant and useful genes which facilitate their exploitation through hybrids. These include dominance of spiny fruits over non-spiny fruits, tall over dwarf plant, purple over green fruit colour, elongated over round fruit shape, round over oval fruit shape, clustering over solitary fruit habit, bacterial wilt resistance over susceptibility.

STRATEGIC CONSIDERATIONS IN DEVELOPMENT OF HYBRIDS IN INDIA

Several vegetable seed companies have made significant progress in development, seed production and large-scale marketing of hybrids in major vegetables. These are given in Table 1.

Now it would be appropriate to point out a few examples from the above list to demonstrate how an F1 hybrid is developed based on some strategic considerations. In watermelon, the female of NS 295 and Tambola are somewhat oval and striped and the male is uniformly light green and round. By crossing these two, we get the hybrid where the fruits are oblong, striped and the hybrid has better yield than the better parent. In watermelon Madhubala, the female is light green, oblong and the male is striped but round, and both parental lines are not liked by the consumers. However, the hybrid is striped, oblong and liked by the consumers.

In bottle gourd, Warad and Santosh, the female has almost bottle shaped fruits and the male has short cylindrical fruits with bulging at stem end. The hybrid combines the desirable features from both the parents and produces cylindrical, medium-long fruits. In ridge gourd Murali, the female parent has short fruits and the male parent has too long fruits. The hybrid has medium long fruits liked by the consumers.

In tomato, the hybrids combine the advantages in terms of vigorous growth, higher yield, long shelf life, and what fruit shape. In chilli, the hybrids have been found primarily heterotic showing robust vigorous growth with expanded fruiting and ultimately leading to high yield. The disease resistance advantage available in up-right fruited cultivars is easily exploited in hybrids, as the pendent fruiting habit has been found to be dominant. Further, these types of marker traits become handy tools in distinguishing selfed seeds from the hybrid seeds. With okra, the consumers' choice is usually cultivars with shiny green fruit, which are designed easily by crossing cultivars with light green fruit with the dark green pod cultivars. Additionally, the okra hybrids are high in yield and show high level of tolerance to yellow vein mosaic virus. With brinjal, the main advantage of the hybrids is being robust, vigorous, tall and non-lodging plants, with desirable shape and fruits size. Resistance to bacterial wilt found to be dominant in brinjal has been utilized in the development of bacterial wilt resistant brinjal hybrids.

These examples demonstrate that in developing hybrids of vegetables, strategic considerations play a major role and the hybrids to be developed in future will rely more on value addition in terms of strategic disease and pest resistance governed by dominant genes. It is hoped that the future hybrid breeding projects will give more emphasis on this aspect.

SUMMARY AND FUTURE THRUST

- In general, there are insignificant inbreeding depression and heterosis in cucurbits except few specific combinations where improvement in earliness in what and quality has been noted.
- Hybrid breeding in cucurbits has been in place mostly on account of strategic deployment of dominant genes conferring resistance to diseases, insect-pests and a few other key characters related to sex expression and fruit development, and product designing per demand of consumers and market.
- Hybrid breeding in tomato, hot-pepper, okra and eggplant has been popular due to advantageous traits: high yield, early yield, vigorous, robust and sturdy plant having tolerance to major diseases, and characteristics favoring the strategic designing of products in terms of fruit shape, size, appearance and better shelf-life.
- Hybrid development in vegetables provides an opportunity for public-private-partnership mode of project implementation where public funded institutions can devote their resources and manpower to develop desirable inbreds/parental lines and develop hybrids per demand of consumers/market and once these are found to be acceptable by the seed industry, the seed industry can enter into some kind of mutually beneficial agreement for seed production and marketing of the hybrids using their vast resources, infrastructure, network and business acumen in

this segment. This arrangement shall leverage synergy and open vast possibilities of meaningful interaction between the two organizations, which at the moment appear to be working in isolation.

- The hybrid breeding program shall have to travel in reverse gear from market/consumer feedback to the product designing and value addition by R&D team, which is happening in the private sector but perhaps not to the desired level in the public sector.
- There is an urgent need to diversify the parental lines and the products.
- Deployment of dominant genes conditioning resistance to disease and pests needs to be implemented on war footing through heterosis breeding to reduce cost of extensive use of fungicides and insecticides and ultimately to reduce the cost of cultivation and to minimize the toxic residues on the fresh vegetables.

ACKNOWLEDGEMENTS

The author is thankful to Sri Sushil Karwa, Managing Director, Krishidhan Seeds Ltd., Jalna, Sri Mangesh Kadgaonkar, Director, Krishidhan Vegetable Seeds (I), Pvt., Ltd., Pune and Dr. G. K. Garg, Director, R&D, Krishidhan Seeds Ltd., Jalna for providing necessary support and encouragements to prepare the manuscript and to participate in scientific seminars and symposia.

References

Janick, J. 1998. Hybrids in horticultural crops. In: Lamkey, K.R. and Staub, J.E. (eds.). Concepts and Breeding of Heterosis in Crop Plants, pp. 45-56. Crop Science Society of America.

Jones, H.A. and Clarke, A.E. 1943. Inheritance of male sterility in the onion and the production of hybrid seed. Proc. Amer. Soc. Hort. Sci. 43, 189.

Kumar, Sanjeet., Banerjee, M. K. and Kalloo, G. 2000. Male sterility: Mechanisms and current status on identification, characterization and utilization in vegetables. Veg. Sci. 27, pp. 1-24.

Pearson, O.H. 1983. Heterosis in vegetable crops. In: R. Frankel (ed.). Heterosis. Springer Verlag, Berlin. pp. 133-188.

Ram, H.H. 2005. Vegetable Breeding: Principles and Practices. Kalyani Publishers, Ludhiana.pp 514.

Riggs, T.J. 1998. Breeding F1 hybrid varieties of vegetables-Review article. Journal of Horticultural Science 63, pp. 369-382.

Robinson, R.W. 2000. Rationale and methods for producing hybrid cucurbit seed. In: A. Basra, S. (ed.). Hybrid Seed Production in Vegetables: Rationale and Methods in Selected Crops. Food Product Press, New York, London, Oxford, pp. 1-47.

Takahashi, O. 1987. Utilization and seed production of hybrid vegetable varieties in Japan. In: Feistritzer, W.O. and Kelly, A.G. (eds.). Hybrid Seed Production of Selected Cereal, Oil and Vegetable Crops. FAO Plant Production and Protection, Paper 82, pp. 313-328.

Wehner, T.C. 1999. Heterosis in vegetable crops. In Coors, J.C. and Pandey,S. (eds.). Genetics and Exploitation of Heterosis in Vegetable Crops. Amer. Soc. Agron., Crop Sci. Soc. Amer., Soil Sci. Soc. Amer., Wisconsin, pp. 387-397.

Yardanov, M. 1983. Heterosis in the tomato. In: Frankel, R. (ed.). Monograph on Theoretical and Applied Genetics, Vol. 6. Heterosis, pp. 182-219. Springer Verlag, Berlin, Heidelberg.

Yu, C.H. 1982. Breeding for hybrid varieties and production of F1 seeds in vegetables. Proc. Symp. Plant Breed., Taichung, Taiwan, pp. 211-217.

contd...

Table 1: Some prominent hybrids developed by private sector seed companies in India

Crop/Hybrid	Company	Chief features
Bottle gourd		
Warad	Mahyco	Cylindrical, good for box packing
Santosh	Krishidhan	Cylindrical, and high yield
Ramdev	Krishidhan	Short-cylindrical (Gutka type)
Shambhu	Krishidhan	Flat-round
Bitter gourd		
Vivek	Sungro	Medium-long-green fruit, compact, firm prickles
Vishesh	Golden Seeds	............ditto................
Abhishek	Seminis	.............ditto...............
Chaman	Nunhems	Spindle shape, green, compact-firm prickles
MBTH-102	Mahyco	White, long, firm prickles
Cucumber		
Gypsy	Namdeo Umaji	Greenish-white fruit, about 15 cm long, good keeping quality
Malini	Seminis	Green, cylindrical, good yield
Ridge gourd		
NS 3	Namdhari	Light–green-shiny, medium long fruits
Naga	East-West Seeds	Light-green-shiny, medium long fruits, tolerant to downy mildew
Murali 9	Krishidhan	Green, medium long
Sponge gourd		
Harita	Mahyco	Green, attractive shiny fruits
Java	Seminis	White-shiny fruits, 25-30 cm long
Watermelon		
NS 295	Namdhari	Oblong, striped fruit, better shelf-life, 8-10 kg, red-flesh, 10-12 percent TSS
Tambola	Krishidhan	Oblong, striped fruit, better shelf life, 10-12 kg fruit, red crispy flesh
Madhubala	Nunhems	Oblong, striped fruit, 10-12 kg fruit, red flesh, vigorous vine
Black Magic	Seminis	Black round fruit, 6-8 kg, red flesh,
Muskmelon		
Keshar	Seminis	Netted, round, orange flesh, sweet, tolerant to wilt and powdery mildew
Pumpkin		
MPH 1	Mahyco	Flat round, medium size fruit, variegated skin

Contd...

Tomato		
NS 2535	Namdhari Seeds	Determinate growth, square-round fruit, heavy yield
Shaktiman	Namdhari Seeds	Determinate growth, square-round fruit, tolerant to leaf curl virus
Abhinav	Syngenta	Determinate growth, square-round to oval fruit, good for long distance transportation, tolerant to LCV
Syn 1389	Syngenta	Determinate growth, open plant, square-round large fruit, tolerant to LCV
US 1080	US Agri	Deteminate growth, square-round fruit, early, heavy yielder
JK Desi	JK Seeds	Determinate growth, bushy, globe fruit shape with green shoulder, heavy yield
Laxmai	Nunhems	Determinate growth, flat round fruit with sweet and sour taste, tolerant to tomato leaf curl virus
Mahalaxmi	KVSIPL	Determinate growth, square-round fruit, tolerant to ToLCV, heavy yielder
Kundal	KVSIPL	Determinate growth, square-round fruit, tolerant to ToLCV, heavy yielder
Chilli		
Ujala	Nunhems	Sturdy and vigorous plant, excellent fruit setting, long, light green and wrinkled fruits with high pungency
Devnur Delux	Nunhems	Vigorous plant growth, strong stem, attractive glossy green fruits at fresh stage, heavy wrinkles, high pungency, blood dark red skin after drying
Sitara	Seminis	Strong plant, 12-14 cm long plods, 1.0-1.2 cm diameter fruit, smooth skin light green fruits, medium pungency
Shama	Choice Seeds	Pale green colour 16-18 cm long fruits, medium pungency
BSS 273	Bejo Sheetal Seeds	Strong tall plant, dark green fruits turn into deep red on drying, highly pungent
BSS 378	Bejo Sheetal Seeds	Tall vigorous plant, 12-15 cm long fruits with 1.2 cm diameter, pale green highly pungent fruits
Wonder Hot	Seminis	Strong dark green colour robust plant, 14-16 cm long thick fruits, smooth skin, medium pungency, fruits turn to dark red on drying
Okra		
US 7109	US Agri	Dark green pods, 12-14 cm long, tolerant to YVMV
Syn 152	Syngenta	Green pods, 10-12 cm long, highly tolerant to YVMV
Syn 016	Syngenta	Shiny green pods, 12-14 cm long, tolerant to YVMV
Mahyco 10	Mahyco	Dark green pods, 10-12 cm long, tolerant to YVMV
Tulasi (NP 1104)	Nunhems	Glossy green pods, early, 10-12 cm long fruits, tolerant to YVMV
Makhmali	Nunhems	Dark green pods, 11-12 cm long, tolerant to YVMV
Anjali Gold	KVSIPL	Dark green pods, 10-12 cm long, tolerant to YVMV

Contd...

Brinjal		
Mridula	KVSIPL	Oval-round fruits, purple colour with white stripes, spiny
Green Beauty	KVSIPL	Oval green fruits with white stripes towards the base of the fruit, cluster bearing, heavy yield
Chaman	KVSIPL	Oval round purple fruits in clusters, heavy yielder
Ajay	Ankur	Oval-round fruits with purple and white streaks
Kalptaru (MHB 10)	Mahyco	Erect, bushy, compact and spiny plants, round to slightly oval fruits, greenish purple with white streaks (bi-colour), green spiny calyx, 60-70 g fruit,
Ravaiya	Mahyco	Compact, non-spiny plants, fruits oval, shining, reddish purple and non-spiny, bearing in clusters, early hybrid,
(MHB 39)	Sungro	Large bulb shape dark purple fruit
Kanhaya	Nunhems	Large, oval, purple fruits
Sandhya BE 706	Nunhems	Medium long, purple fruits
Chhaya (3715)	Nunhems	Medium long, purple fruits
Abhisek (256)	Nunhems	Small, oval-round, purple fruits
Nisha-Improved	Century	Medium long, thick, light purple fruits
Mahadeva	Century	Round, purple fruits

❑❑❑

2

Indian High Altitude Horticulture Development

BRAHMA SINGH
Word Noni Research Foundation and World Wellness Forum, New Delhi, India

HIGH ALTITUDE ENVIRONMENT

India is a vast and diverse country. It has sizeable arable area under high altitudes. High altitudes have sizeable concentration of tribal population. People of these areas have fascinating time-tested knowledge worth exploiting and refining through modern scientific value addition. Area above 9000 ft above mean sea level has been considered as high altitude. Arable lands at high altitudes have two broad categories. First one is cold but humid receiving lot of precipitation in the form of rain during summer and snow during winter. The humidity in these areas is very high through out the year. Low temperature and high humidity are to be manipulated to have quality horticulture produce in such areas. The second category is low temperature and none to very low precipitation that too during winter months only. These areas are mostly sandy, have few months agriculture or horticulture season. Such areas are popularly known as cold deserts or high altitude deserts. Cold deserts occupy sixteen percent area on the earth. India has sizeable area, approximately 74,809 sq. km, in the states of Jammu and Kashmir, Himachal Pradesh and

Uttrakhand. Altitude of Indian cold deserts varies from 2900 to 5900m. Some of the salient features of Indian cold deserts relevant to horticulture development are given in below:

Wind and water erosion prone sandy soils have adequate depth

Infra red and ultraviolet radiations are high

Air density is low

Low concentration of carbon-di-oxide in the atmosphere

Five to seven months frozen winters

Large variation in diurnal (0-30° C) and seasonal (+40 to -40° C) temperatures

Precipitation is low mainly in the form of snow during winters

Relative humidity is low

Irrigation is possible with snow melt water

During winter soils are frozen

Cropping season is 3-5 months during summer season only

This article pertains to horticulture development in such environment.

HIGH ALTITUDE HORTICULTURE

Horticulture is a vast subject. It deals with cultivation of fruits, plantation crops, spices, vegetables, foliage and flower crops, medicinal plants, bio-fuel trees, algal crops, mushrooms, bee keeping and value addition to these. This article covers fruits, vegetables, medicinal plants, foliage and flower crops, and their post- harvest handling, mainly shelf life extension. Cold deserts have been found suitable for these horticultural crops, particularly the temperate crops/varieties.

Horticulture Scenario Fifty Years Back in High Altitudes

In 1960, the following horticultural crops were grown in arid areas of high altitudes of India.

Fruit Crops: apricot, apple

Vegetable Crops: turnip, radish, carrot, Mongol palak (beet leaf)

Flowers: Wild rose and marigold.

Medicinal Plants: number of them but in nature-No cultivation

It is hard to believe but it is a fact that above horticulture crops were the only source of nutrition and health to the inhabitants of the area. Apricot fruit is perhaps the only source of sugar in the area.

The trade in horticulture crops was limited to solar dried apricot and its kernels, some wild medicinal plants and rare spices.

Jammu and Kashmir State Horticulture Department had apricot nursery and vegetable production demonstration centre. The visitors to the area and elite local farmers families visiting other parts of the country and abroad have tried to introduce certain vegetables like cabbage, pea, broad bean, knol-khol and beet root.

Thrust to horticulture development in the area started with the establishment Field Research Labòratory now Defence Institute of High Altitude Research (DIHAR) at Leh in 1962 at the instance of Pandit Jawaharlal Nehru, the first Prime Minister of India.

FRUIT CROPS

Fruit crops grown in cold deserts are as follows:

Prunus armeniaca	Apricot
Malus domestica	Apple
Juglans regia	Walnut
Vitis vinifera	Grapes
Morus alba	Mulberry
Hippophae rhamnoides	Seabuckthorn/ Siberian pineapple
Pyrus communis	Pear
Pyrus persica	Peach
Prunus domestica	Plum
Prunus avium	Cherry

Fragaria sp. (strawberry) has been recently introduced to Leh valley of cold desert.

Among the fruit crops apricot is still a major fruit and source of livelihood of some of the inhabitants in these areas. The fruit is rich source of carbohydrates, vitamin A and C, potassium, calcium and iron. The total soluble solids in apricot cultivars of Ladakh varies from12 to 33° Brix. The deserts have enormous exploitable genetic diversity in terms of fruit morphology, quality and yield for use as table fruit, dehydrated apricot, kernels and oil. Besides having a host of medicinal properties, its oil is potential bio-fuel for the area. DIHAR, Leh has collected and evaluated 84 apricot genotypes. The institute has established a gene bank of 54 apricot genotypes of Ladakh region of Jammu and Kashmir which needs to be strengthened. It has reported prominence of following cultivars of apricot in arid region: Halman, Rakchey-Karpo, Tokpopa, Rogan, Safaida, Nari, Australian, Shakarpara, Charmagz, Kaisi and Suka.

In propagation studies at Leh, chip budding performed in the fourth week of February, cleft/tongue grafting done during the first fortnight of February and shield budding during the month of August have been found suitable with rate success of 85, 90 and 81.6 percent, respectively. Chip budded plants however exhibited better vegetative growth. Top working has been found useful for conservation of apricot germplasm and rejuvenation of old trees. Mulching with organic matter or plastics in apricot has been found beneficial.

Techniques of solar dehydration of apricot have developed which are in vogue there. Sizeable quantity of dehydrated apricot worth several crores of rupees is supplied annually by the region. Lot more of research and development is to be done in promotion of production, processing, packaging and marketing of top class apricots in cold deserts.

Apple is second most important fruit crop of the region. The region has valuable germplasm of varieties like Tha, Mongol, Karkichoo, Shing, Mar, Yangma, Squirmo, Khara and Shamer Kushu. These are small to medium fruit sized varieties with high percentage of juice and very short shelf life but high yielder. Delicious group of varieties introduced to the region are found suitable for the region.

Another important fruit of the region is Siberian pineapple or seabuckthorn which is though not a table fruit but is full of bioactive compounds having amazing health benefits (richest source of vitamin C and store house of A,B, K and E). The three species of it i.e. *Hippophae rhamnoides*-most prevalent, *H. salicifolia* and *H. tibetana* are growing wild covering more than 11,500 hectares in Ladakh alone producing annually huge quantity of fresh fruit, most of which is going waste. DIHAR could identify this nature's gift, create awareness, develop processing and cultivation technologies and bio-prospect them leading to availability of several products including, beverages, cosmetics, health drinks and food products. DIHAR and other life science laboratories of DRDO as well as other research organizations in India have worked extensively on this wonder plant for the benefit of soldiers, society and inhabitants of cold deserts and generated lots of literature on its food, pharmaceutical and nutraceutical importance.

The plant has amazing abiotic (sub-zero temperature, drought and ultra-violet radiation) stress tolerance. This is fantastic fruit, fodder, feed and fuel crop of high altitudes in India where nature nurtured this plant for providing nourishments, medicine molecules, and disease preventive nutrients to inhabitants for effectively combating abiotic stresses of high altitude cold deserts. Seed oil of seabuckthorn has been found to increase high density lipids/cholesterols. This plant has exceptional potential to prevent wind and water soil erosion and maintain soil health by fixing atmospheric nitrogen through frankial root nodulation. *Hippophae* in high altitudes and other colder regions of the world is an important fruit crop to bring prosperity.

Other fruit crops listed above need to be popularized in high altitudes through extensive research.

VEGETABLE CROPS

High altitude arid regions are quite suitable for vegetable crops cultivation. This can be achieved by developing appropriate technologies which are generally area specific. Some of the vegetable production technologies developed for cold desert of Jammu and Kashmir are given below with very brief description.

1. *Early Vegetable Nursery*

It has become possible to grow vegetables like tomato, brinjal, capsicum, cauliflower, cabbage, broccoli and onion by raising early nursery in plant protected structures like soil trench, poly-house and glass-house during March-April and transplanting them as soon as the cropping season starts. The technologies of raising early nurseries of different vegetable crops including potato using true potato seeds have been developed by DIHAR. These technologies are being upgraded continuously since 1963. Nursery production in cold desert has been adopted by many farmers as a business.

2. *Suitable Vegetables and their Varieties*

In limited agricultural season at present, all sorts of vegetables can be grown in Ladakh. This could be possible because of five decades of vegetable research and development by DRDO. Suitable varieties in each vegetable have been identified or developed through introduction, selection, hybridization and mutation. High yielding and suitable varieties and/or hybrids of different vegetable crops are being continuously developed by DIHAR. Production technologies for each of the vegetable crops have been developed, disseminated and documented by DIHAR.

Among vegetable production technologies, protected cultivation and plastic mulch have contributed distinctly to boost vegetable production in cold desert.

3. *Protected Cultivation or Greenhouse Cultivation*

DIHAR had put up first glasshouse during 1964 at its vegetable research unit at Murtse. Since then this laboratory remained a pioneer in research on protected cultivation of vegetables in India. Harnessing solar thermal energy and its conservation during night in protected structures had been main criterion while developing plant protection structures and package of production technologies. Exploitation of available geo-thermal energy for production of vegetables has been experimented upon. Since there was no taker of the technology, this remained limited to experiments. The following areas were researched upon

a. Development of package of cultivation of leafy and other vegetables during frozen winters.

b. Raising early nursery of vegetables in inclement weather

c. Attempting cultivation of new vegetables (which were not possible to grow) in protected structures and gradually making their cultivation possible in open fields. Tomato, brinjal, capsicum, cucurbits etc are examples.

d. Raising seed crops of certain vegetables.

Different types of plant protection structures viz glasshouse, poly house, mud walled poly house and underground poly house (soil trench) and their several combinations have been successfully tried to suit different crops in different situations.

As result of extensive research it is now possible to grow greens leafy vegetables during frozen winters. Farmers now do sell surplus fresh leafy vegetables during winter and earn some money besides enjoying fresh vegetables, something unheard in the area before 1962. It has now become possible to grow crops like green onions, potato, root crops during March, April and May under different structures.

Soil trench or underground solar greenhouse has proved to be most suitable plant protection structure for these areas followed by poly house made of local resources like mud-wall, popular and willow woods. Among the leafy vegetables mustard and lettuce yield better than fenugreek, mangol palak, during sub-zero condition under plant protection structures/solar green houses.

As a result of above it has now been possible to grow vegetables like coriander, mint, vegetable mustard, *Chenopodium*, radish, beet leaf, celery, parsley, lettuce, chinese cabbage, fenugreek round the year in cold deserts of high altitudes in India. Solar green houses contribute much more than the physical return as during frozen winters in cold deserts mere look on growing greens provides moral boost, invaluable joy, sense of achievement and wellness.

4. Plastic Mulch

Favorable response in terms of earliness and productivity in cauliflower, cabbage and tomato has been registered in open field cultivation and in capsicum and brinjal under solar green houses. A remarkable contribution of plastic mulch in cold deserts is in making possible cultivation of all cucurbits and solanaceous vegetable crops like tomato, brinjal and capsicum in open field. Now commercial cultivation of cucurbits has become common in cold deserts. This was considered earlier impossible. Drip irrigation can further boost cucurbits cultivation and their productivity in these deserts. There is need to develop and popularize cultivation of sharda melon (*Cucumis* sp) in these areas using plastic mulch and drip irrigation. The use of biodegradable plastics only needs to be promoted for mulch.

5. Vegetable Seed Production Technology

Seed is a basic and very important input in production of crop. Because of short growing period, seed production in biennial vegetables in cold desert is difficult. DIHAR has succeeded in standardization of technology for commercial production of seed of carrot, radish, onion, cabbage and knol - khol by storing roots, bulbs, heads etc, during frozen winter. It takes two seasons (18 months) to produce seed of these crops in cold deserts. This practice helps the local farmer to get reliable seed of specific variety suitable to the region. Besides this seed production of annual vegetables such as potato, pea, beans, lettuce, palak, coriander, fenugreek, vegetable mustard and tomato has been standardized and popularized. Garlic crop production and its multiplication through locally produced bulbils have been successfully demonstrated by DIHAR.

Seed production in this region is by and large disease free and is therefore of high quality. This area has great potential for commercial production of seeds of several vegetables. True potato seeds (TPS) of good quality is profitably produced here under local natural conditions. Locally produced seeds have been found to perform better than imported seeds in this area. Vegetable seed production technologies developed by DIHAR in cold desert are unique and affordable by the farmers. Since humidity is low, rainfall is nominal, longer photoperiod prevails during seed production period, quality of vegetable seeds produced in these areas is extraordinary.

6. Organic Vegetable Farming

Dependence on synthetic chemical like fertilizers, insecticides, pesticides, fungicides, herbicides and growth regulators is necessitated because of their quick response in crop production. In vegetable production these chemicals play important role. But, besides damaging soil structure and environment, these chemicals are potential health hazards in foods. Moreover, chemicals are costly both to the farmers and society. Organic farming makes positive contribution for the protection of soil, environment and human health.

High altitude areas are following organic farming by default. In vegetable production use of fertilizers is mostly avoided. Use of pesticides and growth regulators is negligible. For enriching soil animal manures, night soil, kitchen waste etc are used. At certain places legumes are included in crop rotations as well as mixed crop. Insects and disease though not many, but are left to natural factor for their control. Organic farming in these areas needs to be promoted by introducing collective/area organic certification.

Natural Medicinal Flora, its Conservation and Multiplication

DIHAR has carried out extensive ethno - medico- botanical survey of different parts of cold deserts Jammu and Kashmir and Himachal Pradesh. It has collected 425 plant species which are used by Amchis (herbal doctors) for various ailments. These have been well documented by the institute. The institute has provided several medicinal plants for detailed biochemical and clinical evaluation to different research centers in the country. The important medicinal taxa are *Podophyllum hexandrum* (Mayapple), *Aconitum heterophyllum* (Aconite), *Saussuera obvallata* (Brahma kamal), *Gentiana aligida, Artemisia* sp, *Oxytropis microphylla, Cremanthodium ellisii, Ephedra,* etc. Detailed phyto-chemical and pharmacological studies are required for positive exploitation and wider application of these Amchi-medico - taxa.

Majority of medicinal flora of cold desert are rare. Some of them are endangered and threatened due to unscientific exploitation. There is need to conserve this rare and precious cold desert's medicinal plants. To achieve this objective herbal gardens in cold deserts are being established.

Cut Flower Production

Cold desert climate is suitable for production of cut flowers like gladiolus, tulips, lilies, gerbera and certain orchids. DIHAR has listed suitable varieties of these crops. It is possible to produce high quality of cut flowers in this region because of prevailing low humidity and long photoperiod. Environmental conditions are quite favourable for development of bright coloured big flowers and long spikes with more number of flowers in gladiolus. There is great scope for dehydrated flower industry with variety of bright and big flowers of marigold, aster, zinnia, wall flower, carnation and others. Low to subzero temperature during October–November facilitates natural freeze dehydration of annual flower crops in full bloom during September.

Extension of Shelf Life of Fresh Fruits and Vegetables

The fresh vegetables and fruits are available in cold deserts up to October. After that great fluctuation in day and night temperature spoil the stored fresh vegetables. During night temperature is subzero and day temperature is 8-12^{0}C making frequent freezing and thawing of produce. To avoid repeated freezing and thawing thereby spoiling fresh produce underground stores are made to harness soil heat and create refrigerated atmosphere for the produce. The fresh produce remains fresh in this condition till February. After that the remaining produce is taken out to natural conditions in shade where it is more or less refrigeration atmosphere. This concept has been given scientific input and is working well in the area on small as well as large (Defence

Forces) scale. Mainly potato, carrot, radish, turnip, beet root, cabbage, knol-khol are being stored for off season utilization. Adequate power supply is not there in cold deserts to have controlled atmosphere storages hence the above technology.

ACKNOWLEDGEMENT

All those who got an opportunity to serve at DIHAR and contribute in development of above appropriate technologies for Indian cold deserts deserve appreciation and thanks.

Reference

Chaurasia and Singh, B. 1995-2000. Cold Desert Plants Vol 1-5. Field Research Laboratory, C/o 56 APO.

Dwivedi, S.K. and Singh B. 2001. Studies on chip budding in apricot (*Prunus armeniaca* L.) in cold arid conditions of Ladakh. Ind J. Hort. 58(3), 215-217.

Dwivedi, S.K. and Singh, B. 2000. Gladiolus - A potential crop in Ladakh. Exploring the gladiolus in India. NBRI - Lucknow, pp. 75-80.

Dwivedi, S.K. and Singh, B. 1999. Topworking of apricot in Ladakh. Prog. Hort. 31 (1-2), 29-31.

Dwivedi, S.K., Kareem, A. and Ahmed, Z. 2007. Apricot in Ladakh. Field Research Laboratory, DRDO, Leh, Ladakh (J&K) India.

Dwivedi, S.K., Kareem, A. and Raut, B. 2005. Some characteristics of apple varieties indigenous to Ladakh (India). Acta Hort. 696 (2).

Dwivedi, S.K., Paljor, Eli., and Attrey, D.P. 2002. Gladiolus in Ladakh. Field Research Laboratory, DRDO, Leh, Ladakh (J&K) India.

Dwivedi, S.K., Singh, B. and Attrey, D.P. 2001. Studies on propagation and utilization of seabuckthorn in Ladakh. In: International workshop on seabuckthorn orgtanised by HPKVV-Palampur, 18-21 Feb. 2001 New Delhi.

Dwivedi, S.K., Singh, B. and Paljor, Eli. 2000. Effect of mulch on establisment and growth of apples saplings in cold arid conditions of Ladakh. Prog. Hort. 32, 42-45.

Dwivedi, S.K., Singh, B. and Paljor, Eli. 2000. Studies on vegetative propagation of apricot through grafting in Ladakh. Ind J. Hort. 57(1), 39-41

Dwivedi, S.K., Singh, R. and Ahmed, Z. 2006. The Seabuckthorn. Field Research Laboratory, DRDO, Leh, Ladakh (J&K) India.

Singh, B. 1992. Societal Contributions to Ladakh. Field Research Laboratory, C/o 56 APO.

Singh, B. 1992. Vegetable production in Ladakh. Field Research Laboratory, C/o 56 APO.

Singh, B. and Chaurasia, O.P. 1998. Medicinal flora of Indian cold deserts. Paper presented in 25th International Horticulture Congress, 2-7 August 1998, Belgium.

Singh, B. Chaurasia, O.P. and Dwivedi, S.K. 1998. Plant biotechnology and herbal flora of Ladakh. In: National Symposium on Herbal Medicines, S B S Postgraduates Instt. of biomedical Sciences and Research - Dehradun, March 27-29, 1997. Proceedings. pp. 1-8.

Singh, B. Dwivedi, S.K. and Paljor, Eli. 2000. Studies on suitability of various structures for winter vegetable production at sub zero temperatures. Acta Hort. (517), 309-314.

Singh, B. et al. 1995a. Vegetable production in Ladakh. Field Research Laboratory, C/o 56 APO.

Singh, B., et al. 1994. Agriculture in Ladakh. Field Research Laboratory, C/o 56 APO.

Singh, B., Jadhav, K. L. and Paljore, E. 1997. In high altitudes of Ladakh – protected vegetable farming potentials. Indian Horticulture. 42(3), 16-17.

Singh, B., Jadhav, K. L.and Paljor, E. 1997. Potential of cabbage seed production in Leh Valley. Indian Hoticulture. 41(4), 36-37.

Singh, N. 2004. Vegetable cultivation in Ladakh (In Hindi). Field Research Laboratory, C/o 56 APO.

Singh, N., Dwivedi, S.K. and Paljor, Eli. 1999. Protected vegetable cultivation in Ladakh (In Hindi). Field Research Laboratory (DRDO) - Leh -Ladakh (J&K) - India.

❑❑❑

3

Enhancing Production of Quality Fruits by Using Phyto-Hormones / Plant Bio-Regulators

L.D. BIST
Department of Horticulture, G.B.Pant University of Agriculture and Technology, Pantnagar, U S Nagar, Uttarakhand, India.

INTRODUCTION

Plant hormones are prominent among those endogenous factors which regulate in a complex manner the environmental and other innate mechanisms influencing growth and development in plants. They function as correlative signals to ensure the orderly development of trees, their flowering and regulate fruit growth, quality and ripening. They are also involved in other physiological processes such as rooting, seed germination, seed production, organogenesis and micro propagation of horticultural plants. Plant bioregulators (PBRs), which include plant hormones and their synthetic analogues, are used for enhancing productivity and quality in fruits and vegetables. There are five classes of plant hormones /PBRs viz. auxins, gibberellins, cytokinins, abscissic acid (ABA) and ethylene.

AUXINS

Auxins promote growth along the longitudinal axis, root initiation, apical dominance, parthenocarpy, fruit size, abscission, and organogenesis. Important auxins are Indole-3-acetic acid (IAA), Indole-

3-butyric acid (IBA), 1-naphthaleneacetic acid (NAA), 2,4 dichlorophenoxy acetic acid (2,4-D), 2,4,5-trichlorophenoxy acetic acid (2,4,5-T), MCPA; Important antiauxins are 2,4 dichloroanizole and DCA.

Significant Uses

i. Control of Fruit Drop and Floral Malformation (Plate-1) in Mango

Spray of 40 ppm NAA thrice at 15 days interval after fruit set for controlling fruit drop in mango. Apply 200 ppm NAA in October followed by a single deblossoming at the bud burst stage for controlling floral malformation (Plate 1).

Plate 1: Floral malformation of mango

ii. Crop Regulation

Two sprays of 600-800 ppm NAA in April-May at 15 days interval at 50% flowering in *Tarai* condition is effective in reducing the crop during rainy season and enhancing quality fruit production during winters in guava.

iii. Root Initiation

IBA increases number of roots in mango stools. Propagation by cutting in pear Mehal/Kainth and Patharnakh is successful with the use of 500-1000 ppm IBA. (Plate 2).

iv. Propagation

GIBBERELLINS

Gibberellins extend shoot growth, regulate dormancy, flowering, fruit-set and seed germination. More than 100 GA_s have so far been identified in plants. Important among them are GA_3, GA_4 and GA_7. Important antigibbrellins are Paclobutrazal, Uniconazal and Daminozide.

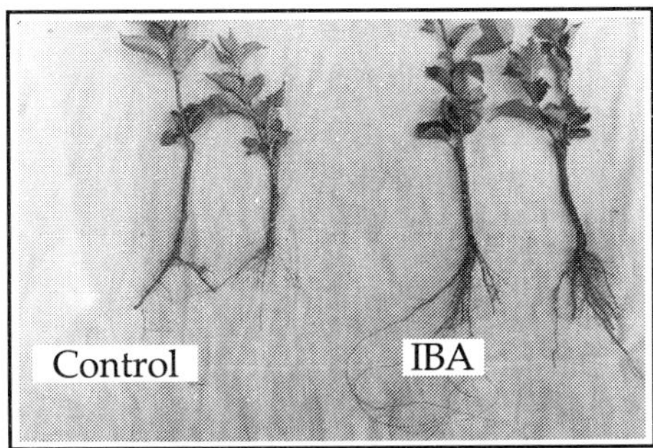

Plate 2: Effect of IBA on rooting of pear cv Patharnakh

Significant Uses

i. Flowering

In most fruit trees and other woody plants GAs generally have no effect or they inhibit flowering. Endogenous GAs are natural inhibitors of flowering in these plants and there is evidence that in deciduous fruit plants GAs move from developing fruit to adjacent apices where they inhibit floral initiation (Plate 3).

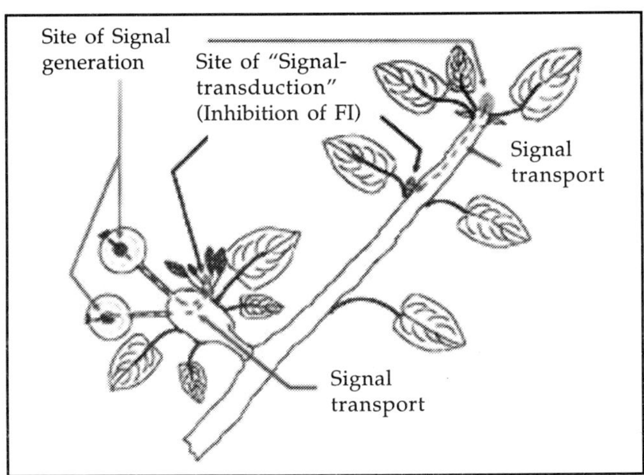

Plate 3: Schematic representation of flower bud inhibiting hormonal signals

GA_4 promote return bloom in spur cultivars of apple while GA_3 and GA_7 inhibit flowering. In conifers such as chilgoja (*Pinus girardiana*), GA_3 or GA_{4+7} is used to promote flowering and thereby significantly increase female cones and seeds per tree. The designer gibberellins such as 16, 17-dihydro GAs generally promote flowering with reduced vegetative growth and in some cases growth is actually inhibited.

The ring D substituted GAs inhibit growth more strongly. The new GAs may be prepared from commercially available fungal GAs. In mango endogenous GA is inhibitory to flowering. GA and cytokinin-like compounds have been involved in causation and control of floral malformation in mango.

ii. Fruit Size and Quality

GA_3 control fruit russeting in apples, GA_{4+7} and BA (Promelin) increase fruit shape, size and quality in Starking Delicious apple. GA_3 is used commercially for rachis and berry enlargement in seedless grape and cluster loosening and flower thinning in wine grape. In citrus, GAs delay fruit senescence and maturity of rind without altering the maturation process, regulate scheduling of harvesting to synchronize with market demand. In cherries, GA_3 controls fruit fly by creating juvenile fruit surface. Peel oils present in juvenile fruit surface are toxic to fruit fly. GA_3 is effective in delaying senescence/fruit ripening. GA_3 is used for stimulation of fruit in blueberries, increase fruit size and quality in sweet cherries and delay senescence/fruit ripening in banana.

In pear, antigibberellin (Paclobutrazol) and Promalin (GA_{4+7} & BA) @250mg/tree by TSLP (Plate 4) is affective in controlling excessive vegetative growth (Plate 5), enhancing flowering and fruit quality. GA applications have been reported to compensate for chilling in several fruits. In pears chilled spurs of high-chill pears grafted on low-chill pears growing in subtropical plains fruited normally (Plate 6).

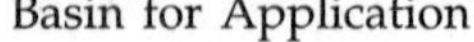

Basin for Application

Tuft of Roots in the Collar Region

Plate 4: Location of application of Paclobutrazol by TSLP method in pear

Control

Paclobutrazol

Plate 5: Effect of Paclobutrazol on growth of Gola pear

Plate 6: Growing high-chill D D Comice pear on low-chill Gola root stock Flowering stage (A), Fruit set stage (B), Fruits at maturity (C)

CYTOKININS

Cytokinins, regulate cell division, dormancy, fruit set and prevent chlorophyll degradation. Cytokinins are also known as juvenile hormones which have major use in plant tissue culture. They are used in seed germination and early seedling development, cotyledon expansion, synthesis of chlorophyll, cell division and organ differentiation, control of leaf senescence and root shoot interaction. Important cytokinins are Zeatin, Kinetin, 6-benzyl adenine (BA, BAP), N-(2-chloro-4-pyridyl)-N-phenylurea (CPPU)

Significant Uses

i. Lateral Branching

BA (Paturyl 10 WSC) is effective as branching agent in young apple trees. BA, causing feathering of apple nursery, is helpful in early training of the trees. In pear, BA is effective in regulating canopy habit and return bloom.

ii. Flowering, Fruiting and Fruit Quality

Promalin (GA_{4+7} and BA) @200 ppm or BA (Paturyl 0.2%) (10WSC) is effective in tree shaping, increased branching, productivity and quality in Gola pear. BA has been shown to improve shooting, flowering and fruiting in Gola pear grown in Tatura Trellis and Penn State low Trellis Hedgerow systems (Plate 7 & 8). GA_{4+7} enhance fruit quality in Red Delicious apple fruits. BA with 2-chloropyridyl phenyl urea (CPPU) or with GA increases fruit size and weight in apple.

Plate 7: Effect of cytokinin on shooting & fruiting in Tatura Trellis grown Gola pear

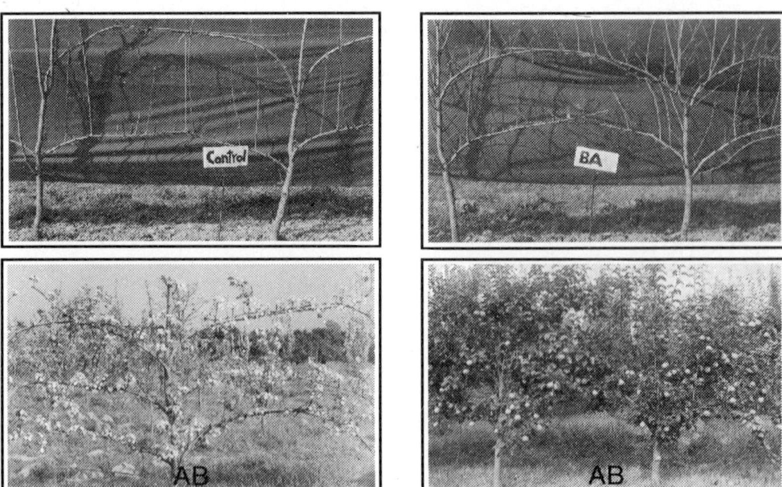

Plate 8: Effect of cytokinin on shooting, flowering and fruiting in Penn State Low Trellis Hedge Row System grown Gola pear

iii. Tissue Culture

Production and mass multiplication of pathogen free elite plant material in pear, peach, banana, strawberry etc. can be achieved by propagation of plants by tissue culture in which cytokinins (BA, Kinetin etc.) along with auxins (IAA, IBA, NAA, 2,4-D etc.) and gibbrellins are used in different combinations. Tissue culture protocols were developed in pear (Plate 9).

Shoot Proliferation

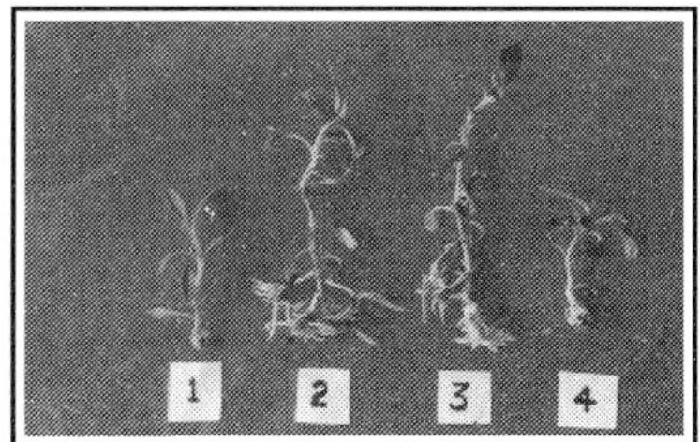

Shoot Proliferation

Plate 9: Micropropgation of Hawthorn

ABSCISSIC ACID (ABA)

Abscissic acid (ABA) is involved in reduction of growth, abscission of plant organ, induction of vegetative buds and regulation of fruit ripening. ABA is involved in stomatal closure and water relations, photosynthesis, root growth, bud and seed dormancy, abscission, stress management, senescence and resistance to frost damage. Endogenous IAA and ABA have been involved in fruit quality of pears. In a number of fruits (pear, avocado, apple, citrus, grape, cherry) the level of free ABA is constant during maturation and increase during ripening.

Significant Uses

No practical application has so far been reported.

ETHYLENE

Ethylene is known as ripening hormone. It is also involved in inhibition of root growth, abscission and other growth processes. It is also known as stress hormone and used to enhance abscission, chlorophyll degradation, stimulation of germination, induction of female flower (cucumber and melon), male sterility (wheat) and growth inhibition by inhibiting GA biosynthesis. Important ethylene releasing compounds are Ethrel, Ethephon (CEPA), Etacelasil, ACC (1-amino-cyclopropane-1 carboxylic acid). Important inhibitors of ethylene synthesis are amino venyl glycine (AVG) , aminoxy acetic acid (AOA)

Significant Uses

i. Stimulation of Flowering in Pineapple

Ethrel, ethephon and precursor of ethylene (ACC) are being used for flower initiation in pineapple.

ii. Fruit Ripening

Used in ripening of number of climacteric fruits and decolouring non-climacteric citrus fruits e.g. spray of 1000 ppm ethephon during fruit development in apple, banana, grape, mango, strawberry and cranberry.

iii. Growth Control

In grapes vigorous vegetative growth interferes with reproductive growth, berry size, coloration and ripening of fruits. Ethephon may be used for inhibiting vegetative growth of grape vines resulting in higher yield and better quality.

GROWTH RETARDANTS

These are synthetic organic compounds causing retardation of cell division by inhibiting biosynthesis of plant hormones without evocating substantial growth distortions. Important retardants are Alar (B-Nine, SADH), MH, paclobutrazol (PP333, cultar, Bonzi, flurprimidol, uniconazol and CCC.

Significant Uses

i. Control of Excessive Vegetative Growth

Alar 1000 ppm has been effective in controlling excessive vegetative growth in low-chill peaches. Alar, CCC and Promalin have been shown to affect micro nutrient status in pear leaf.

ii. Flowering and Fruiting

Alar (SADH) has been in use for flowering and fruiting in high density apple orchards. TSLP applications of paclobutrazol (@1g/tree/year) reduce excessive vegetative growth and increase flowering and fruiting in Gola pear with residual effect for 3-4 years.

References

Bist, L.D. and Kaul, G.L. 1977. Studies on the control of excessive growth in sub-tropical Peach (*Prunus persica* Batch.) under Tarai conditions of Uttar Pradesh. I Effact of various treatments on shoot growth. Indian J. Hort. 34, 237-242.

Bist, L.D. and Ram, S. 1985. Effect of malformation on changes in endogenous gibberellins and cytokinins during floral development of mango. Scientia Hortic. 20, 235-241.

Bist, L. D.1990. Influence of PP 333, Alar, CCC and Promalin on micronutrient status of pear leaf. Acta Hortic. 274, 43-50.

Bist, L. D. and Rai, N. 1994. Residual effect of paclobutrazol and promalin on tree shaping and productivity of 'Gola' pear. Indian J. Hort. 51, 41-48.

Bist, L.D. 1994. Long term effect of paclobutrazol and promalin on growth, yield and fruit quality of pear *(Pyrus pyrifolia* (Burm.) Nakai) cv Gola. Prog. Hort. 26 (3-4), 125-132.

Bist, L.D. and Buban, T. 1997. Effect of Paturyal 10 WSC on vegetative growth, owering, fruit-set and yield of young trees of 'Gola' pear. Hort. Sci. (Hungary). 29, 36-38.

Bist, L.D., Yadav, Ashish and Chander Prakash. 2003. Performance of low-chill pear cultivars under submontanus Tarai region. Prog. Hort. 35, 20-24.

Dwivedi, S.K. and Bist, L.D. 1997. *In vitro* micropropagation of Mehal pear (*Pyruspashia* Buch Ham.). Indian J. Hort. 54, 223-228

Dwivedi, Deepa H. and Bist, L.D. 2003. Correlation between endogenous indole acetic acidlevels in developing fruit of low-chill pear (*Pyrus pyrifolia*) cultivars and fruit quality at maturity. Indian J. Ag. Sc. 73 (5), 272-275

Kumar, R. and Bist, L.D. 2002. Micropropagation of hawthorn (*Crataegus oxyacantha* Linn.) through shoot tip culture. Indian J. Hort. 59(4), 435-439.

Looney, N.E. 1995. Plant bioregulator usage in fruit production: some notable achievements and future prospects. Acta Horticulturae. 394,15-23.

Mander, L.N., Camp, D., Evans, L.T., King, F.W., Pharis, R.P., Sherburn, M. and Twitchin, B. 1995. Designer gibbrellins:The quest for specific activity. Acta Horticulturae. 395, 45-55.

Prang, L., Stephan, M., Schneider, G. and Bengerth, F. 1997. Gibberellin signals originating from apple fruit and their possible involvement in flower induction.(Personal Communication).

Rai , N. and Bist, L.D. 1991. Effect of promalin, SADH and Chlormequat on tree growth flowering, fruit-set, yield and fruit quality of Gola pear. J. Hort. Sci. 66, 443-447.

Rai, N. and Bist, L.D. 1992. Effect of soil and foliar applied paclobutrazol on vegetative growth, flowering, fruit set and yield of oriental pear. Scientia Hortic. 50, 153-158.

Ram, S., Sirohi, S.C. and Bist, L.D. 1981. Interaction of diphenolics with auxin in the stooling of mango. Punjab Hort. J. 21, 46-52.

Shukla, G.S. and Bist, L.D. 1994. Studies on the efficacy of IBA and NAA on clonal propagation by cutting in low chilling pear rootstock. Indian J. Hort. 51(4), 351- 357.

Yadav. A and Bist, L.D. 2001. Growing high-chill pears under subtropical plains in high density system. South Indian Hort. 49, 50-54.

❑❑❑

4

Sustainable Vegetable Research for Nutritional Security and Socio-Economic Upliftment of the Tribal's in Indian Cold Desert

NARENDRA SINGH, RAGHWENDRA SINGH, HITESH KUMAR, MANISH S. BHOYAR AND SHASHI BALA SINGH
Defence Institute of High Altitude Research, DRDO, Leh-Ladakh, J&K, India

INTRODUCTION

Cold arid zones are usually confined to high altitude and circumpolar region. Sixteen percent of total land mass is under cold arid zones. Indian cold arid region comes under the Trans-Himalayan zone and covers about eighty thousand square kilometer land mass. More than 90 percent cold desert of India is situated in Jammu and Kashmir State as Ladakh region. This region is characterized as snow cladding mountain peaks, extreme harsh climatic and geographical conditions which makes the life and agriculture difficult. Due to extreme long winter, the agriculture season is short which extends from May-September in general and in case of particular location its depend upon altitude (Table 1).

Wide variation in daily and seasonal climate makes it difficult to standardize package of practices for cultivation of vegetable crops in the region. Among the climatic factors, temperature is the most important factor which has great impact on cropping pattern and production technique. Temperature decreases with increasing altitude.

Therefore, cropping season and duration depends upon the altitude (Table1). It is depicted from the table1 that area lies between 2400-2700 m altitudes has longer cropping duration and potential to achieve 200 percent cropping intensity by cultivating short duration vegetable crops. After every 300 m height, there is change in cropping season and duration, therefore, there was need to evolve crop specific techniques to obtained optimum yield. Although, army deployed up to 6600 m altitude in the region but human settlement and agriculture is confined between 2400 to 4200 m altitudes where above 2700 m altitude single cropping system is practiced. For enhanced production of vegetables to ensure the nutritional security in high altitude cold desert, a comprehensive research programme was carried out at Defence Institute of High Altitude Research (DIHAR), Leh-Ladakh, Jammu & Kashmir, India on various aspects of vegetable production during last four decades.

Table 1: Cropping Season and Duration at Different Altitude

Altitude (m AMSL)	Cropping season	Duration (month)
2400 - 2700	April – October	7.0
2700 - 3000	May - Mid October	5.5
3000 - 3300	Mid May - September	4.5
3300 - 3600	June - Mid September	3.5
3600 - 4000	Mid June – August	2.5

Cold Desert Ladakh and the Vegetables

Ladakh, the land between earth and sky is considered virtually unfit for vegetation. But long ago some plant species were able to establish themselves in this difficult terrain after natural introduction and with the reference of Charles Darwin's theory "the origin of species" may be the plants of these species modified in their morphological and taxonomical traits in due course of time not only for survival but also complete there life cycle for multiplication in adverse conditions of the region. Majority of such plants are perennial herbs and some of them were suitable to use as vegetable (Table 2) which were identified by local populace through over the years experience may be by hit and trial method. The plants of these species start their growth and quite a good number of such plants are in full vegetative state earlier than the introduced cultivated vegetables in the beginning of summer season. Before introduction of some cultivated vegetable crops such as turnip, radish, carrot, onion, potato and certain leafy vegetables in late nineteen fifties by the Government organization (Foreigner and Indian) and missionaries, the people of this region were dependent on these wild plants species as vegetable and this practice can still be seen in far flung villages. After establishment of Defence Institute of High Altitude Research (1962), a systematic experimentation was carried out and some vegetables were introduced and production technique of these vegetables were evolved and demonstrated, after that, the farmers under took cultivation of further more number of vegetable crops such as cabbage, cauliflower and knolkhol. Up to the nineteen seventies mostly vegetables were grown through direct seed sowing in the month of June and fresh vegetables were available only for 2 months (August- September). The availability period of vegetables cultivated in open field during summer season has been enhanced from 2 to 5 months (June-October) by protected nursery raising techniques and introduction of more number of suitable vegetables/ varieties/ hybrids. After introduction of cucurbitaceous, solanaceous and exotic crops in last decade, total number of introduced vegetables has been increased and agro-techniques have been evolved for more than 60 types of vegetables. Techniques have also been standardized to produce more than 20 types of vegetables even during severe winter months (October-April) through greenhouse technology.

Table 2: Production Potential of Natural Leafy Vegetables of Cold Desert Ladakh

Vegetables	Common name	Leaf yield/cutting (Kg/m^2)
Chenopodium botrys	Sagani	0.8
Fagopyrum esculentum	Buckwheat	0.7
Lactuca dalichophylla	Khala	1.4
Lepidium latifolium	Dittander	0.7
Oxyria digya	Mountain sorrel	0.7
Rumex patientia ssp tibeticus	Shoma	1.0
Urtica hyperborea	Zazut	0.9

Vegetables, Nutritional Security and Health

Vegetables are potential source of vitamins, minerals, carbohydrates proteins and fibres. They are essential ingredients of a balanced human diet to meet daily requirements. It has also been recognized that most of the vegetables are more nutritious than other crops such as one kilogram of spinach provides more nutrition than one kilogram of rice. High nutrient value of leafy vegetables produced in trench under cold desert condition of Leh was also reported. Vegetables are usually higher in productivity than other crops. These crops can provide more nutrition per unit time and area of land. Solanaceous vegetable like tomato can yield over 200 tons per ha in protected conditions and more than 60 tons per ha in open field conditions in cold desert. Likewise cabbage has much higher yield potential (120 t/ha) than cereal crops. Therefore, more nutrition per unit area can be produced by cultivating the diversified vegetables and it can reduce the pressure on cereals also for food stability. Based on the research findings, certain green leafy vegetables considered for the prevention of some serious disorders like night blindness and anaemia particularly in children and women. Garlic can be used to cure cough and cold. Even serious diseases can also be prevented by particular element rich designer product from the vegetables.

Vegetables Requirement in Cold Desert

To fulfill the requirement it is difficult to import the vegetables from other parts of the country to the cold desert region mainly due to high perishable nature and involvement of long distance transportation. The region remains land locked every year from November-May due to heavy deposition of snow on the road passes. Air lifting of the fresh vegetables results heavy expenditure which is unaffordable to the people. Army also requires huge quantity of fresh vegetables to feed their troops deployed in this region and spends a substantial amount to import the vegetables from other part of the country. Including army and floating population, vegetables are required for about 3.0 lakh population in the region. If it is calculated based on the ration scale provided to the army then total of 37200 MT/ year vegetables are required (Table 3). Out of this quantity approximately 17705 MT vegetables are produced locally and remaining quantity is either imported from other places of the country or the people replace it by other food ingredients during non-availability of vegetables.

Table 3: Total Requirement and Local Production of Vegetables in Ladakh

Items	Requirement (MT)	Production (MT)
Vegetables (more than 30 types)	17000	8160
Potato	13000	8450
Onion	5400	1080
Garlic	1800	15
Total	**37200**	**17705**

Sustainable Vegetable Research and Cold Desert

To solve the diversified problems faced by the farmers in vegetable production, techniques are developed and being employed elsewhere, they do not hold good for the region may be due to segregated differentiation in physiographic conditions. Therefore, to enhance the productivity systematic research work has been carried out by DIHAR on following priority aspects of vegetables during last four decades.

a) Induction of Vegetable / Variety / Hybrid

Selection of vegetable/variety/hybrid is most important factor to maximize the yield in the region. Every year large number of new varieties/ hybrids of vegetables developed by public/ private Institutions or imported from the different countries are coming in market. Therefore, there is a requirement of continuous screening of the suitable variety/hybrid of the vegetables for desirable traits because the different varieties/hybrids have significant impact on yield and productivity of the crop (Fig. 1). Army is a substantive market for local farmers to sale their produce besides civil market. Army has their own specifications to purchase the vegetables. To consider the nutrition aspect, variation in meals and seasonal availability, Army requires more than 30 types of vegetables. The farmers had been producing vegetables like, potato, radish, onion, turnip, carrot, cauliflower, cabbage, knolkhol, karamsag and beet leaf upto late nineteen eighties and these were provided in limited quantity to the Army deployed in the sector. Up to the middle of nineteen nineties about 40 types of vegetables were introduced and during last decade further more than 20 types of vegetables have been introduced in the region. During this period more than 60 types of vegetables and their high yielding varieties and hybrids have been evaluated and techniques have been standardized to obtain optimum yield.

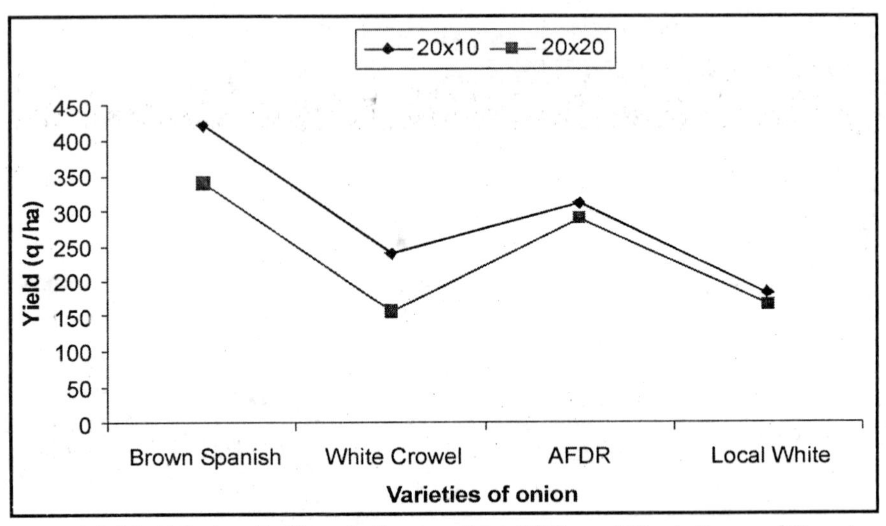

Fig. 1: Yield of Onion Bulb as Influenced by Different Varieties and Spacing

b) Development of Vegetables Varieties

In high altitude cold desert usually single crop is grown in a year during summer months (May-September) therefore, higher production desired as priority aspect. High production can be achieved by growing suitable varieties with scientific management practices. Variety that takes long duration to grow but give higher yield is found suitable for lower altitudes however short duration varieties of vegetables are found suitable to be grown in higher altitudes. These varieties were selected after evaluation and simultaneously certain varieties of different vegetables were developed during the period. The variety Sindhu-1 and Sindhu-2 of tomato were developed through mutation breeding. To induce the mutation, radiation treatment of gamma-rays to the tomato seed was given by Bhaba Atomic Research Centre, Mumbai. These are short duration varieties with bigger size fruits and higher yield than their parent (Pusa Early Dwarf) and recommended to grow up to 4200 m altitude. The cabbage variety developed by hybridization and selection method is resistant to head splitting which is a major problem in cold desert. Another Sindhu Summer variety of cabbage is a short duration and was also evaluated in plain condition. The study revealed that the variety has wide adaptability. It produced head even in the month of June and able to produce seed in single winter season (annual). Likewise, during the period different indigenous varieties have also been developed i.e. Sindhu Yellow (pumpkin), Sindhu Brahma (broad bean) and Sindhu Red (potato).

c) Protected Vegetable Nursery Raising

Nursery raising is a highly technical aspect and has great impact on planting time, availability period, yield and quality of the produce. Successful production of the certain vegetables like solanaceous fruit crops and cucurbitaceous crops in field conditions depend on early preparation of seedlings. Therefore, to get the early produce of the vegetables, practices have been standardized for healthy and vigorous seedlings raising of more than 40 types of vegetable crops in different type of protected conditions like specific polycarbonate greenhouse, trench, polyench and seasonal greenhouse. During the period DIHAR has developed the capacity to raise more than 8.0 lakh seedlings of different vegetables annually in protected condition. The nursery raised in the greenhouse is ready for translation in first week of April (45 days early) for protected vegetable cultivation and this helps to get the produce about 60 days early than open condition with increased yield up to 60 percent depending upon the crop.

d) Package of Practices

Fluctuation in seasonal and diurnal temperatures is very high in cold desert region. Sudden drop in temperature during early stage (April-May) and later stage (September-October) of crop growth is a common phenomenon. This wide variation in climate makes it difficult to standardize common package of practices for vegetables production in the region. Therefore, in due course of time the Institute has developed crop/location specific practices for optimum productivity. To get early crop with higher yield, timely plantation is essential especially for the crops which are grown through transplantation. Protected vegetable nursery raising techniques of different vegetables have been standardized to raise the vegetable seedlings for early transplantation. For nutrient management of the crops, a study on soil analysis revealed that the type of soil in the region depends upon weathered rocks in the particular location. Majority of the soil are sandy to sandy loam in texture and poor in fertility and water holding capacity. The doses of fertilizers and organic manure affect the yield of the vegetable crops. The study reveals that the combined use of fertilizers and organic manure gave higher yield. Therefore, to obtained higher yield dose of fertilizers and organic manure have been standardized for different vegetable crops.

The local white variety of onion grown by the farmers has low yield and character of splitting. After splitting the bulbs are smaller than the specifications fixed by the army therefore difficult to introduce. To find out the solution large number of varieties were evaluated and identified Brown Spanish for higher yield (Fig. 1) and low splitting rate (Fig. 2). It is also clear from the fig. 2 that splitting rate also affects with spacing and closure spacing discourage the rate of splitting. Further higher yield with enhanced availability period can be achieved by cultivating the crop through onion sets.

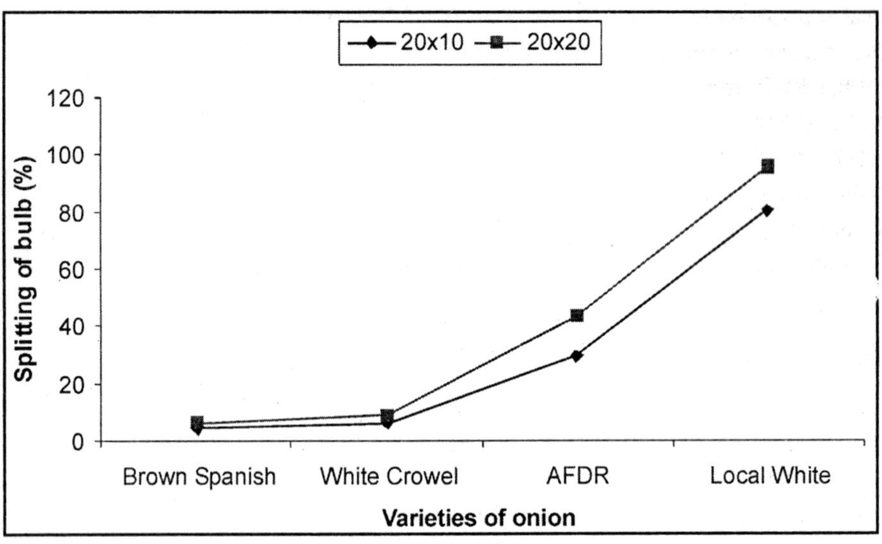

Fig. 2: Splitting Percentage of Onion Bulb as Influenced by Different Varieties and Spacing

e) Mulching

Low temperature and limited availability of water due to the geographical constraints have adverse effect on vegetable production in the region. Being a desert, the region also suffers from low humidity during cropping season which leads to more evapotranspiration and number of irrigation to the crops. To over come these problem innovative methods of vegetable cultivation are required to increase the production and productivity. Therefore, techniques have been standardized to grow the vegetable crops with plastic mulching. The study revealed that mulching of black polythene of 200 micron thickness conserve moisture and reduce upto 30 percent requirement of irrigation and it increases up to 6.5°C soil temperature depending upon the crop which is beneficial for early and healthy emergence of the crop during early stages. It creates a physical barrier to emergence of the weeds. Mulching stimulates soil microorganisms which results in rapid breakdown of soil organic matter and release of plant nutrients for better crop growth and yield (Table 4).

Table 4: Effect of Polythene Mulch on Yield and Earliness of Vegetable Crops

Crop	Yield (q/ha)		Earliness (days)
	Mulched	Nonmulched	
Tomato	617	428	16
Zucchini	560	415	18
Bottle gourd	385	235	12
Capsicum	240	140	22
Egg plant	121	58	25
Chilli	57	22	25
Okra	128	67	17
Cauliflower	516	389	18
Cabbage	780	691	20

f) Greenhouse Technology and Cold Desert

Greenhouse effect is being used since long in many places of the world to grow various plants/crops under controlled environmental conditions by using electricity. But in cold desert region production and throughout year availability of electricity in required quantity is either difficult or not economical and this may not be eco-friendly. Without using greenhouse effect it is difficult to think about vegetable production during winters when temperature goes below - 20°C. Therefore, research work was started on greenhouse technology after establishment of DIHAR and during the period various greenhouses have been designed, evaluated and demonstrated for different aspects of vegetable production. Under the study nine types of greenhouses were evaluated and identified suitability for various aspects of vegetable production such as round the year vegetable production, protected nursery raising, seed production and particular vegetable/variety production at particular time. Design of polycarbonate and fibre reinforce polyester sheet clad greenhouses are found most suitable for production of Indeterminate hybrids of tomato, capsicum, chilli, brinjal , gynoecious hybrid of cucumber and zucchini during summer months from 2400-3600 m altitude and leafy vegetables i.e. Swiss chard, beet leaf, kale, lettuce etc. during winter. Above 3600 m altitude, selection of the crops suitable to grow in the greenhouse is changed as cabbage, cauliflower, radish, turnip and short duration varieties like Pusa Sadabahar and Sindhu-1 of tomato. The polyench greenhouse made of locally available material except cladding material is found suitable to raise early nursery and round the year production of leafy vegetables. However, the trench greenhouse is most suitable to produce off season potato, vegetable nursery and certain leafy vegetables during March-May. The early protected nursery especially of cucurbitaceous crops can be produced in seasonal greenhouses as they provide better scope for hardening of the seedlings before transplantation.

Economics is also an important factor to utilize the greenhouse for different aspects of vegetable production. The Institute has evaluated and recommended economically viable crop rotations and production techniques of the crops for different greenhouses. Tomato (indeterminate hybrid) during summer months (April-October) and Swiss chard during winter months (November-March) give highest return (1:1.8 benefit ratio) in FRP greenhouse, however, leafy vegetables like beet leaf, Swiss Chard and kale during winter months (September-May) and radish during summer months (June-August) found most viable crop rotation for polyench (1:3.2 benefit ratio). The data presented in fig. 3 revealed that due to low construction cost and high production, single and double wall polyench have low production cost followed by trench.

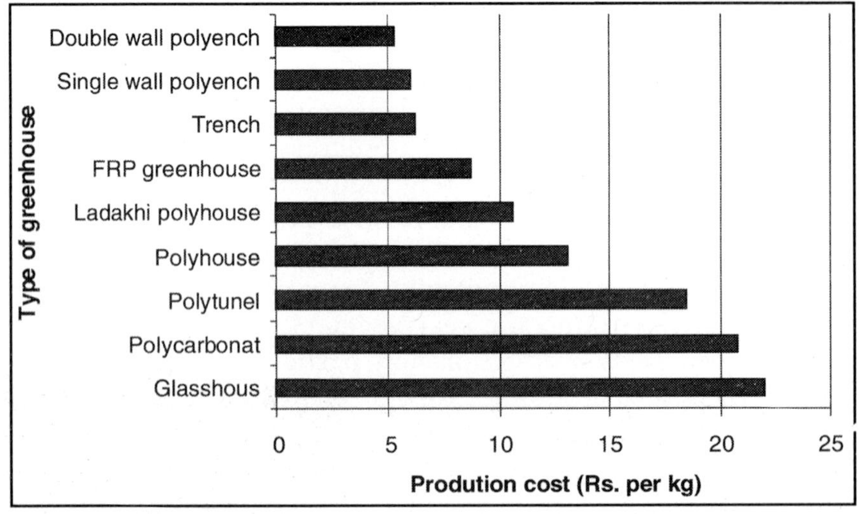

Fig. 3: Cost of Vegetable Production in Various Greenhouses

g) Post Harvest Management / Storage

The availability of the vegetables for longer period of a year can be maintained by giving interval in sowing/planting, harvesting and by storage. Almost all the vegetables are sown/ transplanted in the month of May and harvested up to October from the field. They have to be stored to maintain further availability. The vegetables are stored in refrigeration cold stores at about 4^0C temperature. But unlike plains vegetable storage facilities (cold store or control environment store) are not available in cold desert regions. Therefore, zero-energy based underground storage techniques has been standardized by the Institute. Certain vegetables like, carrot, radish, turnip, cabbage, knolkhol, onion, pumpkin etc can be stored even during frozen winter (November-April) with this technique. The Institute has also developed technology for storage of locally produced potato up to 10 months (November-August) in geothermal energy based underground store with Clorpropham treatment.

Plant Protection

The long prevailing severe winter of the region is not conducive to survive/spread of many pathogens and insect-pests. But some insect-pests and diseases encountered in cultivation of vegetable crops. Therefore, health management practices of vegetable crops have been standardized after identification of the major insects pests like cutworm, aphids, cabbage butterfly caterpillar and diseases i. e. damping off, black rot fusarium wilt, powdery mildew, purple blotch and common scab. The intensity of the diseases and insect-pests decreases with the increasing of altitude therefore more than 3600 m altitude recommended for healthy seed production of certain vegetables.

Vegetable Seed Production and Quality Assessment

Seed is an important input in production of any crop. Non availability of quality seed is also an adverse factor for vegetable production in cold desert. However, the climatic conditions such as long photoperiod, low incidence of major diseases and insect-pests, bright sunshine, low humidity and negligible rainfall are the favorable features for seed production of vegetable crops in the region. Therefore, agro-practices have been standardized for seed production of certain biennial and annual vegetable crops. Seed of biennial crops like cabbage, knolkhol, brussels sprouts, broccoli, carrot, radish, onion and beet leaf can be produced by storing the heads, knobs, stems with sprouts, roots, bulbs etc during frozen winter and by replanting them in the field in subsequent cropping season. However, crops like pea, beans, tomato, capsicum, true potato seed (TPS), lettuce, coriander, fenugreek, vegetable mustard, zucchini, pumpkin and bottle gourd produced seeds in same season and practices for quality seed production of these crops have been standardized and popularized. Yield potential of different vegetables/ varieties obtained from the experiments is given in Table 5.

Table 5: Seed Yield Performance of Different Vegetable and Varieties

Vegetable	Variety	Seed yield (q/ha)
Cabbage	Golden Acre	9.9
	Early Drums Head	12.6
Knolkhol	White vienna	8.0
Broccoli	Packmen	6.3
Brussels spouts	Hills Ideal	5.6
Carrot	Nantes	19.6
	Yamdagni	17.3
Radish	Local	18.2
	Himani	14.3
Turnip	Purple Top Whit Globe	7.4
Beet root	Detroit Dark Red	24.2
Swede	Selection -1	5.7
Beat leaf	Sindhu Harit	12.6
Parsley	Masscurled	5.6
Celery	Timmuf	6.8
Onion	Local White	6.2
	Agri Found Dark Red	8.6
	Brown Spanish	6.8
	White Crowel	6.6
	Red Pinoy	7.4

In a study on seed quality, it has been recorded that seed produced in cold desert were superior i.e. tomato 21.0 percent, coriander 47.8 percent, packchoi 22.3 percent, pea 23.5 percent and snow pea 8.8

percent in boldness than the seed produced in plain condition. Similarly, packchoi seed produced at Leh has less (4.8 percent) of under size seeds in comparison to 22.8 percent in the seeds produced at Chandigarh. The study revealed that cold desert regions are suitable to produce good quality seeds of certain vegetable crops.

Organic Farming

The Physical and chemical properties of the soil of Ladakh is very poor. The use of chemicals such as fertilizers, insecticides, weedicides may harm this fragile ecosystem. The organic farming is the only means by which we can get sustainable high productivity, which only provides essential plants nutrients for growth and development of plants but also leads to build up of organic carbon and improve the soil physical and chemical condition as well as the soil biotic life. Decomposition of organic matter is a serious problem in cold desert because of low humidity and low temperature. Therefore, the technique for decomposition of the biowaste through vermicomposting has been standardized. The earthworms breakdown the animal and plant waste to release of nutrients for better crop growth and yield. Crop like beet leaf, Chinese cabbage, turnip, karam sag, bean, peas and summer squash do not require inorganic fertilizers in the soil supplied with 30-40 t/ha of well decomposed organic manure.

Information and Documentation

Data base is a prior requirement to make any policy/ project for development of a particular sector. For development of vegetable sector in cold desert this Institute has generated lot of valuable data and information in term of publication of more than 50 research papers in the Journals of national and international repute. During this period 5 books, 04 technical bulletins and several chapters in different books have been published/written on various aspects of vegetable research in cold desert. Out of the research carried out during the period, four patents have been filed and one is in pipe line.

Vegetable Research and Socio-Economic Upliftment of Local Farmers

The significant findings of the research on vegetables were demonstrated and disseminated to the farmers through various extension programmes. With the adoption of these technologies the farmers are able to produce enhanced quantity of vegetables every year. They sell their produce to Army deployed in the region besides civil market. Where army was getting only about 15 percent of its total requirement of the fresh vegetable in 1999 through farmers Cooperative Marketing Society, it has been raised up to 54 percent in the year 2007 (Fig 4) and the farmers are getting a substantial amount per annum by selling the vegetables to army other than the civil market after household consumption.

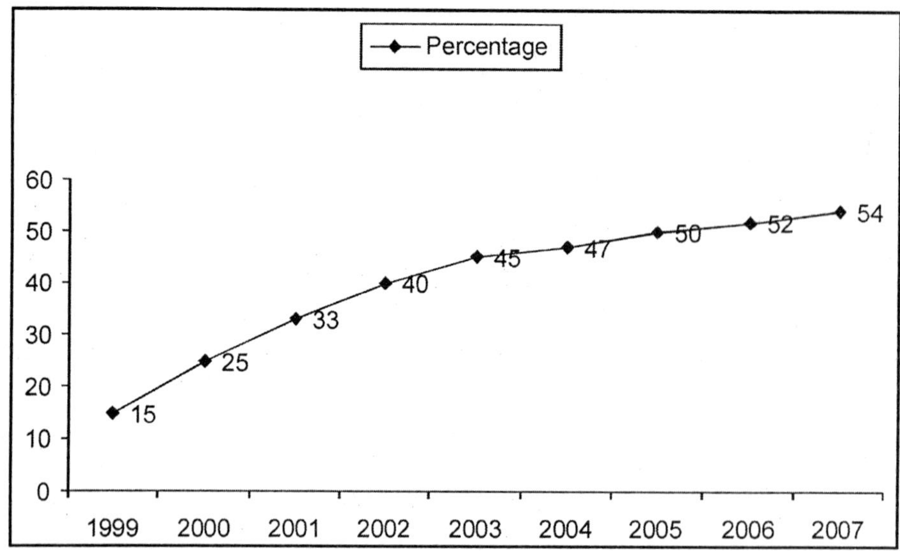

Fig. 4: Supply of Fresh Vegetables (Percent of Total Requirement) to Army

CONCLUSIONS

It can be concluded from the study carried out during last several decades on different aspects of the vegetables, providing scientific inputs to the farmers are helpful in safeguarding nutritional security for army deployed in the sector and socio-economic upliftment of the peoples in high altitude cold desert regions.

References

Anonymous. 2004. Closure report of the project entitled "Pilot scale research cum demonstration unit for round the unit vegetable cultivation in 'L' Sector" SL-P1- 2001/FRL-99.

Bilab. K. Nandi and Bhattacharjee, L. 2004. Emerging issuses in nutritional security. Food security and vegetables A global prespective. Dr. Prem Nath Agricultural Science Foundation, Banglore, India.

Chaurasia, O.P., Singh, Narendra and Singh, Brahma. 2000. Potential wild leafy vegetable from Ladakh. Indian Horticulture. 4-5.

Energy efficiency policy and the environment, Elsevier Science Ltd (UK), pp. 2311-2314.

Garg, I.D., Chandla, V.K., Paul Khurana, A.M. and Singh, N. 2003. Potato cultivation and health status of the crop in Leh-Ladakh (J&K). J. Indian potato Assoc. 30(1-2), 121-122.

Singh, B. 1999. Vegetable production in cold desert of India: A success story on solar greenhouse. Acta Horticulture. (534). 205-212.

Singh, B. and Dhoulakhandi, A.B. 1998. Application of solar greenhouse vegetable production in cold desert. In saying AAm (Ed) Renewable energy:

Singh, B., Dwivedi, S.k., Singh, N. and Paljor, Eli. 1998. Sustainable Horticulture practices for cold desert areas. Lead paper "International Symposium on sustainable Agriculture in Hill Area" 29-31 Oct 1998 at Himachal Pradesh Krishi Vishwavidyalya, Palampur, H.P. India.

Singh, B., Singh, N. and Singh, D.V. 2001. Perspective of vegetable production in Ladakh. Dynamics of cold arid agriculture. Kalyani Publishers, Ludhiana (Pb.), pp. 85-98.

Singh, N. and Ahmed, Zakwan. 2007. Double walled polyench greenhouse. Indian Horticulture. 8-9.

Singh, N. and Singh, B. 2003. Ladakh Men Sabji Utpadan (vegetable production in Ladakh), Field Research Laboratory, Leh-Ladakh, India.

Singh, N., Ahmed, Zakwan and Kumar, N. 2006. Production Technology of Garlic for cold desert. Field Research Laboratory, Leh, Ladakh, (J&K), India.

Singh, N., Dwivedi, S.K. and Paljor, Eli. 1999. Protected vegetable cultivation in Ladakh, book in Hindi. Field Research Laboratory, Leh- Ladakh, India.

Singh, N., Raut, B. and Kumar, N. 2003. Cultural tips for growing vegetable in Ladakh, Field Research Laboratory, Leh-Ladakh, India.

Singh, N., Raut, B. and Kumar, N. 2003. Techniques for cucurbits cultivation in cold desert Ladakh. Field Research Laboratory, Leh-Ladakh, India.

Singh, N., Raut, B. and Kumar, N. and Mohanty, J.N. 2003. Off season potato production in trench in cold arid conditions of Ladakh. J. Indian potato Assoc. 30 (1-2), 95-92.

Singh, N. and Ahmed, Zakwan. 2005. Solar greenhouse technology for vegetable production in high altitude cold arid region. Field Research Laboratory, Leh-Ladakh, India.

Yadav. J., Singh, N. and Kumar, Narendra. 1999. Nutritional values of off-season leafy vegetables grown in trench under cold arid conditions of Leh. Agri. Sci. Digest. 19(2), 103-104.

□□□

5

Fruit Production in Cold Arid Regions of India: Challenges and Opportunities

SANJAI K. DWIVEDI[1], TSERING STOBDAN[2], ASHISH YADAV[2] AND SHASHI BALA SINGH[2]
[1]PEACE, Defence Research and Development Organisation, Metcalfe House, Delhi
[2]Defence Institute of High Altitude Research (DRDO), Leh-Ladakh, India

INTRODUCTION

Ladakh constitutes almost 2/3 of geographical area of J&K state and comprised of Leh and Kargil district which are at an altitude between 2900-5900m above msl. However, only a little area is presently under the cultivation of various agricultural crops. Agriculture is the main source of livelihood in Ladakh. The major agronomical crops include gram and wheat while horticultural crops like vegetables, apricot and apple are playing major role in supplementing the income to the farmers and hence assured great importance in Ladakh. Despite development of tourism industry and deployment of armed forces at China and Pakistan borders, agriculture remains on the forefront in providing income, food/ nutritional security and employment to the tribal farmers especially the ladies in the region. Fruit crops like apricot, apple are one of the major sources of income. This area is characterized by uneven land, barren mountains, arctic desert and extremely cold weather for 7 months of the year and thus called cold arid region.

Cold arid regions in India are located in the high altitude zone which is described as a treeless world and an environment of low atmospheric pressure. All other environmental conditions are only chain effects of this reduced atmospheric pressure. Significant physiological, structural and ecological modifications are observed in plants and animals, all of which can be distinctly correlated with altitude. The entire area of the Indian cold arid region is approximately 74,809 sq. km of which a major part is confined to Ladakh in the state of Jammu and Kashmir (68,321sq.km) followed by Lahaul and Spiti in Himachal Pradesh (6,488sq.km) and only a very minimal part being found in Uttaranchal.

The Indian cold arid regions are mainly characterized by barren mountains and uneven terrain with sandy soils. The climate is extremely harsh with sub zero temperature prevailing during the winter months and the region remains frozen from December to February. The summer days are extremely hot where temperature reaches up to 40°C but falls during night time. Since the altitude is very high and the air is thin consequently larger ultra violet radiation are received.

The atmospheric humidity in the region is low throughout the year (45%) and annual precipitation is very low (80-300 mm) primarily in the form of snow during winter. The soils in this region are coarse and sandy with poor organic matter and thus, the vegetation in this region is sparse. The characteristics of cold arid region are given below:

Climatic Features of Cold Arid Zones

- Extreme temperatures (-40°C in winter and +40°C in summer)
- Very low annual precipitation (80-300 mm) mostly in the form of snow
- Low relative humidity (20-40 %)
- High wind velocity (8-10 km/hr)
- Low atmospheric pressure (493 mm Hg)
- Very low partial pressure of oxygen
- Thin atmosphere results in heavy influx of ultra violet and infra red radiations
- Immature, coarse textured and highly permeable soil
- Sparse plant density

- Short growing period (120-150 days)
- Long frozen winters (5-7 months)
- Maximum sunny days (300-310 days per annum)

Soil

Soils of cold arid zones are immature, highly porous with a high proportion of sand. High permeability, poor nutrient status, poor biological properties and high salt concentration, wind/water erosion in certain patches are some basic soil properties. The pH was found to range from 5.94 -10.12, electrical conductivity was found to vary from 0.05 -1.55 mmhos /sec and organic carbon ranged from 0.07-3.58 per cent. A study conducted by DIHAR (DRDO), Leh, reports that available nitrogen is low in category because of the slow mineralization process. Available phosphorus is generally low in the soils of Ladakh (2.35 - 137.4 Kg/ha), available potassium was high in the soils of Leh and Kargil districts of Ladakh (11.2-496.15 kg/ha), while the extractible macronutrients like Zinc, Copper, Iron and Manganese are generally deficient in this part of the cold arid region (Table 1 & 2). Soil depth is another problem especially for the cultivation of fruit crops. Presence of stone/rocks under the soil surface poses serious problem in the plantation of fruit and forest plants.

Fruit Production : The Present Scenario

Fruit industry in the cold arid region is very limited and still in their infancy. Although abiotic stresses like poor soil conditions, low atmospheric humidity, poor irrigation facilities, short growing period, low temperatures and isolation from the mainland for about period of 7 months in the year are major limitations for development of Horticulture in the region, at the same time long day length with high light intensity, high day temperature followed by low night temperature and low atmospheric humidity make conditions congenial for quality production of certain fruits. The major fruit crops grown in the region are apricot (*Prunus armeniaca*), apple (*Malus domestica*), walnut (*Juglans regia*), Pear *(Pyrus communis*), peach (*Pyrus persica*), plum (*Prunus domestica*), cherry (*Prunus avium*), mulberry (*Morus alba*), oleaster (*Elaeagnous hortensis),* grape (*Vitis vinifera*) and seabuckthorn (*Hippophae rhamnoides*). Recently strawberry (*Fragaria sp.*) has been successfully introduced.

Apricot is the major fruit crop grown in the region on commercial scale. However, apricot and apple is scattered plantation found along

Table 1: Comparative Study of Physico-Chemical Properties of Soils of Leh and Kargil Districts

Soil Properties	Leh district #						Kargil district ##					
	Per cent sample under classes of nutrients			Range	Nutrient Index	Nutrient Index class	Per cent sample under classes of nutrients			Range	Nutrient Index	Nutrient Index class
	A	B	C				A	B	C			
pH (1:2.5)*	15.6	76.04	8.3	5.94-10.12	-	-	22.0	72.0	6.0	6.57-9.47	-	-
Electrical conductivity (dsm^{-1}) **	87.5	7.29	5.2	0.05-1.41	-	-	72.0	20.0	8.0	0.08-1.55	-	-
Organic carbon (%) ***	50.0	34.38	5.63	0.07-1.41	1.66	Medium	54.0	14.0	32.0	0.14-3.58	1.78	Medium
Available P (kg/ha) ***	59.38	39.58	1.04	2.35-25.66	1.42	Low	48.0	30.0	22.0	2.5-137.4	1.74	High
Available K (kg/ha) ***	17.70	37.50	4.79	11.20-496.15	2.27	Medium	6.0	38.0	56.0	103.00-861.3	2.50	Medium

Total villages/samples analysed were 96.0 ## Total villages/samples analysed were 50.0

* A B C of pH (A= < 7.5, B= 7.5-8.5 and C= > 8.5) ** A B C of EC (A= < 0.5, B= 0.5-1.0 and C= >1.0)

*** Available Organic C, P and K (A= low, B= medium and C= hi the agricultural fields. Other fruit crops are very few in number and found in limited patches, especially in the altitude below 3,000 m above msl. Besides its use for table purposes, apricot is also consumed after drying with or without stone. The sweet kernel is used as dry fruit and bitter kernel varieties are used for extraction of oil. The cold arid regions of India have excellent genetic diversity in apricot.

Table 2: General Distribution of DTPA Extractable Micronutrient under soils of Leh and Kargil Districts

Micronutrients	Leh district			Kargil district		
	Percent Total samples analysed	samples deficient	Range	Percent Total samples analysed	samples deficient	Range
Iron (Fe)	96	82.89	054-33.79 (4.214)**	50	66	0.20-38.01 (7.668)
Copper (Cu)*	96	-	0.43-3.52 (0.843)	50	-	0.50-3.41 (0857)
Zinc (Zn)	96	63.54	0.06-16.3 (1.026)	50	48	0.14-5.17 (1.199)
Manganese (Mn)	96	79.16	0.20-32.17 (3.994)	50	94	0.34-4.74 (1.010)

* All samples under different village of both the district are sufficient in DTPA extractable Copper

** Figures in parentheses indicate mean values

Apple is the second major crop of the region only after apricot. It has been found growing 3,600 m above msl where even apricot failed to grow successfully. Study conducted by DIHAR (DRDO), Leh has also highlighted the presence of valuable indigenous apple genotypes in the region. This institute has identified Tha, Mangol, Karkichoo, Shing, Mar, Nas, Yangma, Squirmo, Khara, Shamer Kushu (apple) genotypes which have not been reported so far from Leh and Kargil districts of Ladakh. However, detailed studies on conservation and exploitation need to be initiated. The local varieties are generally small in size, highly juicy but have poor shelf life. However, their abiotic stress tolerance can be exploited.

Strawberry varieties are mostly short day in nature and thus, cannot be grown successfully under the long day conditions of the cold arid region. Introduction/development of long day or day neutral varieties has opened up avenues for successful cultivation. Varieties recommended for Ladakh conditions are as under:

(i) *Long day type :* Addie, Confetura, Hera

(ii) *Day neutral type:* Brighton, Fern, Douglas, Selva, Tieoga,

Strawberry can be grown mainly in the open and also in greenhouse or in pots for domestic consumption. Flowering generally occurs from May to June under open conditions while in the greenhouse it starts by end of March to April. Fruits are harvested 5-6 weeks after flowering and generally 3-4 pickings are available at weekly intervals.

Seabuckthorn : An Emerging Fruit

Seabuckthorn, a naturally growing thorny, deciduous bush, indigenous to the region, an unexploited fruit crop of the region, is fast emerging as an important crop due to its nutritional and medicinal properties. Seabuckthorn, (*Hippophae rhamnoides*) belonging to the family Elaeagnaceae, is a dwarf (1-5 m tall), much branched, thorny, dioecious shrub which grows naturally in approximately 30,000 ha area in Ladakh. It grows in a wide range of soil conditions from rocky, sandy and marshy to saline soils. The plant is able to tolerate moisture as well as temperature stress (-30^0 C). Flowering generally occurs in early June and ripening sets in by mid August. The fruits are small, round to oval in shape with a diameter of 6-8 mm, yellow to scarlet in colour and sour to highly acidic in taste. Fruits of seabuckthorn are a rich source of vitamins having vitamin C (800-1600 mg/g), vitamin E (162-255 mg/100g), vitamin A (11 mg/100g), vitamin B_1 (0.04 mg), vitamin B_2 (0.56 mg/100g) and organic acid (2-4 %). They also contain lysine and microelements such as iron, cobalt, selenium and molybdenum. Presence of antioxidants and flavonoids make it ideal for health. However, because of the perishable nature and highly acidic taste of fruits they can not be consumed fresh.

Varieties / Genetic Diversity

The indigenous fruit wealth of the Ladakh posses wide variability in tree and fruits parameters. Out of various germplasm collected, valuable morphotypes have been identified. The most important among all are apricot (84), apple (9), seabuckthorn (19) (Table 3, & 4) mulberry (5), and oleaster (3), on the basis of physico-chemical properties of fruits. The need of the hour is to develop modern methods of propagation of these crops/ varieties for their quick multiplication and exploit the genetic wealth for its inherent potential under fruit breeding programmes. The commercially growing promising apricot and apple cultivars and introduced varieties of strawberry have also been evaluated and documented (Table 5, 6 & 7). There are limited numbers of cultivars of cold arid fruit crops which can meet the requirement of fruit industry of the region.

Table 3: Physico-Chemical Characteristics of 84 Genotypes of Apricot from Cold arid Region of Ladakh

Parameter	Range		Mean
	Min	Max	
Moisture (%)	70.4	82.5	75.0
Fruit weight (g)	8.7	35.5	19.04
Fruit volume (ml)	9.3	34.6	20.4
Specific gravity	1.02	1.14	1.10
Fruit length (mm)	26.5	44.0	34.00
Fruit width (mm)	24.9	41.8	33.20
Cheek diam (mm)	22.9	38.8	31.00
TSS (°B)	12.0	33.4	19.35
Pulp weight (g)	2.5	31.2	16.97
Stone weight (g)	1.0	20.0	2.37
Pulp:stone ratio	1.5	17.86	9.42
Kernel oil (%)	27.0	45.0	35.0
Av. Kernel weight (g)	0.3	0.5	-

Table 4: Physico-Chemical Parameters of Fruits Morphotypes of Seabuckthorn (*Hippophae rhamnoides* ssps *turkestanica*)

Parameters	Range	Mean Value	SEm
Fruit Length(mm)	5.42- 9.27	7.34	0.0889
Fruit Width (mm)	4.67 - 7.16	5.91	0.0736
Weight of 100 fruits (g)	7.61 - 20.54	14.07	10.69
Volume of 100 fruits (ml)	7.9 -24.1	16.0	0.314
TSS (°B)	10.2 -18.4	14.3	0.189
pH	1.50 -2.3	2.15	0.016
Acidity (%)	1.46 -3.621	2.54	0.103
Sugar (%)	3.65- 5.23	4.44	0.097
Vitamin 'C'(mg/100g)	138 -1400	769	26.36
Juice (%)	68.18 – 80.85	74.5	0.655
Seed (%)	4.44 -8.65	6.54	0.377
Pomace (%)	11.4-27.5	19.45	0.585
Seed weight(g/100 seeds)	1.13-1.35	1.24	0.021

Table 5: Physico-Chemical Characteristics of the some Common Apricot Cultivars of Ladakh

Cultivars	Length (cm)	Breadth (cm)	Chick diameter (cm)	Fruit weight (g)	Pulp weight (g)	Stone weight (g)	Pulp - stone ratio	TSS (°B)	Titrable acidity (%)
Afghani	2.63	2.64	2.63	13.2	12.0	1.2	10.0	19.6	-
Australian	4.15	3.86	3.67	33.2	30.8	2.0	15.4	15.0	0.92
Charmagz	3.44	3.76	3.42	24.8	23.0	1.8	12.8	19.0	-
Halman	2.43	2.43	2.39	11.6	10.2	1.2	8.5	18.4	0.67
Raktsey karpo	3.85	3.82	3.63	34.6	32.0	2.4	13.3	21.2	0.61
Safeda	3.35	3.17	3.42	19.5	18.1	1.4	12.9	18.0	0.64
Shakanda	3.05	2.89	2.78	11.5	10.0	1.5	6.6	19.3	0.95
Shakarpara	3.73	3.58	3.64	28.0	25.3	2.7	9.3	12.0	0.13
Tokpopa	3.30	3.34	3.39	14.5	12.8	1.7	7.5	12.2	1.00
CD (0.05)	1.51	1.41	1.15	-	-	-	-	3.93	0.09

Table 6: Physico–Chemical Properties of Major Apple Varieties of Ladakh

Genotypes	Fruit length (cm)	Fruit diameter (cm)	Fruit weight (cm)	Fruit volume (cm)	T.S.S. (°B)	Acidity (%)	Vit-C (mg/100g)	Reducing Sugars (%)	Non Reducing sugars (%)	Total Sugars (%)
Mongol	5.37	5.93	89.42	77.00	3.46	0.16	50	2.28	1.66	3.94
Tha	4.29	4.95	48.06	54.17	6.66	0.28	50	3.53	2.05	5.24
Bong Kushu	4.63	5.42	60.47	75.00	5.20	0.24	50	3.62	3.29	6.91
Shing Kushu	3.79	4.55	36.25	37.74	5.50	0.32	50	4.04	2.58	6.62
Mar Kushu	4.17	4.84	43.88	53.33	5.33	0.16	50	1.88	0.23	2.11
Khara Kushu	4.40	5.70	62.50	54.00	5.88	0.46	50	1.24	0.63	1.87
Squirmo	4.07	5.60	66.25	56.70	6.10-	0.46	50	6.53	2.36	8.89
Karkichoo	4.44	5.09	51.20	49.76	7.30	0.16	50	2.99	2.66	5.66
Phamer Kushu	4.34	4.70	40.67	41.67	12.21	0.46	50	3.60	2.89	6.49
Sanker Kushu	4.14	5.04	44.33	46.67	16.09	0.41	50	4.33	2.68	7.01

Table 7: Physico-Chemical Properties of Fruits of Important Strawberry Cultivars grown in Cold arid Conditions of Ladakh.

Cultivars	Fruit length (cm)	Fruit diam (cm)	Av fruit weight (g)	Fruit volume (ml)	T.S.S. (°B)	Titrable acidity (%)	No. of fruits/ plant	Yield/ plant (g)	Estimated fruit Yield (t/ha)
Addie	3.32	2.98	9.09	11.03	16.76	0.69	11.6	100.54	7.45
Brighton	2.76	2.33	6.37	8.81	10.03	0.56	15.3	97.46	7.22
Chandler	3.22	2.48	5.88	7.98	17.12	0.73	18.4	108.19	8.01
Confitura	2.79	2.32	5.76	7.65	10.28	0.44	20.3	116.93	8.66
Douglas	2.80	2.57	6.22	8.03	9.91	0.56	13.7	85.21	6.31
Gorella	2.87	2.26	7.47	8.58	8.98	0.56	17.9	133.71	9.90
Fern	3.52	2.86	9.09	11.50	18.68	0.49	13.1	119.08	8.82
Pajaro	3.26	2.84	7.14	9.15	17.32	0.61	12.8	91.39	6.77
Selva	3.16	2.84	7.69	8.95	18.68	0.60	13.9	106.79	7.91
CD at 0.05 P	0.38	0.15	1.07	1.30	2.41	0.09	3.54	10.67	0.76

Introduction / Development of New Fruits / Varieties

There is tremendous potential for introduction of new fruit crops like Kaphal (*Marica nagi*), pomegranate (*Punica granatum*), cape gooseberry (*Physalis peruvianum*), quince (*Cydonia spp*),raspberry (*Rubus spp*) blue berry (*Vaccinium spp*), chilgoza (*Pinus jeradiana*) in the region. The initial studies conducted by DIHAR have shown some rays of hope for introduction of such crops (Table 8). However, there is need to evaluate these crops on long term basis under different agro-climate zones of the region. The major points to be kept in mind while introducing new fruit crops/varieties include, the high chilling requirement, suited for long day conditions, short fruit growing period (from flowering to ripening), long dormancy period and inherent ability to tolerate abiotic stresses of the region. Fruit breeding programme for the development of ideal variety may also be initiated, keeping above requirements in mind.

Propagation and Nursery Development

Non-availability of genuine planting material is a major problem since there is no operational nursery which can meet the demands of local orchardists. To meet this demand the State Horticulture department, imports about 30-40 thousand apple plants from Shimla and Srinagar which are airlifted to Ladakh, since transport by road is restricted in winter months. This enterprise has not proved very successful since survivability of the plants is very poor, perhaps because of fluctuations in temperature during the transit. One major problem is that in the hills of Uttaranchal, Himachal etc. generally, temperate fruit trees are planted before spring i.e. February, for which these are uprooted in January. However, the optimum time for planting in Ladakh is mid March-April when the ambient temperatures become conducive for growth. In the interim period the plants are either in transit or they are stored in cellar houses. Thus, the time gap between procurement and planting is a major problem which affects survival of these trees.

Seabuckthorn growing wild under natural conditions has been planted by local farmers or by Forest department for protection of flood/ soil erosion on river banks, sand dunes fixation or for fencing the new plantation or agricultural fields where fruit yield and its quality have no importance. However, for commercial cultivation, it is imperative that the crop can be multiplied from the best material which has been selected for its fruit yield and other desirable parameters.

Table 8: Performance of some New Fruit Crops at DIHAR (DRDO) Leh During the Year 1998-2006

Name of the Crop	Botanical Name	Variety	Observations/Remarks
Grape	*Vitis vinifera*	Local Ladakhi	Fruiting, short growing period
Pomegranate	*Punica granatum*	Kandhari, Kabuli	Vegetative growth/flowering, Cold tolerance var needed
Capegoose Berry	*Physalis peruviana*	Punjab Selection	Short day plant, Late fruiting
Raspberry	*Vaccinium spp*	Kumaoni Local	Grows in Green House, Fruiting awaited
Mehal	*Pyrus pashia*	Kumaoni Local	Growing successfully, Fruiting awaited
Kaphal	*Myrica nagi*	Kumaoni Local	Vegetative growth, Cold hardy variety needed
Kiwi fruit	*Actinidia chinensis*	Monti,Alison,Abbot Heyward,	Failed to grow

Standard Propagation Techniques

Traditionally seedling plants of apricot, apple, walnut, etc are found in Ladakh. No vegetative propagation was practiced by the locals which often resulted in long gestation period and wide variation in fruit quality. It was felt that in order to overcome this problem vegetative propagation techniques suited to the cold arid region needed to be standardized. Thus, vegetative propagation techniques in apricot were standardised, which is the most important fruit crop of the region, using seedling rootstock. It was concluded that different methods of vegetative propagation viz. chip budding, t budding and cleft grafting are successful at different periods. However, chip budding was found to be the best method. Old seedling apricot plants were converted to productive plants by top working using the shield method of budding which was performed during mid June-early July. This has shown 90.0 per cent success (Table 9). This technique can also be used for germplasm conservation by adopting the orchard on a tree concept.

Seabuckthorn can be propagated successfully by seed which are viable for a period of two years. May is the best time for sowing of seeds in nursery beds. Before sowing the mature seeds should be stratified for about 25 days by keeping in a box in alternate layers of moist sand. These boxes are kept under cool and shady condition for about 25 days. This method gives over 85 per cent germination. Seabuckthorn has profuse suckering habit and has been reported to yield 13-40 suckers per plant. These suckers are true to the type which can be separated along with the roots from the mother plants and planted directly at the planting site. The best time for separation of suckers and planting is in March before bud break. However, the major disadvantage is difficulty in operation due to thorny plant structure.

Seabuckthorn can be propagated successfully by stem cuttings and this method has been recommended for commercial propagation. One to two year old stem cuttings having pencil thickness are ideal for its propagation. The stem cuttings are taken from identified male and female plant to get elite planting material. Seabuckthorn cuttings are not easy to root on their own, thus indole butyric acid (IBA) is applied @ 500 ppm before planting it in the nursery beds. The best time for planting of cuttings has been found to be the month of April. About 85 percent success has been recorded in rooting of stem cuttings under Ladakh conditions. These cuttings are ready for plantation in the main field in March of the following year. DIHAR has successfully standardized asexual method of propagation of seabuckthorn through

Table 9: Performance of Different Methods of Vegetative Propagation in Apricot in Cold arid Region of Ladakh

Method/ Techniques	Time	Sprouting (%)	Bud take (%)	Days taken in sprouting	Saleable plant (%)	Linear growth (cm)	Radial growth (cm)
Grafting - tongue/cleft	First week of February	95.0	90.0	73.5	82.5	164.0	0.55
Budding - chip	Fourth week of February	90.0	85.0	71.4	77.5	28.4	0.58
Budding - shield	Second fortnight of August	88.3	81.6	29.5	76.5	17.0	0.45
Top working - shield	Second fortnight of July	93.3	80.0	31.5	-	76.5	0.80

soft wood cuttings using 200-300 ppm NAA. The technique has yielded over 90 percent rootability in cuttings with excellent root quality under mist condition having course, sand rooting media. The plantlets get ready for transplantation in poly bags within 90 -100 days after planting of cuttings.

Orchard Management

Local farmers do not follow orchard management practices viz. manuring, fertilisation and irrigation etc. Since the fruit trees are generally planted around the agriculture fields and receive nutrients and water from field crops. It is essential to highlight that no scientific study has been undertaken to work out the nutrient and water requirement of fruit crops for the cold arid region. The practice of pruning and training of fruit crops is also uncommon. Trees are allowed to grow freely which is why substantial fruit damage occurs during harvesting. Effort should be made to evaluate modern training systems for adopting the high density planting system. The training method which ensures maximum exposure of fruits to sunlight and enables the branches to bear maximum crop load should be preferred, since there are no hail storms in the region.

Seabuckthorn should be pruned regularly for ease in harvesting and exposure of plant to sun for production of quality fruits. The best time for pruning of seabuckthorn in Ladakh is the month of March, before sprouting of the plant in spring. Intensity of pruning depends upon the age and growth of plant and soil conditions. However, 50 per cent growth of 3-4 year old shoots should be removed. Seabuckthorn, being a thorny bush, should be restricted to a height of less than 2 meter for ease in harvesting. Normally properly managed seabuckthorn plants develop a crown of 2-3 m diameter at the age of four years after planting.

Insect / Pest / Diseases / Disorders

Incidence of pests and diseases in the cold arid region is low due to low temperature and relative humidity. This makes the region ideal for producing organic fruits which has arising market the world over.

Codling moth (*Cydia pomonella*) is a major threat to the fruit industry of Ladakh. Apple is the primary host for the insect which destroys the fruit by feeding on the fruit pulp. The pest has shifted even to apricot, pear and walnut. In India the problem of codling moth is restricted to the Ladakh region, while rest of the apple growing belts like Kashmir, Himachal Pradesh, Uttaranchal, etc, have not reported significant incidence of the pest. It is for this reason that the government has imposed restrictions under SRO 397 dated 8th September 1981, issued under the Jammu and Kashmir plant and disease act, 1973, on the movement of apple and apricot from Ladakh. Other pests reported in fruit crops are aphids, leaf minor, pin hole borer, white grubs, blossom thrips, two spotted spider mite, gypsy moth, leaf curling aphid etc. However, their incidences are almost negligible and are restricted to certain areas in the region. Diseases and disorders like blight, gummosis and sunscald etc., in apricot have been observed which need to be studied in detail. Bird damage of ripe fruits of apricot is one of the major threats, which needs to be managed properly.

Post Harvest Management

Development of techniques for post harvest handling of fruit crops is the immediate requirement for the cold arid region. Lack of knowledge regarding maturity indices, harvesting, grading, packaging methods and poor storage facilities result in 40-60 per cent loss in apricot alone. Apricot is generally preserved by open sun drying which affects the fruit quality. Solar driers viz. Polyhouse, tent and tunnel driers have been developed for hygienic dehydration of surplus fruits and vegetables in the region. Methods for value addition viz. osmotic dehydration, intermediate moisture, apricot bar, hurdle technology, squash, nectar, jam, etc., have also been developed. Another potential area is extraction and processing of apricot kernel oil which is of medicinal and cosmetic use and is found in abundance in the region.

Studies on value addition and processing of seabuckthorn have also been initiated. A multivitamin herbal beverage as squash and nectar, without using any artificial colour, flavour and stabilizers, etc., has been developed in which vitamins and nutrients have been conserved in their natural form. This technology has been patented and subsequently transferred for manufacture on commercial scale. The beverage has also been inducted in the Indian army as special ration for Siachen troops since it has special ability to help in acclimatisation in the high altitudes. The process for preparation of seabuckthorn jam, sauce, pickle and *chyavanprash* has also been standardised. Studies have been undertaken to develop procedures for extraction and evaluation of seabuckthorn oil having immense medicinal value for its use in the cosmetic industry.

Green Houses in Fruit Production

Ladakh remains cut off from rest of the nation from November to June every year due to heavy snowfall which causes blockage of roads. Subzero temperature prevails from December to February throughout Ladakh whereas, zero degree is experienced during the rest of the winter months. Various greenhouse structures viz glasshouse, polyhouse, Ladakhi polyhouse, trench and tunnel for cultivation were studied. Of these, the trench has been found to be the most suitable and economic structure for the region followed by Ladakhi polyhouse. This is because it retains the increased day temperatures even during night. The trench, as proposed by DIHAR, is an underground structure of convenient variable dimensions covered with UV stabilized transparent polyethylene (400 gauges) sheet. At night another covering of black polyethylene is given to harness the soil temperature which increases during the day and to minimize the heat loss during the night. Thus, greenhouses suited to local needs were developed and are called 'Ladakhi polyhouses'. Three sides of this structure are made of unbaked bricks and the fourth side, which harvests sunlight, is made of polythene. This has been found to be very economical and functions efficiently even in the winters.

However, another structure especially suited for the region is the underground trench. This is an underground structure 30′ x 12′ and 3′ deep covered on the top with transparent polythene with the support of wooden poles. The polythene is removed on bright sunny days and the trench is covered when there is snowfall. This has been found to be the most effective structure for propagation of plants through runners (strawberry), cuttings (seabuckthorn, grape, mulberry, oleaster), grafting/budding (apricot, apple) and the production of crops like strawberry, grapes etc., since it regulates the microclimate inside the trench most effectively as compared to any other protected structure.These structures can be effectively used for winter cultivation, raising early quality planting material, conservation of valuable germplasm and extending the period of growth of crops.

Mulching

Water scarcity is a major limitation in the area and use of mulches has been found to enhance water retention by the soils by decreasing loss of soil moisture through evapo-transpiration under high light intensity and low environmental humidity. Besides mulching also regulates the soil temperature which subsequently improves plant growth. It decreases weed population in the vicinity of the main crop thereby decreasing competition for nutrients and water and thus, productivity is enhanced. It has been observed that use of organic as well as synthetic mulches viz. straw, hay, dried leaves and paper, pebbles, polythene, asphalt sheets has immense advantages. Organic mulches improve soil organic carbon levels and also enhance water holding capacity of the soil.

Black polythene of 150 gauge thickness has been found to enhance yield and also reduces frequency of irrigation by 50 per cent and completely negates the need for weeding. Among the fruit crops mulching has obvious benefits. It enhances establishment of plants, accelerates plant growth and reduces the need for frequent irrigation in crops like apricot, seabuckthorn and apple. In strawberry, mulching has been found effective to improve the runner production and fruit yield and its quality. Use of dry grasses and hay has also been found to increase fruit yield and summer production in strawberry.

Organic Farming

The cold arid region in India is alienated from the rest of the country because of its geographical topography. One major advantage of this is that the farming society here remained untouched by the use of inorganic fertilizers, pesticides, fungicides etc. This part of the world still follows traditional methods of organic farming and manures such as night soil, farm yard manure, etc., are used for enriching soil. Legumes are included in crop rotation. Another feature is that the land remains fallow for 5-6 months in the year during winter when it is able to rejuvenate itself. Low atmospheric humidity and temperatures are effective in preventing disease and insect pest infestation in most of the crops. This ensures minimum use of pesticides in the region. The natural advantage of the region can be economically exploited by cultivation of organically grown fruits on a commercial scale and their export to foreign and domestic market.

Problems of Fruit Industry in Ladakh

The climatic conditions prevailing in the region, inaccessibility and lack of technical know-how are the major bottlenecks in the development of the fruit industry. Despite the maiden efforts being made the existing problems are listed below:

- Non availability of quality planting materials through local nurseries

- Lack of location specific scientific cultivation/orchard management techniques
- Improved cultivars of fruit crops for various purposes / locations
- Crisis of irrigation water and modern scientific irrigation methods
- Lack of post harvest technology and processing methods of fruit crops
- Poor knowledge/infrastructure & linkage for packaging, grading, storage, transportation and marketing etc.
- Neglection of fruit crops by policy makers/local administration/ scientists /research institutions /government agencies working in the region.

Scope of Fruit Production in Ladakh

- Development of organic farming of fruits.
- Production of export quality fruits meeting International standards
- Off season production of crops like strawberry under natural conditions
- Exploiting fruit crop (apricot, seabuckthorn) for oil production
- Conservation and utilization of highly valuable germplasm of fruits,
- Identification of gene responsible for plant growth under extremely stress conditions due to low temperature, moisture stress, high altitude.
- Production of organic fruit pulp/value added products for domestic/ global markets.

References

Anonymous. 2002. Annual Report 2001. Field Research Laboratory (DRDO). C/o 56APO.

Dwivedi, Deepa H., Kareem A., Dwivedi, S.K. and Mir, A.A. 2002. Physico-Chemical characteristics of important apricot (*Prunus armeniaca* l.) cultivars of the cold arid region of Ladakh. Ind. J. Hort. 59(2), 118-121.

Dwivedi, S.K. and Singh, B. 1999. Topworking of apricot in Ladakh. Prog. Hort. 31 (1-2), 29-31.

Dwivedi, S.K., Singh, B. and Paljor, Eli. 2000b. Studies on vegetative propagation of apricot through grafting in Ladakh. Indian J. of Hort. 57(1), 39-41.

Dwivedi, S.K., Attrey, D.P. and Paljor, Eli. 2000c. Standardisation of budding in apricot in cold arid conditions of Ladakh. Prog. Hort. 32, 117-120.

Dwivedi, S.K. Singh, B. and Paljor, Eli. 2000d. Effect of mulch on establisment and growth of apples saplings in cold arid conditions of Ladakh. Prog. Hort. 32, 42-45.

Dwivedi, S.K. and Singh, B. 2001a. Studies on chip budding in apricot (*Prunus armeniaca*)arid conditions of Ladakh. Indian J. Hort. 58(3), 215-217.

Dwivedi, S.K., Singh, B. and Attrey, D.P. (2001 b). Studies on distribution, propagation and ultilisation of seabuckthorn in Ladakh. Seabuckthorn-a resource for health and environment in 21st century. Eds. V. Singh and p. K. Khosala. Proceedings of international workshop on seabuckthorn, Feb. 18-21, 2001. CSKHPKVV-Palampur, India, pp. 20-24.

Dwivedi, S.K., Attrey, D.P. and Kareem, A. 2002a. Post harvest management of seabuckthorn (*Hippophae* spp.) In cold arid regions of Ladakh. In: national workshop on post-harvest management of horticultural produce organized by national productivity council, 7-8 Feb. 2002 at hotel shivalikview, Chandigarh, pp. 89-98.

Dwivedi, S.K., Attrey, D.P., Paljor, Eli. and Singh, B. 2002c. Studies on conservation of genetic diversity in apricot in cold arid region of india. In: 26th international horticultural congress, Toronto (Canada), 11-17 august, 2002. Abstract , pp. 524-25.

Dwivedi, S.K., Paljor, Eli., Attrey, D.P. and Singh, B. 2002d. Propagation of common seabuckthorn through hard wood cutting in ladakh. Indian J. of Agril. Sciences. 72 (4), 228-9.

Dwivedi, S.K., Singh, R. and Ahmed, Z. 2006. The Seabuckthorn. Field Research Laboratory, DRDO, Leh, Ladakh (J&K) India.

Dwivedi, S.K., Singh, R. and Ahmed, Z. 2007a. Apricot in Ladakh. Field Research Laboratory, DRDO, Leh, Ladakh (J&K), India.

Dwivedi, S.K., Hemraj and Ahmed, Z. 2007b. Standardisation of a fast method of vegetative propagation of Sea buckthorn through soft wood cutting in India. Paper presented in the 3 rd International Seabuckthorn Association conference held at Quebec, Canada, August, 12-16, 2007.Paper No 83.

Singh, B. 1995a. Vegetable production in Ladakh. Field Research Laboratory, C/o 56 APO.

Singh, B. 1995b. Green house development in Leh. Paper presented in national workshop on commercialization of tissue culture and greenhouse technology, Banglore, India. 5-6 may 1995.

Singh, B., Dwivedi, S.K., Singh, N. and Paljor, Eli. 1998a. Sustainable horticulture practices for cold arid areas. Paper presently in international symposium on sustainable agriculture in hill areas. 29-31 oct. 1998. HPKVV Palampur. Abstract, pp. 35-36.

Singh, B., Chaurasia, O.P., Dwivedi, S.K. and Paljor, Eli. 1998b. Effect of plastic mulch on vegetable production in Ladakh. In: national symposium on emerging scenario in vegetable R&D. PDVR, Varanasi (UP), 12-14 december 98. Abstract, pp 137.

Singh, B., Dwivedi, S.K. and Sharma, J.P. 2000a. Greenhouse technology for winter vegetable cultivation in cold arid zones. In: dynamics of cold arid agriculture. Eds. J.P. Sharma and A.A. Mir, Kalyani publishers-Ludhiana, pp. 279-293.

Singh, B., Dwivedi, S.K. and Paljor, Eli. 2000b. Studies on suitability of various structures for winter vegetable production at sub zero temperatures. Acta Hort. 517, 309-314.

□□□

6

Research on Seabuckthorn (*Hippophae rhamnoides* L.) in India

TSERING STOBDAN[1], GIRISH KOREKAR[1], SANJAI K DWIVEDI[2], ASHISH YADAV[1], KONCHOK TARGAIS[1] AND SHASHI BALA SINGH[1]
[1]Defence Institute of High Altitude Research, DRDO, Leh-Ladakh-194101, India
[2]PEACE, DRDO, Metcalfe House, Delhi-110054, India

INTRODUCTION

Seabuckthorn (*Hippophae rhamnoides* L.) is a unique and valuable crop for the cold desert of India. It has silvery deciduous leaves and colourful red, orange or yellow berries that remain on the shrub most of the winter. The plant is hardy and usually cold tolerant. It can withstand extreme temperatures from -43°C to +40°C and considered to be drought resistant. The shrub develops extensive root system having ability to fix atmospheric nitrogen. It is therefore an ideal plant for soil erosion control, land reclamation, wildlife habitat enhancement, and farm stand protection. Traditionally every part of the plant is being used for a variety of purpose in cold desert of Ladakh. Seabuckthorn berries are among the most nutritious of all the fruits and have immense medicinal properties. The shrub serves as a storehouse for researchers in the field of biotechnology, neutraceutical, pharmaceutical, cosmetic and environmental sciences. In India, seabuckthorn has become popular after Defence Institute of High Altitude Research (DIHAR), DRDO has developed and commercialised technology for preparing beverage from its highly acidic fruit. In recent years a number of research institutes in India are working on geographical distribution, genetic diversity, conservation, product development, neutraceutical and medicinal properties of the lesser know shrub of the Trans-Himalayas. As a result of R&D efforts, few products are available in Indian market and several others are at different stages of commercialization. Importance of seabuckthorn for its medicinal properties has been recognised by Defence Research & Development Organisation (DRDO), which has carried several R&D programme. Extensive studies have been carried on antioxidant, immunomodulatory, anti-inflammatory and radioprotective properties of seabuckthorn. EST database of seabuckthorn leaves has also been initiated in recent years. Department of Biotechnology (DBT), Govt. of India and Indian Council of Agriculture Research (ICAR) have initiated mega research programme on seabuckthorn involving several R&D institutes across the country. Research programme in India has already started benefiting the people living in Trans-Himalayas. Several processing units have been set up generating rural employment. The outcome of the ongoing research programme will further add to its commercial and medicinal values.

Geographical Distribution

Hippophae rhamnoides, *H. salicifolia* and *H. tibetana* are the predominant seabuckthorn species in India. Of these, *H. rhamnoides* is widely distributed in the country. Seabuckthorn (*H. rhamnoides*) is naturally distributed in six valleys namely Leh, Nubra, Changthang, Suru, Lahaul and Spiti in Trans-Himalayan cold deserts of India. It is also found in few pockets of Nathula in Sikkim. Remote sensing and GIS have been used to map seabuckthorn in Lahaul and Spiti in Himachal Pradesh and Ladakh region in Jammu and Kashmir. The high altitude cloudless sky, arid and sparse vegetation cover provides an ideal condition for mapping in the hostile and inaccessible terrains. Studies on satellite images by DIHAR and Defence Electronic Application Laboratory, DRDO, have shown that 11,500 ha of land are under pure natural seabuckthorn plantation in Ladakh. The shrub is mainly distributed in Nubra, Indus and Suru valleys. It is also found in Changthang valley at an altitude of 4,690 m above MSL. Natural plantation is mainly distributed along Nubra, Shyok and Indus River. Plantation is done around agricultural fields to serve as wind break and biofence.

Genetic Diversity

Seabuckthorn is a cross-pollinated crop and hence there exists high level of genetic diversity. Studies conducted by DIHAR (DRDO) on 17 morphotypes of seabuckthorn collected from Nubra valley revealed large genetic diversity. RAPD markers show five major groups in the natural population. None of the morphotypes shares a similarity more than 25 percent indicating high level of genetic diversity between the morphotypes. Extensive survey in Ladakh region has shown presence of more than 200 genotypes in the natural population with wide variation in physico-chemical properties. It has been recorded that the shrub height varied from 1.8-7.2 m. Inverse relation has been found between the shrub height and the altitude of the region in which the shrub grows. Morphological studies in natural population of seabuckthorn in Lahaul, Spiti and Kinnaur have also shown a similar trend. Plant height is greatest in Kinnaur and lowest in Spiti. Based on physico-chemical properties of the berries and the leaves, 19 morphotypes with distinct variation have being maintained at DIHAR, Leh. A set of morphological descriptors for the genus *Hippophae* has been developed for germplasm obtained from various parts of India. The pattern of variability in morphological characteristics associated with habitat, plant growth habit, stem, leaf, fruit, flower, seed and biochemical parameters are taken into consideration. The suitability of these parameters in characterization of *Hippophae* germplasm has been successfully demonstrated. Using the descriptors, it is found that *H. rhamnoides* and *H. salicifolia* are genetically closer than with *H. tibetana*.

Products

Seabuckthorn is rich in nutritional and bioactive substances. Berries and leaves have attracted interest of researchers to develop products for preventive and curative purposes. A wide array of products is possible from seabuckthorn for neutraceutical, cosmetic and medicinal purpose. DIHAR, Leh has developed a process for preparation of beverage from the highly acidic berries. The patented process has been transferred to several commercial firms, local entrepreneur and NGO. The beverage is commercially available in Indian market under the brand name 'Leh Berry', 'Ladakh Berry', 'Power Berry' 'Sindhu Berry' etc. Squash prepared from the berries does not freeze even at -22°C in Ladakh condition. It does not contain any added colouring and flavouring agent. The beverage has found wide acceptance among consumer in terms of colour, taste, flavour and overall quality. The institute has also developed and patented seabuckthorn jam, which is rich in vitamins and minerals. Semi-ripe berries have also been used for preparing pickle. Dried seabuckthorn pulp is used for preparing seabuckthorn powder. DIHAR has formulated and patented herbal tea with seabuckthorn leaves as the major constituents. Tender and healthy leaves from male plant is mixed with other high altitude herbs like salam panja (*Dactylorhiza hatageria*), local tea (*Bidens pilosa*), local caraway (*Carum carvi*), black caraway (*Bunium persicum*), oregano (*Origanum vulgare*), local mint (*Mentha longifolia*), yarrow (*Achillea* sp.) and rose root (*Rhodiola* sp) etc. The tea is rich in flavonoides, vitamins and has medicinal values. The technology has been transferred to commercial firms and the product is available in market under trade name 'Si-beri'.

Defence Food Research Laboratory (DRDO), Mysore has developed biscuits containing seabuckthorn pulp and leaves. The biscuit is rich in dietary fibre, flavonoides and other phyto chemicals. Other bakery products like buns, bread and cakes have also been developed with antioxidant rich seabuckthorn leaves. The Laboratory has also developed seabuckthorn mixed fruit jelly by blending seabuckthorn with papaya, watermelon and grapes. The seabuckthorn-grape jelly exhibits good organoleptic characteristics with high sensory score. Apart from these products other products like skin cream, animal feed, nutritional supplements, soap, yogurt, wine, colourant etc. have been developed in India. Popularity of seabuckthorn for product development can be judged from the fact that pice of berries has raised from Rs 8 per kg in 2001 to Rs 22 per kg in 2007 in Ladakh.

Propagation

Propagation of seabuckthorn by various methods has been studied. The seeds remain viable for two years. Freshly harvested seeds have a short physiological dormancy and thus do not germinate immediately after harvesting. Seed stratification for 20-25 days in alternate layers of moist sand yield over 85 percent germination. In Ladakh condition, direct sowing of seeds in November also results in high germination percentage. Cold water soaking of seeds for 6 days yields 93.33 percent germination and 73.33 percent survival rate. Hard wood cutting is a method of choice for large scale propagation. One or two year old pencil thickness cuttings dipped 2/3 to 1/2 of lower portion in water for 2-3 days followed by 500 ppm IBA treatment results in 85 percent success in Ladakh condition. Planting of the cuttings in greenhouse / trench / polyhouse results in even higher success rate in cold desert condition. Micropropagation of seabuckthorn has been standardized in Murashige & Skoog media. For shoot proliferation 0.01 ppm IBA is found optimum while 2.0 ppm BA and 1.0 ppm NAA is optimum for rooting. Seabuckthorn has profuse suckering habit. A single plant produces 13-65 suckers which can be separated from the mother plant. The individual suckers have been used as true to type planting material.

Nutritional Composition

Seabuckthorn is an important source of bioactive phytochemicals such as carotenoids, tocopherols, vitamin C, organic acids and polyphenols. Physico-chemical characteristics of ripe berries and oil of Ladakh region, Uttarakhand and Himachal have been studied. The composition varies considerably because of origin/subspecies, climate and geographical condition of the growing areas, and agronomic practices. Seabuckthorn growing areas in Trans-Himalayas host unique geoclimatic condition with extreme temperature variation, low precipitation, high UV radiation, high wind velocity and highly permeable immature coarse textured soil. The plants in cold desert therefore have to experience conspicuously different environment as compare to their traditional niches. Seabuckthorn in India are therefore expected to have distinct phytochemical profile. A novel triglyceride, 1,3-Dicapryloyl-2-linoleoyl glycerol has been isolated and its structure elucidated. Nutritional composition of pulp and oil of *H. rhamnoides* from Ladakh, Uttrakhand and Himachal Pradesh are summarized in Table 1.

Medicinal Properties

Seabuckthorn has been studied extensively for its antioxidant properties. Defence Institute of Physiology and Allied Sciences (DIPAS) has done extensive studies on antioxidant properties in different model systems. Leaf extract of seabuckthorn has been evaluated on chromium induced oxidative stress. Male albino rats feed with potassium dichromate equivalent to a dose of 30 mg/kg body weight of chromium (VI) for 30 days results in decrease body weight and increase organ to body weight ratio. Leaf extract of seabuckthorn at a concentration of 100 and 250 mg/kg body weight protects the animals from the chromium induced oxidative injury. Similarly, cytoprotective and antioxidant properties of alcoholic leaf extract of seabuckthorn against

Table 1: Composition of *H. rhamnoides Growing* in Cold Desert of Himachal and Ladakh (n=3)

Geographical location	Seed (%)	Moisture (%)	Pulp oil (%)	Total ash (%)	Total protein (%)	Vit. C (mg/kg)	Flavonols (mg/kg)
Lahaul 1	10.4	74.4	3.08	1.05	2.25	2924	308
Lahaul 2	8.1	74.4	2.99	0.79	1.86	2304	285
Lahaul 3	9.7	72.7	3.04	0.93	2.11	4877	212
Spiti 1	7.9	76.9	3.38	0.60	2.17	1866	122
Spiti 2	7.4	74.6	3.14	0.73	2.15	1444	156
Spiti 3	9.6	67.2	3.38	0.99	2.44	1666	180
Ladakh 1	9.8	72.3	3.60	0.89	2.12	2145	251
Ladakh 2	8.6	72.1	3.52	0.79	1.99	2568	291
Ladakh 3	8.5	73.4	3.61	0.92	2.34	2425	428

hypoxia induced oxidative stress in C-6 glioma cells have been demonstrated. Exposure of cells to hypoxia for 12 hours resulted in a significant increase in cytotoxicity and decrease in mitochondrial transmembrane potential compared to the controls. Appreciable increase in nitric oxide and reactive oxygen species (ROS) production and increase in DNA damage has been observed. Pre-treatment of cells with alcoholic leaf extract of seabuckthorn at 200μg/ml significantly inhibits cytotoxicity, ROS production and maintains antioxidant level similar to that of control cells. Seabuckthorn extraction procedure has significant effect on antioxidant activity. Microwave assisted extract has significantly higher antioxidant capacity than those obtained by ultrasound-assisted extraction and maceration.

Immunomodulatory activity of seabuckthorn leaf extract has been evaluated in adjuvant induced arthritis (AIA) rat model. Inflammations induced by injecting Complete Freund's Adjuvant (CFA) in rats is reduced significantly by administrating seabuckthorn extract on the same day or 5 days prior to inflammation insult. The result suggests potential use of seabuckthorn for treatment of arthritis. Effect of seabuckthorn leaf extract on nitric oxide (NO) production induced by lipopolysaccharide (LPS) in murine macrophage cell line RAW 264.7 has been studied. The extract showed significant scavenging of NO radicals released by the NO doner. Treatment of macrophage with the extract also causes significant inhibition of inducable nitric oxide synthase. Results suggest significant anti-inflammatory activity of seabuckthorn extract and have potential for treatment of inflammatory diseases. Alcoholic leaf and fruit extract of seabuckthorn at a concentration of 500μg/ml have significant cytoprotective effect against sodium nitroprusside (SNP)-induced oxidative stress. These extracts inhibit SNP-induced cytotoxicity, free radical production and maintain anti-oxidant status identical to than of control cells. The alcoholic fruit extract has been found to have significantly higher anti-oxidant activity than leaf extract against SNP-induced cytotoxicity in murine macrophage.

Topical administration of flavone of seabuckthorn on cutaneous wound healing in rats results in enhanced wound healing activity as indicated by improved rate of wound contraction, decrease time taken for epithelialization and significant increase in hydroxyproline and hexosamine content. The flavone treatment also results in significant decrease in lipid peroxide levels and thus suggests that seabuckthorn flavone promotes wound healing activity. Seabuckthorn ointment is found effective in treatment of infected cutaneous wounds in bovines. Application of 5 percent povidone iodine or seabuckthorn ointment is found better in wound healing of infected cutaneous wound in calves than liquid paraffin.

In a recent study, seabuckthorn has been evaluated for hepatoprotective activity on carbon tetrachloride (CCl_4)-induced liver injury in male albino rats. Pretreatment of leaf extract at a concentration of 100 and 200 mg/kg body weight significantly protect the animal from CCl_4-induced increase of glutamate oxaloacetate transferase, glutamate pyruvate transferase, alkaline phosphatase and bilirubin and better maintain protein level in serum. The result suggests use of seabuckthorn leaf extract as neutraceutical or food supplement against liver disease. Possible mechanism of adaptogenic activity of seabuckthorn during exposure to cold, hypoxia and restraint (C-H-R) stress induces hypothermia and post stress recovery in rats has been elucidated. Results suggest that seabuckthorn extract treatment causes a trend for shifting anaerobic metabolism to aerobic during C-H-R exposure and post stress recovery. Adoptogenic activity of the extract might be due to its antioxidant activity, maintained body glucose level, better utilization of free fatty acids and improved cell membrane permeability. Seabuckthorn oil has significant anti-atherogenic and cardioprotective activity. Therapeutic value of seabuckthorn against subchronic arsenic toxicity in mice has shown that aqueous extract of seabuckthorn significantly protects against arsenic-induced oxidative stress but does not chelate arsenic. Ethanolic and water extract of seabuckthorn have shown protective effect against the toxic effect of mustard gas, a chemical warfare agent.

Seabuckthorn extract has also been used against Dangue virus infection in human blood-derived macrophages. Infected cells treated with seabuckthorn extract are able to maintain cell viability of Dangue infected cells at par with commercially available anti-viral drug, Ribavirin. Seabuckthorn seed extract possess antibacterial activities against *Bacillus cereus, B. coagulans, B. subtilis, Listeria monocytogenes,* and *Yersinia enterocolitica* suggesting use of seabuckthorn as food preservative.

Radioprotection studies at Institute of Nuclear Medicine and Allied Sciences, DRDO, using herbal preparation of seabuckthorn, RH-3, against whole body lethal irradiation in mice has shown that a dose of 30 mg/kg body weight of RH-3 rendered 82 percent survival as compare to non-survival in irradiated control. RH-3 inhibits Fenton reaction and radiation mediated regeneration of hydroxyl radicals in vitro, superoxide anion mediated Nitroblue tetrazolium reduction and $FeSO_4$ mediated lipid peroxidation in liver. RH-3 administration protects spermatogenesis by enhancing the spermatogonial proliferation, enhancing the stem cell survival and reducing sperm abnormalities. Seabuckthorn renders chromatin compaction and significantly inhibits radiation induced DNA break. Possible contribution towards radioprotective efficiency of RH-3 is maintenance of chromatin organisation, induction of hypoxia, hydrogen atom donation, free radical scavenging and blocking of cell at G2-M phase by interfering with topoisomerase I activity. Polyphenol/flavonoids present in seabuckthorn might be responsible for mitochondrial and genomic DNA protection from radiation-induced damage.

Research Programme

Potential of the lesser-known shrub has been recognized by several R&D organizations. Defence Research & Development Organisation (DRDO) is undertaking several R&D projects on seabuckthorn. Biotechnological potential has been recognized by Department of Biotechnology (DBT), Govt. of India, which has recently called for project proposal exclusively on seabuckthorn. Similarly, Indian Council of Agriculture Research (ICAR) has approved a mega programme 'A Value Chain on Seabuckthorn (*Hippophae* L.)' in which research institutes of ICAR, DRDO, ICMR and NGOs are collaborative partners. The following institutes are involved in major R&D on seabuckthorn in India:

Table 2 : Comparative Account of Nutrient Composition in Fruit Pulp and Seeds of *H. rhamnoides* Population from Different Valleys of Uttarakhand and Ladakh

Valley/Country	Moisture (%)	TSS (%)	Acidity (%)	Reducing Sugar (%)	Starch (%)	Lignin (%)	Fibre (%)	Fat (%)	Carbohydrate (%)	Total Protein (%)
Yamunotri	88.4	9.08	6.6	5.0	37.04	21.67	12.0	9.00	0.39	6.38
Bhyundar	85.3	9.72	6.6	2.3	48.87	26.33	12.0	9.33	0.30	5.42
Mana	86.4	8.86	6.8	5.0	85.17	17.67	13.0	10.00	0.37	5.83
Niti	84.9	9.28	6.3	2.1	62.55	18.00	13.0	8.33	0.34	5.96
Gangotri	97.6	9.28	6.5	6.0	29.42	21.33	14.0	10.33	0.40	7.13
Ladakh	74.6	14.3	2.54	0.96	-	-	3.45	1.54	-	2.64

Table 3: Fatty Acid Composition of Seabuckthorn Pulp Oils from Cold Desert of Himachal and Ladakh (n=3)

Geographical location	FA						
	14:0	16:0	16:1	18:0	18:1	18:2	18:3
Lahaul 1	0.5	28.4	50.3	0.6	11.3	8.1	1.3
Lahaul 2	0.6	29.4	50	0.7	9.7	7.9	1.6
Lahaul 3	0.5	28.3	49.7	0.5	11.5	8.5	1.7
Spiti 1	0.6	35.8	45.6	0.5	9.8	8.2	0.8
Spiti 2	0.3	35.8	45.6	Trace	9.8	8.2	0.8
Spiti 3	0.6	30.7	51.8	Trace	12.3	5.9	0.4
Ladakh 1	0.4	33.8	46.4	1.0	13.4	5.4	1.3
Ladakh 2	0.9	31.2	49.1	0.8	13.9	4.6	Trace
Ladakh 3	0.8	30.9	45.6	0.9	16.8	5.8	0.8

Table 4: Comparative Account of Mineral Composition in Fruit Pulp and Seeds of *H. rhamnoides* Population from Different Valleys of Uttarakhand and other Areas

Valley/Country	Fe (ppm)		Mg (ppm)		Cu (ppm)		Zn (ppm)		As (ppm)	
	Pulp	Seed	Pulp	Seed	Pulp	Seed	Pulp	Seed	Pulp	Seed
Yamunotri	1.127	0.593	0.72	2.487	0.133	0.097	0.92	0.817	0.06	0.099
Bhyundar	0.727	0.46	0.62	2.996	0.11	0.043	1.3	0.827	0.077	0.145
Mana	0.753	0.36	1.703	1.806	0.133	0.023	0.817	0.497	0.16	0.78
Niti	0.703	0.573	0.78	30.04	0.09	0.057	1.14	2.83	0.064	0.063
Gangotri	0.81	0.647	1.92	2.843	0.103	0.043	2.74	1.943	0.231	0.095

Table 5: Comparative Account of Nutrient Composition in Fruit Pulp and Seeds of *H. rhamnoides* Population from Different Valleys of Uttarakhand and other Areas

Valley/Country	Sodium (%)		Potassium (%)		Phosphorous (%)		Nitrogen (%)	
	Pulp	Seed	Pulp	Seed	Pulp	Seed	Pulp	Seed
Yamunotri	0.63	0.39	11.62	10.38	0.67	0.61	1.02	3.65
Bhyundar	0.47	0.41	10.12	10.90	0.63	0.65	0.89	3.74
Mana	0.51	0.49	12.07	10.12	0.66	0.66	0.93	3.73
Niti	0.59	0.31	13.21	9.33	0.60	0.63	0.95	4.09
Gangotri	0.53	0.05	14.84	13.42	0.63	0.69	1014	4.53

Table 6: Carotenoid, Tocopherol and Tocotrienol Composition of Pulp Oils of Berries of Indian *H. rhamnoides* (mg/kg of pulp oil) (n=3).

Geographical location	Carotenoids	Tocopherols/Tocotrienols						
		α T1	α T3	β T1	γ T1	γ T3 + δ T1	δT3	Total
Lahaul 1	266	859	293	16	152	362	12	1694
Lahaul 2	3420	1068	227	12	165	288	17	1788
Lahaul 3	2940	773	148	19	149	344	16	1564
Spiti 1	2576	1006	183	18	170	228	10	1615
Spiti 2	2820	1046	164	9	120	117	16	1494
Spiti 3	2706	679	247	11	121	241	17	1318
Ladakh 1	2400	954	95	27	126	187	14	1394
Ladakh 2	2350	858	120	22	82	201	15	1301
Ladakh 3	2650	903	109	21	90	208	17	1348

α T1, α-Tocopherol; α T3, α-Tocotrienols; β T1, β-Tocopherol; β T1, β-Tocopherol; γ T3, γ-Tocotrienols; δ T1, δ-Tocopherol; δ T3, δ-Tocotrienols.

- Defence Institute of High Altitude Research (Formerly FRL), Leh Ladakh-194101, India
- Defence Institute of Physiology and Allied Sciences, Lucknow Road, Timarpur, Delhi, 110054, India
- Defence Food Research Laboratory, Siddhartha Nagar, Mysore, 570 011 Karnataka, India
- Defence Research and Development Establishment, Gwalior 474 002, India.
- Institute of Nuclear Medicine and Allied Sciences, Brig S.K. Majumdar Marg, Delhi-110054,
- CSK Himachal Pradesh Agriculture University, Palampur, 176062, India
- Institute of Himalayan Bioresource Technology, Palampur, India
- Dr. Y.S. Parmar University of Horticulture and Forestry, Nauni, Solan 173230, Himachal Pradesh, India.
- Dr. B.R. Ambedkar Centre for Biomedical Research, University of Delhi, Delhi 110007, India
- Central Food Technological Research Institute, Mysore 570020, India

- G B Pant Institute of Himalayan Environment and Development, Srinagar Garhwal 246174, India
- Regional Research Laboratory (CSIR), Industrial Estate (PO), Thiruvananthapuram Trivandrum, 695019 Kerala, India
- Guru Gobind Singh Indraprastha University, Kashmiri Gate, Delhi
- Forest Research Institute, Dehra Dun, India

CONCLUSIONS

Seabuckthorn has immense nutritional and medicinal properties. Initial studies especially in DRDO laboratories have brought significant benefit to the people in Trans-Himalayas. Several products have been developed and commercialized. Laboratories studies on prevention and cure of several diseases have shown the potential of the shrub as a storehouse of bioactive compounds with a variety of medicinal properties. Seabuckthorn is now recognised as a storehouse for researchers in India as evident from mega programme announced by several R&D organisations. Research outcome of the ongoing programme will immensely benefit the modern society from the lesser known Trans-Himalayan shrub.

References

Basu, M., Prasad, R., Jayamurthy, P., Pal, K., Arumughan, C., Sawhney, R.C. 2007. Anti-atherogenic effects of seabuckthorn (*Hippophaea rhamnoides*) seed oil. Phytomedicine. 14(11), 770-777.

Dhyani, D., Maikhuri, R.K., Rao, K.S., Kumar, L., Purohit, V.K., Sundriyal, M., Saxena, K.G. 2007. Basic nutritional attributes of *Hippophae rhamnoides* (Seabuckthorn) population from Uttrakhand Himalaya, India. Current Science. 92(8), 1148-1152.

Dwivedi, S.K., Attrey, D.P., Chaurasia, O.P. 2004. Studies on preparation and preservation of seabuckthorn (*Hippophae rhamnoides* L.) squash beverage. Ind. J. Horticulture. 61(1), 78-80.

Dwivedi, S.K., Singh, R., Ahmed, Z. 2006. *The Seabuckthorn*. Field Research Laboratory (DRDO), Leh-Ladakh, India.

Ganju, L., Padwad, Y., Singh, R., Karan, D., Chanda, S., Chopra, M.K., Bhatnagar, P., Kashyap, R., Sawhney, R.C. 2005. Anti-inflammatory activity of Seabuckthorn (*Hippophae rhamnoides*) leaves. International Immunopharmacology. 5(12), 1675-1684.

Geetha, S., Sai Ram, M., Mongia, S.S., Singh, V., Ilavazhagan, G. and Sawhney, R.C. 2003. Evaluation of antioxidant activity of leaf extract of Seabuckthorn (*Hippophae rhamnoides* L.) on chromium (VI) induced oxidative stress in albino rats. J. Ethanopharmacology. 87(2-3), 247-251.

Geetha, S., Purushothaman, J., Karan, P., Shweta, P., Ratan, K., Sawhney, R.C. 2008. Hepatoprotective effects of sea buckthorn (Hippophae rhamnoides L.) against carbon tetrachloride induced liver injury in rats. J. the Science of Food and Agriculture. 88(9), 1592-1597.

Goel, H.C., Samanta, N., Kannan, K., Kumar, I.P., Bala, M. 2006. Protection of spermatogenesis in mice against gamma ray induced damage by *Hippophae rhamnoides*. Andrologia. 38(6), 199-207.

Goel, H.C., Kumar, I.P., Samanta, N., Rana, S.V.S. 2003. Induction of DNA-protein cross-links by *Hippophae rhamnoides*: Implications in radioprotection and cytotoxicity. Molecular and Cellular Biochemistry. 245, 57-67.

Goel, H.C., Prasad, J., Singh, S., Sagar, R.K., Kumar, I.P., Sinha, A.K. 2002. Radioprotection by an herbal preparation of *Hippophae rhamnoides*, RH-3, against whole body lethal irradiation in mice. Phytomedicine. 9(1), 15-25.

Gupta, A., Kumar, R., Pal, K., Singh, V., Banerjee, P.K., Sawhney, R.C. 2006. Influence of sea buckthorn (*Hippophae rhamnoides* L.) flavone on dermal wound healing in rats. Molecular and Cellular Biochemistry. 290, 193-198.

Gupta, R., Flora, S.J.S. 2005. Therapeutic Value of *Hippophae rhamnoides* L. against subchronic arsenic toxicity in mice. Journal of Medicinal Food. 8 (3), 353-361.

Jain, M., Ganju, L., Katiyal, A., Padwad, Y., Mishra, K.P., Chanda, S., Karan, D., Yogendra, K.M.S., Sawhney, R.C. Effect of *Hippophae rhamnoides* leaf extract against Dengue virus infection in human blood-derived macrophages. Phytomedicine (in press).

Mahajan, A., Sharma, S.K., Asrani, R.K., Varshney, A.C., Tyagi, S.P., Kumar, A. 2005. Treatment of infected cutaneous wounds with seabuckthorn (*Hippophae* sp.) in bovines: a histopathological study. Ind. J. Veterinary Surgery. 26(1), 31-33.

Mathew, D., Parimelazhagan, T., Gomez, S., Ahmed, Z. 2007 Characterization of seabuckthorn (*Hippophae spp.*) genetic resources in India using morphological descriptors. Plant Genetic Resources Newsletter. 149, 22-26.

Narayanan, S., Ruma, D., Gitika, B., Sharma, S.K., Pauline, T., Sai Ram, M., Ilavazhagan, G., Sawhney, R.C., Kumar, D., Banerjee, P.K. 2005. Antioxidant activities of seabuckthorn (*Hippophae rhamnoides*) during hypoxia induced oxidative stress in glial cells. Molecular and Cellular Biochemistry. 278, 9-14.

Negi, P.S., Chauhan, A.S., Sadia, G.A., Rohinishree, Y.S., Ramteke, R.S. 2005. Antioxidant and antibacterial activities of various seabuckthorn (*Hippophae rhamnoides* L.) seed extracts.J. Food Chemistry. 92(1), 119-124.

Padwad, Y., Ganju, L., Jain, M., Chanda, S., Karan, D., Banerjee, P.K., Sawhney, R.C. 2006. Effect of leaf extract of Seabuckthorn on lipopolysaccharide induced inflammatory response in murine macrophages. International Immunopharmacology. 6(1), 46-52.

Ranjith, A., Kumar, K.S., Enugopalan, VV., Arumughan, C., Sawhney, R.C., Singh, V. 2006. Fatty acids, tocols, and carotenoids in pulp oil of three sea buckthorn species (*Hippophae rhamnoides. H. salicifolia* and *H. tibetana*) grown in the Indian Himalayas. JAOCS. 83(4), 359-363.

Roy, P.S., Porwal, M.C., Sharma, L. 2001. Mapping of *Hippophae rhamnoides* Linn. In the adjoining areas of Kaza in Lahul and Spiti using remote sensing and GIS. Current Science. 80(9), 1107-1111.

Saggu, S., Kumar, R. 2007. Possible mechanism of adaptogenic activity of seabuckthorn (*Hippophae rhamnoides*) during exposure to cold, hypoxia and restraint (C–H–R) stress induced hypothermia and post stress recovery in rats. Food and Chemical Toxicology. 45(12), 2426-2433.

Saggu, S., Kumar, R. 2008. Effect of seabuckthorn leaf extracts on circulating energy fuels, lipid peroxidation and antioxidant parameters in rats during exposure to cold, hypoxia and restraint (C–H–R) stress and post stress recovery. Phytomedicine. 15(6-7), 437-446.

Sankhyan, H.P., Sehgal, R.N. 2005. Morphological variations in the sea buckthorn populations at nine places of Himachal Pradesh. Bionotes. 7 (2), 67-68.

Sankhyan, H.P., Sehgal, R.N., Bhrot, N.P. 2003. Effect of growth regulators and seasons on the propagation of different Seabuckthorn species through cuttings in cold deserts of Himachal Pradesh. J.Indian Forester. 129(10), 1300-1302.

Sankhyan, H.P., Sehgal, R.N., Bhrot, N.P. 2005. Standardization of presowing treatments for different Seabuckthorn species in cold deserts of Himachal Pradesh. J. Indian Forester. 131(7), 931-938.

Selvamuthukumaran, M., Khanum, F., Bawa, A.S. 2007. Development of sea buckthorn mixed fruit jelly. International Journal of Food Science & Technology. 42(4), 403-410.

Sharma, U.K., Sharma, K., Sharma, N., Sharma, A., Singh, H.P., Sinha, A.K. 2008. Microwave-Assisted Efficient Extraction of Different Parts of *Hippophae rhamnoides* for the Comparative Evaluation of Antioxidant Activity and Quantification of Its Phenolic Constituents by Reverse-Phase High-Performance Liquid Chromatography (RP-HPLC). J. Agriculture Food Chemistry. 56 (2), 374–379.

Shukla, S.K., Chaudhary, P., Kumar, I.P., Samanta, N., Afrin, F., Gupta, M.L., Sharma, U.K., Sinha, A.K., Sharma, Y.K., Sharma, R.K. 2006. Protection from radiation-induced mitochondrial and genomic DNA damage by an extract of *Hippophae rhamnoides*. Environmental Mol. Mutagen. 47(9), 647-656.

Singh, R., Mishra, S.N., Dwivedi, S.K., Ahmed, Z. 2006. Genetic variation in seabuckthorn (*Hippophae rhamnoides* L.) population of cold arid Ladakh (India) using RAPD markers. Current Science. 91(10), 1321-1322.

Stobdan, T., Angchuk, D., Singh, S.B. 2008. Seabuckthorn: an emerging storehouse for researchers in India. Current Science. 94(10), 1236-1237.

Swaroop, A., Sinha, A.K., Chawla, R., Arora, R., Sharma, R.K., Kumar, J.K. 2005. Isolation and Characterization of 1,3-Dicapryloyl-2-linoleoylglycerol: A Novel Triglyceride from Berries of *Hippophae rhamnoides*. Chemical & Pharmaceutical Bulletin. 53(8), 1021.

Vijayaraghvan, R., Gautam, A., Kumar, O., Pant, S.C., Sharma, M., Singh, S., Kumar, H.T., Singh, A.K., Nivsarkar, M., Kaushik, M.P., Sawhney, R.C., Chaurasia, O.P., Prasad, G.B. 2006. Protective effect of ethanolic and water extracts of sea buckthorn (*Hippophae rhamnoides L.*) against the toxic effects of mustard gas. Ind. J. Experimental Biology. 44(10), 821-831.

□□□

7

Producing World Class Apples in Cold Deserts for Health and Economic Security – A Case Study of Regional Horticultural Research Sub-Station, Tabo in Himachal Pradesh

K.R. DHIMAN., D.R. BHARDAWAJ., R. BAWA AND AMIT VIKRAM
Dr. Y.S. Parmar University of Horticulture and Forestry, Nauni (Solan), HP, India

INTRODUCTION

The cold deserts in India covering approximately 79,278 km^2 pose a greater developmental challenge due to rugged terrain, isolation, inhospitable climate and lack of communication facilities.

The inhabitants of the area are mainly farmers. These areas have very difficult terrain with perpetual snow-covered peaks and hostile climate. The altitudes of this zone in the valley range from 7,000 to 14,000 ft above msl. The occurrence of heavy snow during winter is regular feature of the area with long freezing winters that continue for five to six months. As a result, there is a short growing season of five to six months only. There are extreme variations in seasonal temperature at high altitudes. The rainfall is almost negligible.

Precipitation is mainly in the form of snow. The area receives a high intensity of UV and IR radiation followed by high velocity winds (40-60 km/hr). Glacial melts serve as sources of irrigation during the summer. The soils are shallow, sandy loam in texture, low in organic matter with neutral to slightly alkaline in reaction.

Constraints in Apple Growing

- Lack of locally grown nursery plant material
- Small landholding and poor economic conditions of the people
- Difficult area with respect of approach and steep in slopes
- Changing weather conditions, especially the decreasing trend of snowfall
- Lack of cold tolerant varieties and rootstock
- Lack of scientific knowledge of training and pruning
- Lack of awareness regarding the use of chemical fertilizers
- Poor knowledge of insect pest management
- Totally dependent on irrigation
- Far distant from the commercial marketing centres of the country
- Cultural hindrances

Despite these constraints, because of higher productivity and income per unit, horticulture is playing a vital role in improving the socioeconomic conditions of the rural masses in cold deserts. More fruit production and additional area brought under it are liable to bring vital changes in the structure of our rural economy.

The Tabo Experience

Regional Horticultural Research Sub Station, Tabo is situated in Spiti Division of Lahaul & Spiti district of Himachal Pradesh at an elevation of 10,672 feet above MSL. The mandate of the station is introduction, evaluation and development of cultivation packages for temperate fruits and vegetable crops suited for cultivation in dry temperate region of Himachal Pradesh. The station came into existence in 1979 but land was allocated in 1981. The station has 8.0 hectares of land wherein plantation of wild apricot and almond was done in 1985. The plantation of apple and apricot was taken up n 1989-92 by introducing 145 plants of apple and 40 plants of apricot comprising different varieties. The high density plantation of apple was taken up from March 1996 onwards.

The Present Scene

Apple (*Malus domestica*) being the most prevalent fruit crops of the dry cold desert region of Himachal Pradesh, serves as the main source of income. The total area under apples in cold desert region is 9,776 ha with a total production of 26,656 tons.

The climatic conditions are highly congenial for the production of export quality apples, owing to a long fruit bearing season, high concentration of UV rays, and clear sky during July and August, and high diurnal temperature variations during the growing season and ensure irrigation facilities.

Apple Nursery Production under Polyhouse Conditions

Perhaps one of the main constraints to the establishment of orchards in cold deserts is lack of locally acclimatized nursery plants. Owing to the short growing season, apple nurseries cannot be produced under open field conditions. Experiments conducted at the Regional Horticultural Research Sub Station, Tabo, Spiti in HP, however, have shown that fruit nursery growing is possible under greenhouse structures. The graftable seedlings are procured from a wet temperate zone in the month of February and directly planted in the greenhouse. In the month of April when these seedlings start sprouting, grafting is performed. Tongue grafting has usually given better results. The weeding, deshooting, hoeing, irrigation and manuring have to be performed at regular intervals. The ventilators and doors of greenhouses are kept open during the daytime. Greenhouses of 10x4x2m size provide better results than longer ones because; they provide better cooling and ventilation without depending on fans. The plants are ready for sale by the end of the winter season when planting conditions are congenial in dry cold deserts.

The station is producing nursery plants of apple, apricot and pear on regular basis. In addition, multiplications of clonal rootstock of M7, M9, MM106, MM111 and M793 under polyhouse conditions are also being undertaken.

Top-working/Rejuvenation of Apple

The station has standardized complete top working technology for rejuvenation of old fruit plants of apple under cold desert conditions.

The method employed is T-budding performed in the months of July-August, by taking extra precautionary measures to prevent bud sprout from first chilly winter.

Installation of Sprinkler and Drip Irrigation System

One of the major problems in protected nursery production under Tabo conditions was high incidence of root-rot. Major cause of the disease is the application of excess of irrigation water through flood system of irrigation. To overcome this problem, sprinkler irrigation system has been installed in all the existing 46 polyhouses.

To avoid the loss of nutrient from the tree basins and other organic inputs due to flood irrigation, the entire demonstration trial comprising of 2600 plants of different fruit crops has been put under drip irrigation system. Other experiments on standardization of frequency of drip irrigation for different fruit crops under cold desert conditions are in progress.

Establishment of Bud Wood Orchard of Apple

Apple is the dominating crop of Spiti and upper Kinnaur areas. The station has successfully demonstrated high density planting of apple. Keeping in view the increasing demand of scionwood from the RHRSS, Tabo, a budwood orchard of important, commercial and economic cultivars of apple has been established during March 2004 for further multiplication and supply of scionwood (Table 1).

Table 1: Germplasm of different cultivars of apple at RHRSS, Tabo.

Category	Cultivars
Standard varieties	Royal Delicious, Red Delicious, Rich-a-Red, Golden Delicious and Red Gold
Colour strains	Vance Delicious, Top Red, Bright-n-Early
Spur type	Red Chief, Oregon Spur, Silver Spur, Coe Red Fuji, Red Fuji, Granny Smith
Others	Scarlet Gala, Mutus, Real Macoy, Lord Lamborn and Criterion
Polliniser	Granny Smith, Snowdrift, Golden Hornet, Tydeman's Early Worcester, Manchurian and Gloster.

Recent Introductions

In April 2006, 581 plants of Super Chief, Scarlet Spur, Gale Gala, Red Fuji, Coe Red Fuji and Granny Smith budded on EMLA-7, EMLA-106 and EMLA-111 were introduced for the first time under cold desert conditions on large scale to evaluate the performance of rootstocks and scion varieties in terms of fruit quality and productivity.

Fruit Quality

In cold deserts, the apple crop takes longer time to ripen, which not only attributed to the good size of the fruit, but also the excellent quality which can compete with the best apples in the world. In most cultivars, fruit setting takes place between second and third weeks of May and fruit matures at the end of October or the first fortnight of November. The fruit quality of different apple cultivars growing at RHRSS, Tabo is presented in Table 2.

Table 2: Fruit quality of different apple cultivars growing at RRSS,Tabo.

S. No	Cultivars	Fruit size (cm)		Fruit weight (g)	Firm-ness (kg)	TSS (%)	Acidity (%)
		Length	Breadth				
1	Royal Delicious	7.8	8.3	258.6	7.5	14.4	0.33
2	Well Spur	7.9	8.3	251.2	9.6	14.8	0.40
3	Silver Spur	8.0	7.9	215.3	8.6	13.2	0.36
4	Red Spur	7.5	8.3	231.7	8.0	14.8	0.40
5	Oregon Spur	7.3	7.8	225.4	7.5	16.8	0.52
6	Red Chief	7.6	8.4	247.4	8.6	16.2	0.49
7	Vance Delicious	7.0	7.5	191.7	7.7	18.2	0.22
8	Golden Delicious	6.5	6.7	144.1	8.7	15.4	0.48
9	Golden Spur	6.8	7.5	180.6	7.5	14.8	0.72
10	Tydeman	6.8	7.9	197.5	9.13	8.0	0.51
11	Stark Spur Golden Delicious	5.9	6.6	132.7	7.5	16.1	0.56
12	Greensleeves	5.6	6.7	134.5	9.0	16.0	0.48
13	Honey Gold	7.8	7.8	171.9	7.2	16.1	0.32

Cultivars Tydeman's Early Worcester and Greensleeves mature in September, with others during October or November depending upon the overall weather conditions of the growing season. The fruits are excellent in colour and taste, mainly because of the high concentration of UV rays and diurnal temperature variation. The TSS and acidity of fruit is higher than apples of the wet temperate zone and they are much more crisp and crunchy – thus making them most popular. The apple produced from this region has made a name for itself in commercial marketing centre of our country, where it is sold on premier price.

❏❏❏

8

Novel Approaches for Food Security in High Altitudes

A. S. BAWA
Defence Food Research Laboratory, DRDO, Mysore, Karnataka, India

INTRODUCTION

Indian army faces one of the biggest challenges in maintaining its troops at one of the world's toughest battle fields at an altitude of more than 18,000 feet. In this process it has to fight not only against the enemy but also against the all odds of nature. At high altitude, troops are exposed to inclement weather, perilous altitude, difficult terrain and psychological problems arising out of being away from the main land as well as their near and dear ones. Under such circumstances if an army has to keep its soldiers physically fit and in high morale to fight against the enemy and protect the integrity of the nation, it has to be supplied with palatable, healthy and adequate nutritious food. The food provided to the troops should also be wholesome containing adequate and balanced nourishment.

Normal Body Changes in High Altitude

Certain normal body changes such as breathlessness during movement, hyperventilation (breathing fast), increase in urination, awakening during the night, strange dreams etc. happen in most people who travel to high altitude. Gaining height slowly while taking time to adjust to new altitudes is the best way to avoid illness. Other measures to help prevent altitude sickness include keeping warm and drinking plenty of water because the body needs more water at higher altitudes. One eat enough even if we don't feel like it. A diet high in complex carbohydrates is generally believed to be the best food for high altitudes.

High Altitude Illness

At high altitude, atmospheric pressure is reduced. The consequent reduction in oxygen pressure can lead to hypoxia (i.e. reduced supply of oxygen to the tissues). High altitude sickness mainly occurs at heights above 2500 metres. The main indication of altitude sickness during ascent is a painful headache with loss of appetite, nausea or vomiting, dizziness, sleeping problems, fatigue or weakness, confusion, unsteady walking. The condition occurs because of the body trying to adjust to lower oxygen levels. In rare cases, the illness can lead to life threatening conditions such as pulmonary oedema which causes fluid in the lungs; symptoms of which include breathlessness while resting, a persistent cough and an unusual chesty sound and cerebral oedema. Symptoms of cerebral oedema, which can be fatal due to fluid in the brain, include irrational behaviour, confusion and a staggering gait. Patients with this condition need immediate evacuation and medical attention.

Ultraviolet Radiation from the Sun

The ultraviolet (UV) radiation from the sun includes UVA (wavelength 315–400 nm) and UVB (280–315 nm) radiation, both of which cause damage to human skin and eyes. Exposure to UV radiation, particularly UVB, can produce severe debilitating sunburn and sunstroke, particularly in light-skinned people. Exposure of the eyes may result in acute keratitis ("snow blindness"), and long-term damage leads to the development of cataracts. Long-term adverse effects on the skin include: development of skin cancers (carcinomas and malignant melanoma), mainly due to UVB radiation and accelerated ageing of the skin, mainly due to UVA radiation, which penetrates more deeply into the skin.

Hypothermia

Hypothermia occurs when the body loses heat faster than it can produce it and the core temperature of the body falls. It is easy to progress from very cold to dangerously cold due to a combination of wind, wet clothing, fatigue and hunger, even if the air temperature is above freezing. Symptoms of hypothermia are exhaustion, numb skin (particularly toes and fingers), shivering, slurred speech, irrational or violent behaviour, lethargy, stumbling, dizzy spells, muscle cramps and violent bursts of energy. It is best to dress in layers; silk, wool and some of the new artificial fibres which are all good insulating materials. A hat is important, as a lot of heat is lost through the head. A strong, waterproof outer layer is essential, as keeping dry is vital. Carry basic supplies, including food containing simple sugars to generate heat quickly, and lots of fluid to drink. A space blanket is something all travellers in cold climates should carry. To treat mild hypothermia, first get the person out of the wind and/or rain, give them hot liquids - not alcohol - and some high-kilojoules, easily digestible food. Allow them to slowly warm themselves. This should be enough to treat the early stages of hypothermia.

A number of measures can be adopted to prevent acute mountain sickness: ascend slowly, have frequent rest days, spending two to three nights at each rise of 1,000 m. If one reaches a high altitude by trekking, acclimatisation takes place gradually and the person is less likely to be affected than if the person flies directly to high altitude. It is always wise to sleep at a lower altitude than the greatest height reached during the day if possible. Also, once above 3000 m, care should be taken not to increase the sleeping altitude by more than 300 m per day. Drink extra fluids. The mountain air is dry and cold and moisture is lost as we breathe. Evaporation of sweat may occur unnoticed and result in dehydration. Eat light, high-carbohydrate meals for more energy. Avoid alcohol as it may increase the risk of dehydration. Avoid sedatives.

Issues Related to Food & Development of Novel Processed Food

Normal preparation of Indian style food is extremely difficult at high altitude due to the lack of modern cooking appliances. Food supplies to high altitude areas encounter several problems. Terrain constraints limit land route and food supplies have to be air dropped. This poses the danger of breakage of fragile packaging materials and loss of supplies. Besides this, inclement weather, rarefied atmosphere and low concentration of oxygen at high altitudes decrease the efficiency of fuel. This in combination with low boiling point of water leads to delayed cooking. Due to the lack of sufficient oxygen and lowered sensory perception at high altitude, troops do not feel like eating and suffer from lack of appetite. In addition to this, they also complain of indigestion on taking foods with higher fat content. As a result troops posted at high altitude show preference towards fresh fruits and vegetables. Dehydrated foods are also preferred by them because of their being light in weight and longer shelf life. Taking all these aspects into consideration, Defence Food Research Laboratory (DFRL) has come up with retort processed meals ready to eat ration (MRE), precooked, dehydrated composite pack rations and emergency survival rations, all packed in flexible polymeric film and their laminates, etc. besides fresh fruits and vegetables for soldiers at high altitudes.

The compo pack rations used by army such as full meal compo pack rations, which provide 4100 k cal and supplementary compo pack rations which provide 1430 k cal encompass a wide variety of convenience food products developed at DFRL. It consists of ready-to-eat (RTE), ready-to-reconstitute (RTR) and ready-to-cook kind of convenience foods. Some such items developed for the purpose are, chapaties, parothas, instant vegetable pulav, instant khichdi, instant curried dhals, instant choley mix, instant rajma, instant aial mix, instant potato peas curry, instant halwa, instant upma, instant neutro cereal mix, instant omelette, scrambled egg mixes, instant soup powder, instant fruit juice, beverage mixes, instant vermicelli kheer mix, energy rich cereal and fruit bar.

The compositions of one man compo pack ration and mini compo pack ration are having a shelf life of one year. The instant food can easily be reconstituted by immersing in hot water for 5-10 min.

The ready-to-reconstitute food products (instant mixes) are based on either flaking technology or freeze thaw dehydration technology. The process of flaking technology is based on precooking, conditioning, flaking and drying under controlled conditions. The flaking process reduces the case hardening and improves sensory characteristics. Freeze thaw dehydration technology involves soaking, pressure cooking, cold conditioning and drying under controlled conditions. The instant food products developed through the above technology could be easily reconstituted within 5-10 min. by putting in hot water and are superior to conventionally dried counterparts in colour, flavour, texture and taste. All the products have undergone large scale user trials by army units based at high altitude. They are highly acceptable and stable for more than a year under ambient conditions.

Retort processed, shelf stable, ready-to-eat foods in laminates of aluminium with polymeric films are suited for use at stringent areas. The flexibility of the pouches, light weight and ready-to-eat nature of food, help the soldiers to accommodate and consume it even during operational situations in treacherous terrain as well as cold weather conditions. It could be consumed as such or after warming the pouch in boiling water. DFRL has sterilized a number of traditional Indian foods in such retort pouches. They include vegetable pulav, vegetable khichdi, sooji halwa, potato pea's curry, mutton curry, mutton pulav, chicken pulav, chicken curry, palak paneer curry, avial, upma, mixed vegetable curry, alu choley, rajma curry, noodles etc. many of the retort processed foods products are the basis of Meals ready-to-eat (MRE) ration. Constituents of MRE rations for Army and Marine Commandoes of Indian Navy provide different menus to avoid monotony. The rations are suitable for high altitude region. About 1,50,000 such rations were supplied to Army during Kargil conflict.

In meals ready-to-eat rations, the main constituents of rations are heated by hexamine fuel tablets. The process is time consuming and cumbersome in field conditions. Hence, simple device to heat the product at different field condition with reduced preparation time is preferable. Self-heating system called "Thermopack" developed at this laboratory is based on a multi step exothermic reaction and is found to be adequate to heat the product from ambient to 65°C within 10-15 minutes. Under sub-zero condition, the system enables to heat the product from -10°C to 40°C. The system consists of heater pad, treated water and RTE food pouch along with amenities such as needle, blade, paper plate, tissue paper, instruction manual etc. The heater pad of the system does not have direct contact with food pouch as they are placed independently. The system is user friendly and easy to operate.

DFRL has designed separate survival rations for Army, Navy and Air Force for use during field operations when normal supply lines of rations are disturbed. The army survival ration consists of two survival bars and two types of long life chikki. The ration weighs 500 g and delivers 2200 k cal. The naval survival ration consists of two survival bars in a lot of 600 g to cater to one person for three days. The survival bar is based on calorie rich products formulated from easily digestible carbohydrates. It has a shelf life of 3 years. Besides this DFRL has also designed and developed emergency flying ration consisting of survival bars weighing 330 g and providing 1320 k cal for 48 hours. These survival rations can also be used in high altitude region.

Supply of fresh fruits and vegetables to high altitude area, their storage and transportation pose real challenge due to the lack of established roads, presence of hazardous terrain and hostile climatic conditions. To extend the shelf life of fruits and vegetables during storage and transportation, DFRL has introduced multipurpose plastic crates, pre-treatments and packaging of fresh fruits and vegetables. The usage of plastic crates is not only hygienic and economical but also minimises the spoilage loss when compared to conventional packaging systems such as mulberry baskets. The chemical pre-treatment of fruits and vegetables by washing with chlorine water, selective use of anti-fungal paste and ethylene scavenging sachets for delayed ripening as well as usage of perforated polyethylene pouches to transport assorted units of fruits and vegetables separately has been found to extend the shelf life of different fruits and vegetables by 3-6 days under ambient conditions. The minimally processed vegetables will be fresh / fresh like products with optimal sensory attributes, have logistic utility and operational convenience to the Armed Forces. The minimally processed and stabilized vegetables and the products have a shelf-life of 10-14 days under ambient conditions (23-36°C) and 6-8 weeks at low temperatures. These methods have been found to be useful for storage and transportation of fruits such as pineapple, papaya, jackfruit, apple, mango and vegetables such as tomato, carrot, turnip, radish, onion and ginger at high altitude region. Besides the above, DFRL has also developed shelf stable ready-to-eat fruit slices such as mango, pineapple, banana, papaya, guava, etc., based on hurdle/intermediate moisture technologies with a shelf life of 6-8 months under ambient conditions. In addition to this ready-to-eat fruit bars such as mango bar have also been developed. These products could be considered for user acceptance trials at high altitude area.

Some of the other fruit products for utilizing in high altitude locations include apricot based products i.e. dehydrated fruit, stabilized fruit slices based on hurdle process, RTS beverage and jam. Fruit juice blended tender coconut wafer (lemon, mango, pineapple) may be a high potential liquid food for the high altitudes. Use of antifreeze containers which had been appreciated by the Units during the field trials may be inducted for regular use to prevent freezing of fruits, vegetables and liquid foods during peak winters.

In high altitude areas barley is consumed as a staple food. Various products based on barley like *Thupka* are prepared and used regularly. These products are prepared by grinding roasted barley grains in the form of the powder in order to supplement barley based foods to the native people of high altitude and mountainous regions. Instant products like khichadi and pulav based on barley were developed which can reconstitute in hot water with in 10 min. The product contains 12.9 - 13.2 % protein, 16.8 - 19.6 % fat, 2.4 - 4.7 % moisture,59.4 - 60.2 % carbohydrates and 5.19 - 5.24 % total ash. The products give 440 - 465 cals per 100 gm and have shelf life of one year at ambient conditions.

Fibres especially soluble ones in the processed products are promising in reducing cholesterol and helps for management of diabetes mellitus. DFRL has developed crispy, nutritious, fibre rich biscuits possessing antidiabetic and hypocholesterolemic effects, which have been evaluated in experimental animals. The biscuits are well accepted with a shelf-life of 6 months. These fibre rich products are easily digestible, have soluble fibres and are best suited for use at high altitude areas as well.

Seabuckthorn grows widely in dry temperate Himalayas. Fruits and leaves are rich in bioactive substances and high potential in health food, pharmaceutical and cosmetic industries. The fruits contain more than 190 bioactive substances, out of these, there are 18 aminoacids. The fruits contain significant amount of carotenoids (including beta-carotene, zeaxanthin, lycopene and polyring-lycopene), flavoxanthin, progestin, cryptoxanthin, violoxanthin, neoxanthin and vitamins C, K and E. The main antioxidant content in leaves are flavonoids, which comprises leucoanthocyanidins, quercetin, sorhamnetin, epicatechin and flavonols. Seabuckthorn leaves are used by DFRL as a major ingredient in herbal tea formulation and baked products viz. biscuits, buns, bread and rusks. In addition, the seabuckthorn based products viz. spiced squash, jam, jelly and yoghurt are rich in polyphenols, flavonoids and also possess stress reducing property.

Rhodiola rosea, a medicinal plant popularly known as golden root or arctic root is an adaptogen, known for stimulating the nervous system, decreasing depression, enhancing work performance, eliminating fatigue and preventing high-altitude sickness and cardiovascular diseases. This plant is found in the Indian Himalays as well, and the the local population has been using its root decoction as a beverage. For an effective exploitation of the Indian variety of this unique phytoadaptogen, DFRL has developed nutraceuticals in the form of pills and chocolates.

Aloe vera, known to possess more than two hundred phytoconstituents has been utilized for various food formulations viz. jam, jelly, squash, etc. These medicinal foods rich in glucomannans, anthraquinones, lignin, saponins, aminoacids, vitamins and minerals exhibit pharmacological properties, viz. anti-inflammatory, antiulcerative, adjuvant, analgesic, etc. best suited for alleviating stress situations. Tender coconut water is a natural drink with gentle taste and flavour. It is rich in minerals particularly potassium and consumed by all age group. After harvesting, the quality of tender coconut water in nuts is found to undergo deterioration after 72 hr. The bulkiness of coconuts adds transportation costs. Since, the availability is limited to southern states; a simple preservation technique has been developed for use at extreme conditions. Tender coconut water has been preserved in natural form packed in flexible polymeric stand-up pouches / aluminium beverage cans by combination preservation technology.

The nutritional cereal, ragi called finger millet (*Eleusine coracana*) is rich in calcium, iron, dietary and functional fibre. Various technologies have been adopted for the development of calcium rich, fibre rich convenience foods. Ragi based nutritional products are in ready to eat, ready to reconstitute or ready to cook form. The biggest advantage of some ragi based products is its ability to reconstitute using cold water which is a boon for people at high altitudes. The products are also rich in easily digestible carbohydrates. A range of ragi based products in the form of convenience mixes have been developed at the laboratory. Some of them especially sweetened millet mix and millet cookies have been highly liked by the Armed Forces during field trials conducted at high altitude. Some of the nutritional products based on ragi developed at the lab are millet ada mix, millet roti mix, millet dosa mix, millet pakoda mix, millet nippattu mix, millet beverage mix, millet sweetened cookie mix, functional millet beverage mix, functional millet cookie mix, sweetened millet mix, millet chapathi mix, millet halbai mix, millet kheer mix, spiced millet mix etc.

One of the common complaints of troops placed at high altitude is the lack of appetite which leads to decrease in body weight and another problem is indigestion. In order to improve their intake, inclusion of appetisers in the ration is considered better. DFRL has developed a number of instant mixes of appetisers/munches which could help the troops to have adequate appetite for food consumption. These have received good response in the high altitude trials.

Appetisers, based on spices, fruits and vegetable developed at the laboratory are spice drink mix, chakota soup mix, spiced tomato mix, appetising mix, ginger munch, fruit munch, pepper munch, jeera munch, lemon munch.

□□□

Post Harvest Management and Value Addition of Selected Fruits for Sustainable Development in High Altitudes

R.T. PATIL, R.K. GUPTA AND D.B. SINGH
Central Institute of Post Harvest Engineering and Technology, Abohar, India

INTRODUCTION

Fruit diversity of high altitudes in India exists to a large extent from north (Jammu & Kashmir) to subtropical plains in the north, and also to Arunachal Pradesh in the east. Genetic diversity is rich in the northwestern Himalayas and to relatively low extent in the northeastern region. Ladakh, which covers the major area of cold arid regions in India, is located between 8500-12000 feet above sea level and is a rich repository of useful fruits.

Important Fruit Crops of High Altitudes (Ladakh)

Fruit Crop	Family
Wild Apricot (*Prunus armeniaca)*	Rosaceae
Sea Buckthorn (*Hippophae rhamnoides*)	Elaeagnaceae
Apple (*Malus domestica*)	Rosaceae
Pear (*Pyrus communis*)	Rosaceae

SEA BUCKTHORN

Sea buckthorn (*H. rhamnoides*) is one of the few potential wild resources in the higher regions of the Indian Himalayas, where three species of this group, i.e. *Hippophae rhamnoides, H. salicifolia* and *H. tibetana* grow abundantly in nature, and are still awaiting their useful exploitation. Seabuckthorn is locally called as 'Tsermang' and the fruits are called as 'Tsestalullu' in Ladakh. Sea buckthorn is generally found in the cold and hilly areas of Ladakh in Jammu & Kashmir, Lahaul-Spiti in Himachal Pradesh and parts of Arunachal Pradesh and Sikkim. Fruits of sea buckthorn are a rich source of vitamins like A, B1, B2, C, D, E and K. Its leaves are also a rich source of protein and antioxidants. Its fruits are now utilized for making juices and other food products. Its oil has anti-radioactive and UV-protective effect. Berries of Sea buckthorn contain 15 plus carotenoids, important among them are Lutein & Zeaxanthin. Thus sea buckthorn is a multipurpose plant of Ladakh. Its natural population varies in colour, shape, size of fruits, leaves, thorns, etc. The Sea buckthorn fruit is a reservoir of good health - a general tonic and energizer that keeps the entire body metabolism in good order and fights common cold. This fruit nectar is the best for all age groups as it contains zero cholesterol. It is perhaps the only fruit juice that claims to be anti-ageing, with additional useful properties such as enhancing body endurance and improving memory. It improves body immunity especially against diseases like cancer and reduces the risk of coronary heart disease on account of the presence of many bioactive agents like carotenoids and anti-oxidants in it. Sea buckthorn juice strengthens memory, helps in proper functioning of the liver, is good for the nervous system, lowers the level of cholesterol, controls blood pressure, improves micro-circulation of blood capillaries and nourishes skin and hair. Production of sea buckthorn in Leh is estimated at around 12,000 tonnes but owing to lack of appropriate harvesting practices almost 80 per cent is lost.

APRICOT

Apricot (*Prunus armeniaca* L.) is the main fruit crop in the cold arid regions of India, mainly Ladakh. Apricot is rich in Protein 0.8 per cent, Fat 0.1 per cent, Vit.C mg/100g 7.0, Vit. B µg/100g 1500 and mineral 14.2 mg/100g. This tree thrives well even in sandy and wasteland conditions and can tolerate water stress and temperatures as low as -30° C. Among the cultivated indigenous and exotic cultivars grown in the region such as Suffaida, Shakarpara, Suffaida Parachinar,

Charmagz, Kaisha, Australian, Afgani, Nari, Rakchey-Karpo, Tok-Popa and Halman. Cultivars like Halman and Shakarpara has highest TSS and hence possesses excellent drying quality while Suffaida, Charmagz, Australian and Rakchey-Karpo are most popular table purpose cultivars.

APPLE

Most of the varieties are small, juicy and have poor shelf life. However, their capability to tolerate moisture and nutrient stress is inherent which makes them valuable gene sources for apple improvement programmes. Lot of genetic diversity has been found in this fruit crop prominent among these are Bong, Karkichoo, Khara, Mar, Mongol, Phamer, Sanker, Shinz, Squirmo, Tha and Yangm which can be utilized for processing and value added products.

PROCESSING POTENTIAL & VALUE ADDED PRODUCTS

1. Sea Buckthorn

Harvesting : Harvesting is difficult due to the densely thorny nature of the shrubs. A common harvesting technique is to remove an entire branch, though this is destructive to the shrub and reduces future harvests. A branch removed in this way is frozen, and then the berries can be easily shaken off. The branches are cut, deep frozen to -32°C. They are slightly defrosted on the surface during the removal of the berries from the branches and afterwards cleaned. The worker then crushes the berries to remove up to 95 per cent of the leaves and other debris. This causes the berries to melt slightly from the surface as the work takes place at ambient temperature (about 20°C). Berries are later stored at -22°C.

Processing : The sea buckthorn can be processed into a wide variety of products.

Juice and its extraction : Scientists in India have reported a new extraction method for sea buckthorn berries giving a juice rich in vitamin C, polyphenols and flavonoids. The research, published in the *Journal of the Science of Food and Agriculture* (Vol. 86, pp. 2345-2353) reports that using continuous high speed centrifugation (spinning) to separate the juice and the solid sludge the resulting juice retains more than 40 per cent of polyphenols, 50 per cent of flavonoids and 70 per cent of vitamin C present in the pulp of the red berries. The new extraction method significantly improves both the yield and nutrients obtained from the fruit compared to current methods, which reportedly give juices of poor quality. Juice extraction is often the first step in processing. The freshly pressed juice separates into sludge, juice and oil. Traditionally, this oil and the sludge are removed using a centrifuge followed by UHT pasteurization to preserve the heat sensitive compounds in the product. The juice is used to supplement health drinks and foods.

Oil and its extraction: Plants used in traditional medicine have the potential to provide pharmacologically active natural products, which can be used to treat various ailments. Sea buckthorn berry oil is especially rich in omega-7 series of fatty acid (palmitoleic acid). Making up around 30 per cent of the total oil, this rare fatty acid is a key component in sea buckthorn berry oil and it can act as a replacement for other fatty acids in cell membranes. The oil of sea buckthorn effectively combats wrinkles, dryness and other symptoms of malnourished or prematurely aging skin and is utilized in anti-ageing skin cream and lotions. Sea buckthorn oil extraction can be done using several methods. Pulp oil exists in the juice pulp and can be isolated as a cream layer during centrifuge treatment as a part of juice production. Countercurrent extraction of the seed or pulp with an organic solvent, commonly hexane is an old time method of recovering the oils. This method is rarely used because of concern over solvent residues. Supercritical fluid extraction (SCE) using carbon dioxide under high pressure is the modern method of oil extraction. It is both efficient and environmentally sound. Carbon dioxide becomes a liquid under certain pressure and temperature conditions making it useful as an extraction solvent. The liquid CO_2 and extracted oil are transferred to a tank where the pressure is released and both the CO_2 for reuse and the product are recovered.

Pigment: A pigment termed "sea buckthorn yellow" can be extracted from sea buckthorn waste material. The pigment can be extracted with alcohol or using supercritical CO_2 methods.

Other products: A wide variety of products from herbal teas to cattle feed and firewood can be produced from Sea Buckthorn.

2. *Wild Apricot*

Apricots are the main resource of the lower part of Ladakh. They are usually sun dried without any hygienic consideration on the roofs of houses or on large stones. The added value of such products is low.

A cottage-scale unit to process and dry this product in the leh market to increase the added value, attention on a high-quality product for the local and organized market is required.

Drying and Quality Maintenance

Improved method of sun drying: It is not customary to wash the fruit because when it is allowed to stand with free moisture on the surface it may undergo rapid deterioration through growth of brown rot fungus. Fruit is cut completely around the suture. The fruit is then separated and pit removed. Apricots can be dried whole, but should be after dipping in lye solution and exposing to prolonged sulfuring. Steam blanching of fruits will improve the quality of product, reduces dry time and higher sulfur dioxide retention and hastens cooking and dehydrations. The cut fruit is placed, cup side up, on wooden trays. It is desirable to sulfur the fruit as soon as possible to prevent the darkening and preserve vitamin C. The sulfured fruits are taken to the drying yard as soon as possible after sulfuring. The fruit is exposed to the sun for the period from 1- 7 days, depending upon the climate. The dried fruit should contain about 18 per cent moisture and should be soft and pliable, not sticky.

The yield of dried material will vary from 15-18 per cent based on the weight of the fresh, unprepared fruit. Before packing dried apricots are graded and processed. Before grading and processing it should be ascertained that fruit is free from insect infestation, mold and dirt. During processing graded fruits are passed over shaking screen. Water is sprayed in to the cylinder for the purpose of cleaning the fruit and moisturizing it to facilitate the better absorption of the sulfur dioxide during the sulfuring. The washed fruit is spread on trays, which are staggered to the depth of 5-8 cm and exposed to fumes of burning sulfur for 4 hours.

Batch Drying: Some techniques and methods to manufacture better quality dried fruits have been developed which have their applicability in commercial scale production for the mutual benefit of the consumers and manufactures. Dry batch dry (DBD) process, is developed for dehydration of apricots to produce a translucent, quick cooking and excellently flavoured dried product. In this process, the fruit is first washed, halved and pitted. It is then sulphured lightly by a dip in bisulphite solution and dried to 50per cent weight reduction in a conventional tunnel dehydrator. After blanching in a steam blancher, the fruit is returned to the tunnel to finish dehydration. The apricots processed by this way resemble sun-dried apricots in appearance and have a pleasing fruity flavour.

Dehydration of apricot: The halved fruit material is steamed for 2 to 4 minutes before sulfuring. One hour of sulfuring is considered sufficient as against 2 to 3 hours if the fruit is sun-dried. Tunnel dryers are used at a temperature not exceeding 65°C. At the start of drying it is desirable to use rather high relative humidity (50 to 55 percent). Drying time is from 15 to 20 hours under these conditions. After drying, the fruit is not re-sulfured, but is ready for grading and packing. When ready for packing, the moisture content vary from 15 to 20 percent.

Utilization of apricot pits: Apricot pits are of commercial value in that they are used for cookie fillers, activated carbon and chemicals. However, in India at the present the pits have little value. The pits are dried by spreading them on trays or on a clean concrete floor. They must be stirred periodically to encourage uniform drying. Apricot pits should not be sulfured.

Processing of kernels: Kernels may also be used in preparing macaroon paste by blanching to loosen the skins, grinding and then heating to about 122°F to hydrolyze amygdalin, steam distilling to remove most of the oil, and then pressing. The residue is then mixed with sucrose and heated in steam-jacketed kettles.

Dried Sulfured Apricots have Essentially the Following Composition

Water %	24
Protein (Nx 6.25) %	5.2
Fat (either extract) %	0.4
Ash %	3.5
Total carbohydrates %	66.9
Crude fiber %	3.2
Calcium, mg/100g	86
Phosphorus, mg/100 g	119
Iron, mg /100g	4.9
Vitamin A, I.U./100 g	7430
Ascorbic acid, mg/100 g	12
Thiamine, mg /100g	.01
Riboflavin, mg /100g	.16
Niacin, mg/100 g	3.3

Exploitation of wild bitter apricot for value addition: "Chuli," the wild apricot growing abundantly in the trans-Himalayan Ladakh in the Jammu and Kashmir states in India, yields bitter kernels called "khante" which are utilized primarily for extraction of apricot oil by the aboriginal communities. The oil extracted by traditional methods is used for cooking, religious and cosmetic purposes and is reported to have medicinal properties. However, no scientific characterization of the oil has been undertaken thus far. Wide variability exists in the region as far as oil content and composition are concerned.

3. *Apples*

India has not made much headway in the export of apples as its apple industry is not export-competitive and the efforts were largely concentrated in expanding the domestic market. Moreover, India lacks suitable varieties of apple for export markets and the proper quality control measures as well. India has not been able to do well in the export of apple despite achieving a large production of apples in the country. The main reason for this poor performance in export marketing is the existence of a large domestic market for apples. Aside, from HPMC, there is no other organized cooperative marketing organization involved in the export of apples. This is particularly necessary as most of the orchard owners are small and are not in a position to export individually. Coupled with this, the absence of suitable varieties of apples for exports and the lack of proper quality control measures have hampered the growth exports from India.

Shrink wrap packaging for extending shelf life: Shrink-wrapping reduces moisture loss, retards softening, and maintains characteristic freshness with reduced colour development during extended periods of storage and marketing. Recent advances in design and manufacturing of heat shrinkable films have stimulated interest in shrink-wrap packaging. Machinery is now commercially available that automatically applies film to farm produce of various sizes and shapes at rapid rates. Shrink wrapping may be considered as a supplement to proper refrigeration. Individual fruit wrapping also provide the optimum gas and humidity condition for maintaining the quality during transit and storage. As a result, it doubles or at times triples the shelf life of fruits and vegetable without any refrigeration. However, the extent of benefit from shrink-wrap packaging depends upon the type of produce, its physiological maturity and initial quality.

Study carried at CIPHET indicated that shrink wrapping and zip-lock envelop packaged apple (Royal Delicious) created a good protection over the fruit during storage under cold and ambient condition. Both the packaging drastically reduced the weight loss due to natural desiccation process. The firmness of the apple was maintained during the period of storage under shrink wrap and zip-lock packaging in comparison to the control sample that is unpackaged. Firmness retained more in cold stored samples as compared to room stored samples. The colour was also found good in case of both packaging as compared to control sample. Overall the results of the individual fruit packaging was found encouraging and it has positive effect on quality by extending shelf life of apples twice as compared to storage at ambient condition. Hence, this technique should also be evaluated for apples grown at high altitudes.

Suitable Harvest Equipments, Technologies of Processing for Post Harvest Management of Horticultural Crops of High Altitudes

1. Harvesting, Grading and Sorting Equipments

Non-Destructive Methods for Determining Maturity of Fruits

Eating quality and post harvest shelf life of ripe fruits depends on its maturity at harvest. Change of peel color and total soluble solids (TSS) are indicative parameters to measure maturity of most of the fruit. A maturity index is defined based on TSS and colour values. The model was developed to predict maturity using colorimeter at CIPHET. Colour and maturity index chart were also prepared to determine maturity. This technique can be employed to sort the fruits of high altitudes at well-organized markets and in processing plant.

Clippers: Fruit and vegetable clippers developed at CIPHET are efficient to harvest small fruits and vegetables besides holding it. The fruits can be saved by holding it for safe harvesting. The tool is modified over available clippers in the market, which not only harvest the fruits but also hole it thus reducing the soft fruits. This tool can be used for harvesting of Sea-buckthorn with modification as per the fruit requirement.

Safe harvesting: During harvesting/plucking of fruits, if it drops on the ground, it may cause damage due to impact. To avoid such damage, fruit saving Gadgets are developed at CIPHET, Abohar/ Ludhiana. Fruits after plucking from tree are thrown in the trough of fruit saver and collected in a box/container and thus inner damage of the fruits could be avoided. Further, fruits are also graded on size basis at appropriate places in these gadgets. Smallest grade fruit is collected first and largest grade is collected in last.

Fruit grader: CIPHET Tomato Grader which grades tomato on the basis of size into four grades i.e. 25-40 mm, 40-55 mm, 55-70 mm and > 70 mm and works on the principle of gravity and has mobility from one field to another used for other round fruits like apple. It has a capacity of 325 kg/hr with overall grading efficiency of 66per cent. While grading tomatoes roll down the pipes due to gravity and fall immediately wherever they find the space of their diameter. The collector is inclined at 10° so that the tomatoes slide directly in crates. The important feature of grader is its ability to adjust the gap between the pipes and inclination of grading table and hopper.

Fruit washing: Washing is the essential primary process unit operations for value addition of the fruit and vegetables at farm as for processors level. Fruits are washed not only to remove field soils, dust, pesticides but the surface microbial load also. A wide range of fruit and vegetables can be mechanically washed and can replace the prevalent practice of washing which involves drudgery and unhygienic conditions. Fruits and vegetables washer developed at Punjab Agricultural University, Ludhiana can be used both by the producers as well as processors.

2. *Pre-Cooling of Fruits*

To preserve quality and prolong the storage life many methods are available to pre-cool horticultural produce to optimum storage temperature. Studies have shown that pre-cooling of horticultural produce greatly increases storage life. Recommended pre-cooling methods commonly used for horticultural produce are forced-air cooling, hydro-cooling and Evaporative Cooling. Evaporative cooling is an inexpensive and effective method of lowering produce temperature. It is most effective in areas where humidity is low. Dry air is drawn through moist padding or a fine mist of water, then through vented containers of produce. As water changes from liquid to vapor, it absorbs heat from the air, thereby lowering the produce temperature. The incoming air should be less than 65 percent relative humidity for effective evaporative cooling. This method would be suitable for warm-season crops requiring warmer storage temperatures (7-12°C).

Ciphet ECS

ECS designed at CIPHET will fulfill the need of short-term storage of perishables like fruits and vegetables especially in hot and dry regions. Normally, fresh harvested fruits and vegetables to be stored in ECS. The fruits and vegetables prior to storage should be pre-sorted for any extraneous materials around their skin, rotting, damage by birds or insects any visible microbial infection, any physical damage/ bruise etc.

CIPHET ECS in no load condition is capable of dropping the temperature up to 20°C lower than outside and increasing relative humidity to 65 to 80 per cent, depending upon the outside conditions. When ECS is loaded at its 90per cent capacity, drop in temperature reduces by 3-4°C and RH reaches to 99 per cent depending upon the outside temperature and relative humidity.

3. Processing Technology and Value Added Products

Minimal processing: Minimal processing techniques like, trimming, dicing, treatment of antioxidants developed at CIPHET as well as by the other organization can be effectively used for quality maintenance and storage life of the fruits grown at high altitudes.

Extruded fruit based items: Local fruits like apricot, sea buckthorn; apple and pear can be processed into extruded products. A low cost single screw extruder, 25 kg/h and 10 HP motor for production of ready to eat snacks from fruits, cereals, millets and other ingredients developed at CIPHET, Ludhiana can be utilized.

Osmo dehydration: Technology for osmo dehydration of fruits developed at CIPHET, Ludhiana and CIAE, Bhopal can be utilized for fruits of high altitudes to develop novel products.

Technology for making fruit bars, candies etc.: Novel techniques for utilizing fruit and vegetables for making fruit bars, candies etc developed at CIPHET, Abohar/Ludhiana could be utilized for the development of novel products from fruits of high altitudes.

Ready to serve beverages; Technology for preparation of ready to serve beverages developed at CIPHET, Ludhiana can be utilized for processing of fruits of high altitudes.

Effective packaging of fruits and their products: Research carried out at CIPHET, has developed effective and economically viable packaging of fruits and their products. These techniques can be adopted for enhancing shelf life of fruits of high altitudes.

Health tonic/beverages/tablets/Golies: Utilization of fruits like Sea buckthorn for making health tonics, digestive tablets etc by using advanced methods of dehydration, mixing and binding can be promoted for diversification of fruits of high altitudes.

4. Drying Methods and Equipments for High Altitude Fruits

Most fruits and vegetables contain enough water to permit their spoilage by the micro-organisms and enzymes as stated earlier. To be a suitable substrate to support growth of micro-organisms, a food must have free water available for the micro-organisms. By reducing the free water content, thereby increasing osmotic pressures, microbial growth can be controlled. The dehydration reduces the amount of available moisture i.e. the water activity (aw) and hence, product becomes shelf-stable and is preserved for quite a long period because of removal of binding of water. The basic principle in the process of drying or dehydration is the removal of sufficient moisture to protect the product from spoilage.

Advantages of drying/dehydration over other methods of preservation are (1) The weight of a product is reduced to $1/4^{th}$ to $1/9^{th}$ its original or fresh weight and thus, the costs associated with its further operations is reduced drastically (2) Due to reduction in bulk of the product, it requires less storage space (3) Cost of processing is very low (4) No preservative is added for its preservation and (5) Per unit weight of fruit or vegetable nutrients concentration is very high.

Sun drying of fruits and vegetables is practiced in tropical and subtropical regions, where there is plenty of sunshine and practically little or no rains during the drying season. It is carried out in the open sun. Equipment essentially consists of drying trays and few other items like knives, lye bath etc. for food preparation prior to drying. Besides, various types of dryers such as solar driers, cabinet, tray and pan dryers, tunnel and continuous belt driers can be used for drying of fruits of high altitudes.

5. Oil Extraction

Oil content in Sea buckthorn seeds varies from 10-20per cent Supercritical fluid extraction (SCE) using carbon dioxide under high pressure is the modern method of oil extraction being used to extract the oil from the kernels. But investment for establishing such facility is very high. Therefore, extraction of Sea buckthorn seeds oil may be tried using expeller with suitable pre-treatments of seeds and process control. This way extraction of oil may be cheaper and economical than liquid CO_2 extraction.

Constraints in Fully Utilization of Fruits of High Altitudes

(a) *Poor infrastructure for harvesting:* Lack of proper harvesting gadgets leads to post harvest losses. Production of sea buckthorn in Leh is estimated at around 12,000 tones but owing to lack of appropriate harvesting practices almost 80 per cent is lost.

(b) *Below par horticultural development:* The problem of low productivity is further accentuated by the absence of suitable varieties used by the processing industry.

The lack of processing varieties is mainly due to their low and uncertain demands as compared with regular commercial varieties.

(c) *Poor domestic demand:* It is a fallacy to imagine that any industry can be developed on the basis of export alone. In fact, particularly in the case of processed fruit and vegetable industry, domestic demand has given a boost to the growth in several countries abroad and exports have been the off shoot. High cost of the processed products can be attributed to expensive raw materials, high cost of packaging practices are, by and large, obsolete except in certain units. Aseptic packing is the latest technology in the food processing industry. Since this system is highly capital intensive, not many units would be capable of is going in for this system. Only through up gradation of our technology can we meet the high quality standards of products required for export.

(d) *Ineffective publicity and promotion:* The impact of publicity on the sale promotion of certain under various trade names is very well recognized to make these products a household world.

(e) *Lack of standardization and quality control:* Sale of processed products inside the country and abroad has suffered considerably due to the poor and inconsistent quality standards of our products.

Recommended Strategy

A new orientation needs to be given to the fruit processing industry to upgrade nutrition, minimize post harvest losses, ensure remunerative returns to the growers, increase employment avenues in the countryside and earn foreign exchange. The attention needed is as follows:

I. *Increasing productivity of raw material*: To ensure higher productivity in fruits at high altitudes, steps to be taken include supply of planting material and quality seeds, high density planting, economizing fertilizer and water use. Staggering production and harvest, using appropriate plant protection measures and better post harvest management.

II. *Contract growing:* Contract growing of fruits, which has been successfully worked out side, has not been properly exploited in our country. Captive plantation around the processing units can be organized which can save on expensive packaging & transport & guarantee supplies of raw material at optimum maturity.

III. *By- products:* The edible portion in most fruits varies between 50-70 per cent. Instead of transporting fruits over long distances, it may be more appropriate to carry pulps and concentrates. This will save in transport cost and also help economic exploitation of residues in to value added by- products.

IV. *Cost distribution*: The cost of distribution of fruits and vegetables, including the cost on preparation, packing transport, marketing, retailing & control systems is 5-9 times for production cost. The type of handling system envisaged for the future by horticulture industry should be integrated distribution which will optimize all, the components of distribution to obtain the optimum benefit.

V. *Production diversification*: There is excessive dependence on mango based products. Export market outlets for these items being limited, the growth of the industry will be adversely affected with a narrow product range. Products from some minor indigenous fruits of Ladakh like apricots, sea Buckthorn should be promoted as "health foods" as the therapeutical properties of some of these fruits have long been exploited by the indigenous system of the Indian medicines, like Ayurveda. We now find herbal beverages as articles of international trade in partial substitution of tea and coffee.

VI. *Post harvest infrastructure:* Suitable post harvest management infrastructure which includes Harvesting methods and tools, Short term storage, collection, minimal processing, packaging, transport, cold storages, IQF, MAP, CAP etc. should be developed and adopted for sustainability of growing fruits of high altitudes.

VII. Plant pigments: With the ban on the use of coal tar dyes in several countries, India offers a great potential for extracting and supplying natural pigments in different hues and colours as the country possesses a large diversity of coloured fruits, flowers and other plant material.

VIII. Public policies: Guaranteed export promotion drive, especially in agricultural products is largely based on sound public policies. Government should keep a continuous watch on the compatibility of Indian products in foreign countries and come forth with instant remedial measures in terms of incentives, subsidies, duly relief and freight concessions.

IX. Organized promotion: Over seas promotional campaign to boost Indian products are very expensive. Publicity blitz needs to be launched by agencies like APEDA to promote trade marks, designs by distributing attractive folders, increasing press relation activities in individual countries through cultural programmes etc. This will go a long way to add an exotic touch to the sales efforts of our products.

Recommendations

1. Appropriate harvesting tools and techniques should be developed for Sea Buckthorn fruits
2. Research on dehydration of Sea Buckthorn leaves, fruits, pulp and seeds should be accelerated for value addition and development of technology for novel products.
3. Economical oil extraction process for Sea Buckthorn seeds should be evolved.
4. Post harvest operations such as grading, pre-treatment; drying, packaging etc. should be optimized for quality end product of apricot.
5. Screening of apple genotypes suitable for processing and value addition and evolve suitable post harvest management practices to enhance their shelf life and post harvest quality.
6. Value addition in terms of minimal processing, intermediate processing, ready to eat convenience foods and health foods should be prepared and promoted using fruits of high altitudes.
7. Technology/process for novel products from fruits such as Sea Buckthorn, apricot and apple available in high altitudes should be evolved for more economic returns.

References

Anonymous. Hand book of FOOD dehydration and drying. Published by Engineers India Research Institute. N. Delhi, pp. 55-57.

Anzueto, C. R. and Rizvi, S. S. H. 1985. Individual Packaging of Apples for Shelf Life Extension. Journal of Food Science. 50(4), 897–900.

CIphet, Show Case/Annual Report. 2007. CIPHET, Ludhiana.

Dhillon, B.S. and Rana, J.C. 2003. Temperate fruits genetic resources management India- issues and strategies. ISHS Acta Horticulturae. 662: VII International Symposium on Temperate Zone Fruits in the Tropics and Subtropics.

Dwivedi, D.H., Ram, R.B. 2006. Chemical composition of bitter apricot kernels from ladakh, India. ISHS Acta Horticulturae. 765: XXVII International Horticultural Congress-IHC2006: International Symposium on Plants as Food and Medicine: The Utilization and Development of Horticultural Plants for Human Health.

Dwivedi, S.K., Attrey, D.P., Paljor, E. and Singh, B. 2003. Conservation of genetic diversity in apricot in the cold arid regions of India. ISHS Acta Horticulturae, 623: XXVI International Horticultural Congress: Plant Genetic Resources, the Fabric of Horticulture's Future.

Sharma, D.P., Sharma, N., Bawa, R., Kaler, R. 2005. Potential of Apricot growing in the Arid-cold Desert Region of North Western Himalayas, Ishs Acta Horticulturae. 696: VII International Symposium on Temperate Zone Fruits in the Tropics and Subtropics - Part Two.

Harris, Charles, C. 1989. Shrink wrap package with venting openings. (U.S. Patent No. 4815603)

Kaushal, Lal, B. B. and Sharma, P. C. 1995. Apple, In: Hand book of fruit science and technology. Salunkhe D. K. & Kadam S. S. EB (ed), Marcel Dekker, Inc., New York, 91.

Loesecke, Harry W. Von. 1998-99. Drying and dehydration of Foods. Allied Scientific Publication. Bikaner, pp. 42-49.

Ranjith A. 2007. Journal of the Science of Food and Agriculture, Vol. 86, pp. 2345-2353.

Singh, R., Mishra, S. N., Dwivedi, S. K., Ahmed, Z. 2006. Current Science. 91(10), 25.

Suri, Vivek. 2008. Ladakh: Farmers demand higher price for farm products.

Verma, L.R. and Joshi, V. K. 2000. Post Harvest Technology of fruits and Vegetables. Vol. 2. pp. 870-873.

❑❑❑

10

Challenges and Prospects for Food and Health Security of Soldier in North East India – DRL Perspective

B. J. GOGOI AND R. B. SRIVASTAVA
Defence Research Laboratory, DRDO, Tezpur, Assam, India

INTRODUCTION

In North East India, due to its strategic importance and close proximity of international borders, large numbers of troops are located at different stations. Due to the peculiar topography, remoteness and lack of infrastructure facilities as well as hot & humid climatic conditions in the region; collection and transportation of food materials for the troops and their health care aspects are hampered. The troops deployed along the hilly combatant terrain of eastern Himalayan region are generally depending on the supplies from plains. Defence Research Laboratory (DRL), Tezpur being the only Defence R&D laboratory in the region is catering to the needs of food & health security for the troops for bringing betterment in their life and living conditions in the entire north eastern region.

Food Security for Soldiers in North East India

An important aspect of the current R&D programmes at DRL, Tezpur is the utilization of natural resources and value addition to the biodiversity in the region, to cater to the needs of the troops. North-

East India is one of the mega hotspots of biodiversity in the world and is a treasure house of unconventional fruits & vegetables viz., *Pyrus pashia, Litchi acuminata, Pyrus pomifera, Dendrobenthamia capitata, Dillenia indica, Citrus species, Microcos peniculata, Averrhoea carambola, Alternanthera sessilis, Basella rubra, Brassica rugosa, Centella rotundifolia, Ipomoea reptans* etc. which are rich sources of vitamins, minerals and antioxidants.

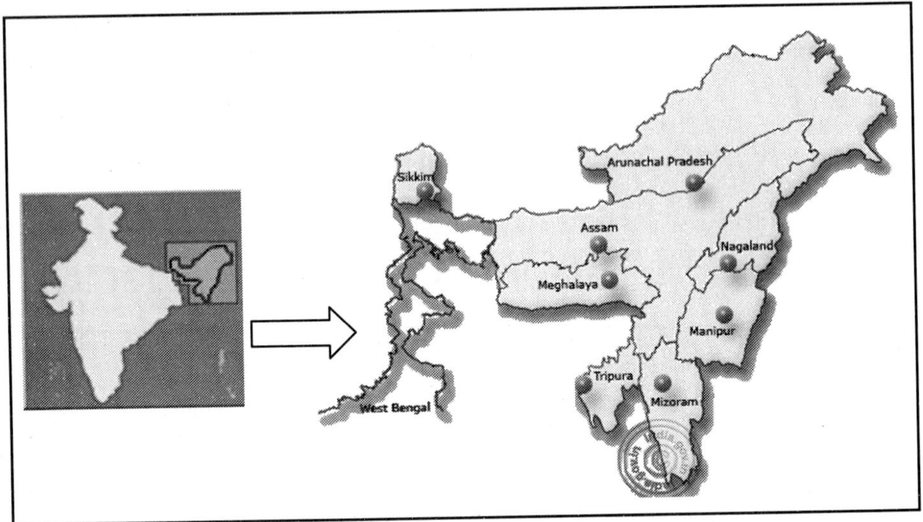

Fig. 1: Map of North East India

Value added products developed from these unconventional fruits of the region have started showing far reaching impact both in defence as well as in civil sectors. In addition to mushroom soft drink, a value added product from the fruit of *Dillenia indica* as health drink has been developed. The *Dillenia* health drink is rich in vitamin C, B_2, B_6 and flavonoids with calorific value of 141.9 Kcal/100g. The LD_{50} of the Herbal Health Drink of *Dillenia indica* is > 5000 mg / kg body weight in rats. The laboratory has also recently initiated work on development of antioxidant based herbal nutraceuticals as adaptogen for hot & humid climatic conditions of North East India.

The hottest chilly of the world i.e., *Capsicum frutescens* var. *nagahari* of Tezpur has also been identified by the laboratory. This variety of chilly contains maximum capsaicin and dihydrocapsaicin contributing a pungency of 8,55,000 Scoville Heat Units. Apart from food supplement to the troops, this has a tremendous potential as a non-lethal weapon for riot control and as an anti-terrorism agent.

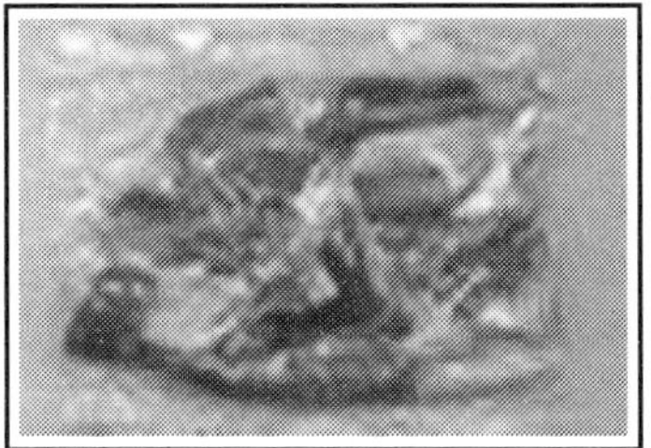

H.T. of *Dillenia*

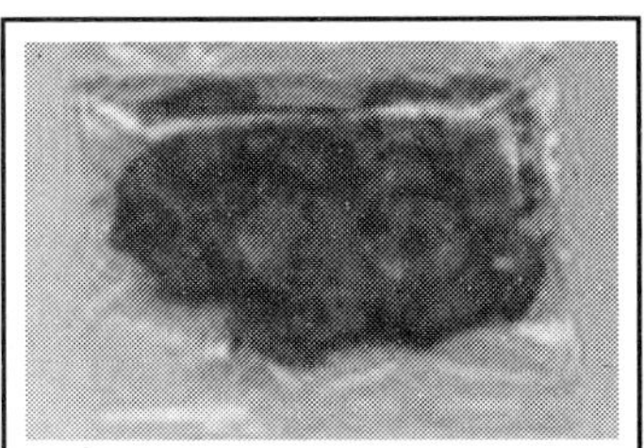

Pickle of *Dillenia*

Health Drink fruit of *Dillenia indica*

Fig. 2: *Dillenia indica*

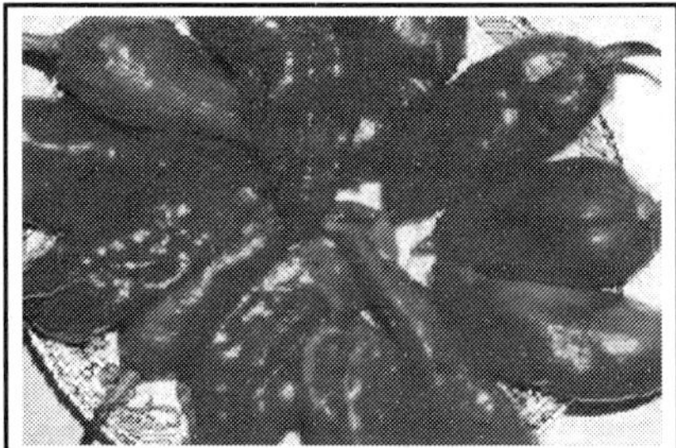

Fig. 3: *Capsicum frutescens*

Fig. 4: Vegetable Production under Protected Environment at 17 Mtn Div.

Work has also been initiated for troops stationed at high altitude areas of Sikkim for production of fresh vegetables through green house technologies. The production of vegetables under protected environment in high altitude areas of North East India will definitely meet partial demand in terms of fresh vegetables for the troops.

Mushrooms are highly prized commodity which is liked by many for their rich protein, vitamins and minerals. It is a good source of nutrition and requires less time, labour and money for their cultivation. The protocol for low cost production of oyster and button mushroom has been developed using locally available agricultural wastes as substrate. A wild edible mushroom "Puffball" was domesticated and spawns production protocol has been standardized for the species. Composting is a process to dispose organic wastes into a valuable fertilizer. Vermicompost is a safe and eco-friendly technique for rapid decomposition of biodegradable wastes using epigamic earthworms. It is a balanced organic fertilizer that helps in development and production of all types of crops. A protocol for low cost vermicompost technology using locally available earthworm *Perionyx excavatus* has been developed for round the year production of vermicompost.

Mushroom Soft Drink

Low cost Mushroom Production

Low cost Vermicompost Technology

Fig. 5

Health Security for Soldiers in North East India

1. Vector Borne Disease & Management

Malaria and other vector borne diseases continue to be a major health problem for troops & civil populace in North East India. The hot and humid climate of North Eastern India is very conducive for proliferation of haematophagous insects and leeches. Mosquito is the most dreaded insect which alone is responsible for transmitting malaria, filariasis, Japanese encephalitis etc. and invites a considerable attention of health services. DRL, Tezpur has, therefore, taken up the issue of combating these insect-vectors in this region by adopting systematic survey of vectors and integrated control measures. A number of malaria awareness camps were organized in the Army cantonments under 4 Corps and 3 Corps from time to time. In these camps, some minimum protective measures as follows were deliberated and demonstrated.

(i) Indigenous cultural practices to keep away mosquitoes.

(ii) Adoption of personal protective measures, viz., use of mosquito repellent, impregnated mosquito net, wearing of full trouser, immediate blood examination of suspected fever cases etc.

(iii) Environment control like residual spray in barracks, draining out of stagnated water body, vector control by ULV fogging, maintenance of neighboring population free from malaria etc.

(iv) Training on impregnation of bed net with K-Othrine was given to jawans.

DRL, Tezpur has developed few products viz., Mosquito Repellent Herbal Vaporizer, Herbal Room Freshner cum Mosquito Repellent, Larvicidal Floating Tablet, Moskill (biolarvicide formulation) for the control of malaria & other vector borne diseases in the region.

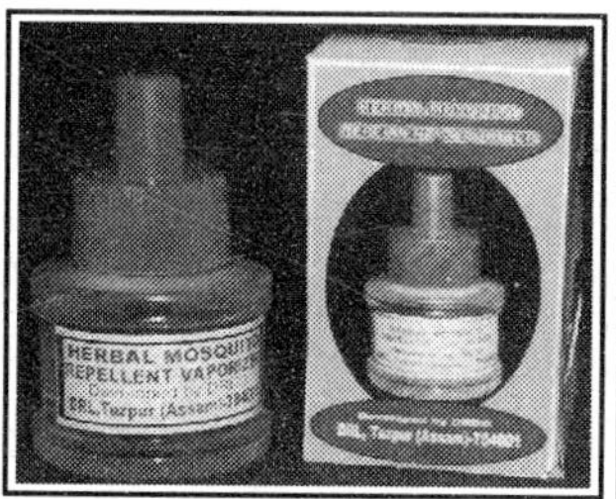

Herbal Mosquito Repellent Vaporizer

Larvicidal Floating Tablet

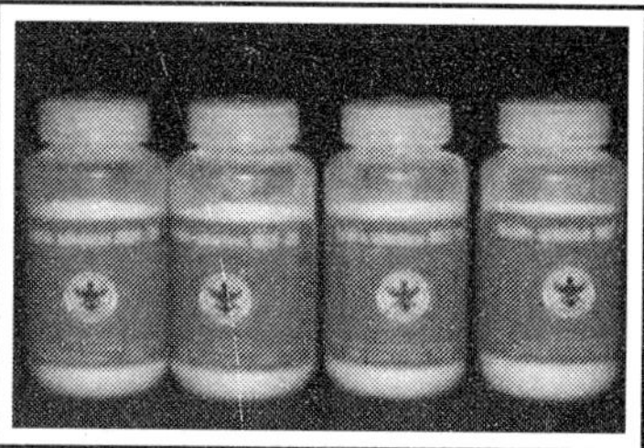

Biolarvicide Formulation

Fig. 6

A large number of antimalarial plants are available in North East India which are being used by the local people effectively to control malaria fever. The laboratory has identified two rare species of antimalarial plants viz., *Gomphostema niveum* and *Gomphostema critinum* which have been found to be very effective against malaria.

Gomphostemma crinitum

Gomphostemma niveum

Fig 7

In the coming years, the laboratory envisages studies on bionomics and distribution of species complexes of disease vectors in North East India, the extent of insecticide resistance in and around Army cantonments using biochemical and molecular markers for better implementation of vector control measures, drug resistance studies on *Plasmodium falciparum* using conventional methods and molecular markers for better implementation of proper treatment policy in Army cantonments, development of insecticide treated Army uniform for prevention of man-mosquito contact and other haematophagus insect bites, macro level mapping of insecticide and drug resistance in North East region with the help of RS-GIS for effective management of vector borne diseases, search for indigenous natural products to control malaria and vectors, distribution of Simulids and their control and prevalence of arboviral diseases and their vector control.

2. *Water Quality Surveillance & Management*

High iron, arsenic & fluoride content in drinking water are major problem in most parts of North Eastern region. Problems related to heavy metal toxicities and water borne diseases are continuously being reported. The troops deployed in remote areas of this region are vulnerable to these contaminations. This laboratory has carried out survey on water quality in the North East India and has acquired expertise in the area of water quality management.

Iron Removal Unit (IRU) developed by DRL, Tezpur has been appreciated by troops, because of its capability to reduce iron content in drinking water to a great extent. The unit is designed for 300 ltrs/hr output and can cater to the need of small size population particularly in army barracks/units. Iron content of treated water through this unit is brought down to below 0.3 mg/ltr with feeding water having iron content upto 40 mg/ltr. Contamination of drinking water with heavy metals like fluoride, arsenic etc. is reported recently in some pockets of the region. Due to this apprehension, the troops are now suggesting for development of an integrated water purifier either to remove these contaminants or to bring it down within permissible limits in drinking water.

The Water Testing Field Kit has been developed for assessing 11 major physico-chemical parameters viz., pH, total hardness, total dissolved solid, chloride, residual chlorine, fluoride, sulphate, nitrate, iron, arsenic and bacteriological test in drinking water. The kit can be used in field condition by a semi-skilled person for testing potability of water.

IRUs delivered to Eastern Command

Hands on demonstration of IRU to Jawans

Dearsenification Unit

Water Testing Field Kit

Arsenic Detector

Fig. 8

Recently, the laboratory has developed a dearsenification unit and arsenic detector for removal and detection of arsenic in water respectively. The laboratory envisages development of an integrated technology for potable water and development of prototype water filtering units for removal of major contaminants.

3. Biotechnological Approaches to Combat Dermatophytes

In North East India, skin diseases are most frequent among the common ailments. Out of various causative agents, microbial infections are the most predominant among the civilians as well as troops posted in North East region. The hot & humid climate prevailing in the region favours the growth of a number of pathogenic fungi. They cause superficial skin diseases in humans and animals. The infection is not fatal but highly contagious and cause intense itching/restlessness and thus lowers the work performance of the troops. Preliminary studies on the prevalence of common skin infections & causative agents have been made in collaboration with Army hospital & local medical practitioners. Potential plants with broad spectrum of activity against various human pathogenic fungi namely *Trichophyton mentagrophytes, T. rubrum, T. tonsurans, Microsporum fulvum, M. gypseum, Candida albicans* and *Trichosporon beigelii* have been identified. *In vivo* efficacy study, toxicity test etc. are under progress. Herbal formulation as skin ointment for all the skin related infections, which is under development in the laboratory, will go in a long way to support troops located in remote jungle areas.

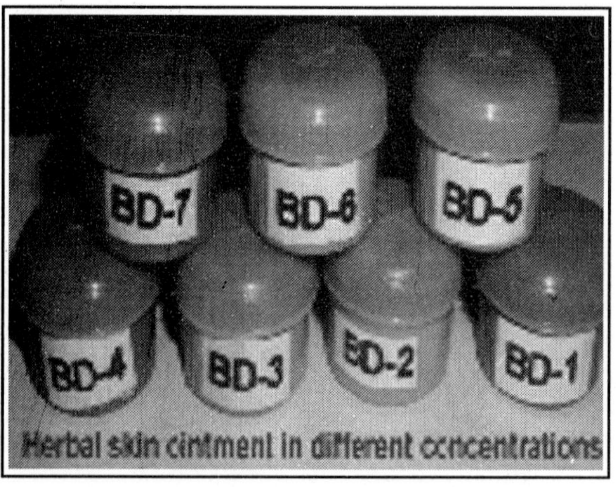

Fig. 9: Antifungal Herbal Skin Ointment

It is increasingly being recognized that simple, rapid and specific diagnosis using molecular tools by characterizing genomic DNA of the causative agents would be helpful in early diagnosis, and proper treatment. On the other hand, treatment of skin ailments is also difficult due to resurgence of resistant strains and may be due to shortcomings in presently available remedies. In this regard, the North East region has immense potential for development of herbal formulations against such skin infections. The laboratory is working towards development of Polymerase Chain Reaction based diagnostic tools for rapid detection of microbes that cause skin ailments and also development of suitable herbal formulations as remedial measures.

New Technology Initiative for Energy Security

The crisis of fossil fuel is increasing day by day and will hit the defence sector in near future. Biofuels have been seen as an environmentally less-damaging alternative to petrol and diesel which would help combat global CO_2 emissions and help to reduce dependence on oil. Biodiesel can be produced from a number of oil bearing plants available in the North East region viz., *Jatropha curcas, Ricinus communis, Mesua ferrea* etc.

Since the region is endowed with high algal diversity, R&D work has also been taken up related to biofuel production technology from microalgae as an alternative energy source. The four most abundant classes of microalgae are diatoms (Bacillariophyceae), green algae (Chlorophyceae), blue green algae (Cyanophyceae) and golden algae (Chrysophyceae). The work on biofuel production technology from microalgae will definitely pave a new horizon in future and will make a strong economic sense.

CONCLUSIONS

The major R&D programmes of the laboratory in the area of water chemistry, medical entomology, agricultural sciences, biotechnology and phytochemistry have provided ample technological support in terms of food & health security to the troops stationed in the plains and high altitude areas of North East India. In continued effort towards the exigencies and need of the troops serving in inaccessible areas and inhospitable environments having severe climatic conditions, the laboratory has dedicated its services for the betterment of their life and living conditions. The scientists of the laboratory are making sincere efforts to achieve innovative R&D products and time bound targets for the troops inspite of many bottlenecks and limited resources. The laboratory has continuous interactions with services, academia and industry people for collaborative R&D programmes on future product requirements of services, its development & productionization and in realising its spin off benefit to the civil populace.

Defence Research Laboratory (DRL), Tezpur in years to come, will be looking forward to meet the future challenges related to food & health security for troops stationed in border & forward areas of North East India through its continuing R&D endeavours.

References

A' Georgii, A., Korting, H.C., 1991. Antifungal susceptibility testing with dermatophytes. Mycoses 34, 193-199.

Bhattacharya, S., Sharma, C., Dhiman, R.C., Mitra, A.P. 2006. Climate change and malaria in India. Current Science. 90(3) 369-375.

Ceccato, P., Connor, S.J., Jeanne, I., Thomson, M.C. 2005. Application of geographical information systems and remote sensing technologies for assessing and monitoring malaria risk. Parasitologia. 47, 81-96.

Chakraboti, D., *et. al.* 2000. Fluorosis in Assam, India. Current Science 78(12), 1421-1423.

Chakraboti, D. *et.al.* 2004. Groundwater arsenic contamination and its health effects in the Ganga-Meghna-Brahmaputra plain. J. Environmental Monitoring 6, 74.

Hudson, B.J.F., Ed., 1990. Food Antioxidants. Elsevier Applied Science, London.

Myers, N., Mittermeier, R.A., da Fonseca, G.A.B., Kents, J. 2000. Biodiversity hotspots for conservation priorities. Nature. 403(6772), 853-858.

Scartezzini, P., Speroni, E.J. 2000. Review on some plants of Indian traditional medicine with antioxidant activity. J. Ethnopharmacol. 71, 23-43.

Singh, G., Singh, H.B., Mukherjee, T.K. 2001. Ethnomedicine of North East India. NISCAIR (CSIR), New Delhi.

Tiwari, A.K. 2001. Imbalance in antioxidant defence and human disease: multiple approach to natural antioxidants therapy. Current.Science. 81, 1179-87.

□□□

11

Soil and Water Salinity in Leh and Ladakh Cold Desert and Management Options

GURBACHAN SINGH
Central Soil Salinity Research Institute, Karnal-132 001, Haryana, India

INTRODUCTION

Leh and Ladakh region is a part of the Indus valley and lies between 30° to 36° east latitude and 76° to 79° North longitude. This is India's highest plateau with most of the area being about 3000 m above the sea level. Historically, this region includes main Indus Valley, the more remote Zanskar (in the south) and Nubra Valley (to the north over Khardung La), the almost deserted Aksai Chin, Kargil and Suru Valley area to the west (Fig. 1). This region is mainly lying in rain shadow of Himalayas which denies entry to the monsoon clouds. This region constitutes about 85% of the Indian cold desert where the annual average rainfall is about 50mm. The main source of irrigation water is the winter snowfall on the mountains. The summer weather is dry and pleasant with average temperatures between 10 to 20°C while in winter, the temperature may dip to -15°C. Summer season is relatively shorter than winter season. Most of the agricultural crops are grown during summer in the lower reaches of valleys. The vegetation is extremely sparse except along stream beds and the wetlands where several wild herbs and shrubs can be seen. For several centuries, this region enjoyed a stable and self reliant agricultural economy based on growing wheat, barley and peas and keeping livestock especially the yak, dzos (yak & cow cross breed), cows, sheep and goats. The Ladakhies developed a small scale farming system where land is irrigated by a system of channels which funnel water from the melted ice and snow of the mountains. Apricots and pashmina are the two commodities for export from this region. Major cultivation constituting lush irrigated farmland is confined in major valleys like Indus Valley, Dras Valley, Nubra Valley, Salt lake Valley, Suru Valley and Shyok Valley (Fig. 2). Wheat, barley, millets, turnip, radish, peas, black peas, grapes, apricots and melons are grown in these valleys. In the salt lake valley, there are two lakes; the fresh water Panluk Lake with an area of about 2 square kilometers and the salt water Tsokar Lake with an area of about 10 square kilometers. Soil salinity is quite wide spread along the Indus River particularly affecting the pastures (Fig. 3). No significant work seems to have been done in the region to identify the source of salts or to develop reclamation and management practices for improved productivity. While traversing to Pangong lake, a number of pastures affected by waterlogging and soil salinity can be observed (Fig. 4). It was also observed that the soil salinity develops along the shores of the Pangong lake as the water line residue due to evaporation (Fig. 5). Since 10 square kilometer lake is salty in the salt lake valley, this raises suspicion that salts are geogenic in nature and might pose several difficulties in reclamation of agricultural lands. In case the present trend of salinity related desertification continued the valley lands may become salinized and thus affecting agriculture sustainability in this region. A special technology package needs to be developed for reclamation and management of salinity affected soil and water in Ladakh valley. A canvas of technologies developed by CSSRI, Karnal for the reclamation and management of salt affected soils in India are discussed in the present paper. Some of these technologies may be relevant to reclaim and manage salinity affected lands in Ladakh valley after incorporating location specific adjustments. The basic idea behind this paper is to create awareness about salinity problem in cold desert and the status of knowledge available around the world to deal with salinity in soil and water.

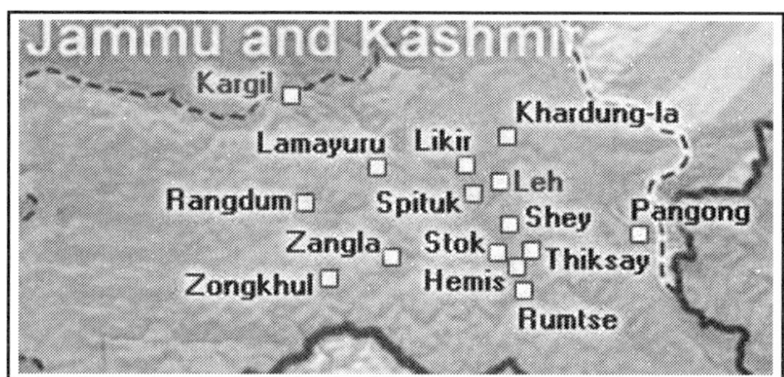

Fig 1: Part of the Leh-Ladakh region in Jammu and Kashmir (*Source:* Website)

Fig 2: A view of irrigated agriculture in the valley of Ladakh

Fig 3: A view of salt affected soil near stoklam village (Leh) on the banks of Indus River

Fig 4. A view of the salt affected pastureland nearly 120 Km from Leh on way to Pangong

Fig 5. A view of the Pangong lake showing salinity development on the shores with receding waterline

Soil Salinity Problem in Leh-Ladakh Region

To ascertain salinity problem in soil and water in Ladakh region, samples were taken at several places between Leh and Pangong lake. It was astonishing to note that salinity of soil near the bank of Pangong lake was more than even the sea water salinity. The soils near Tungtsay and Choglamsor have high pH and very high electrical conductivity. The sodium concentration near the bank of Pangong lake varied from 2520 to 3817 me/l. The Ca+Mg concentration was also very high whereas carbonates were almost absent. Similarly, Cl and S concentration was also very high. The detailed analysis is given in Table 1.

Salinity Issues at Global and National Level

Salt-affected soils are reported to cover a sizeable part of the land surface in more than 100 countries. Based on the FAO/UNESCO Soil Map of the World, a total of 953 million (m) ha, covering about 7-8 percent of the land surface is suffering from this malady (Szabolcs, 1979), more than 50 percent of which is sodic. The salt affected soils are reported to occupy 42.3% area of Australia, 21.0% of Asia, 7.6% of South America, 4.6 % of Europe, 3.5% of Africa, 0.9% of North America and 0.7% of Central America (El-Mowelhey, 1998). Depending upon the source, agency and methodology used estimates about salt-affected areas in India differs. As per reconciled figures of CSSRI, Karnal; NBSSLUP, Nagpur and NRSA, Hyderabad about 6.74 million ha area is affected by salinity and sodicity in India (Table 2, Fig.6).

The salt affected soils reduce the productive capacity of about one-third of the potential arable land in the world that are either lying barren as merely poor grazing ground or have very low biological output due to unfavorable environment for plant growth. In nearly 2.3 percent of the total geographical or about 4% of the total culturable area in India, crops do not grow or yield poorly because of excess salts in the root zone. The problem is becoming more acute with expanding irrigation. The nature and severity of the problem varies from region to region depending upon the topographical situation, hydrological and climatic conditions, drainage availability, land use and cultural practices etc. From the management point of view the salt affected soils are divided into two broad categories - (1) Sodic (alkali) soils, and (2) Saline soils, though at places saline-sodic soils are also met with. However, the management of saline sodic soils will be more similar to that of the sodic soils. Such soil needs extra water for leaching of soluble salts. The sodic soils have higher proportion of sodium in relation to other cations in soil solution and ion exchange complex. The growth of most crop plants is adversely affected because of impairment of physical conditions, disorder in nutrient availability and suppression of biological activity due to high pH exceeding even 10.0 and exchangeable sodium percent upto 90 or so. The solution salts contain preponderance of sodium carbonates and bicarbonates capable of alkaline hydrolysis, thereby saturating the absorbing complex with sodium. The sodic soils of the Indo-Gangetic plain are generally gypsum ($CaSO_4.2H_2O$) free but are calcareous with $CaCO_3$ increasing with depth, which is present in amorphous form, or in concretionary form or even as an indurate bed at about one meter depth. The accumulation of $CaCO_3$ generally occurs within the zone of fluctuating water table. The dominant clay mineral is illite. The processes which target on dissolution of $CaCO_3$ have significant role in reclamation of alkali or sodic soils. These soils are deficient in organic matter, available N, Ca and Zn. Certain micro-nutrients present problems of either deficiency or toxicity. Toxicities of Al, Mn, Fe sometimes pose problems in wheat when over irrigated and results in yellowing of the crop. The major factors responsible for formation of alkali soils in the Indo-Gangetic region include: irrigation with ground waters containing excessive quantities of carbonate and bicarbonate ions, rise in groundwater due to introduction of canal irrigation and salt laden run-off from the adjoining areas/undrained basins. The inland saline lands are widespread in the canal irrigated, arid and semiarid regions. These soils are characterized by the presence of excess neutral soluble salts like chlorides and sulphates of sodium, calcium and magnesium. Sodium chloride is the dominant salt. High soil salinity is often accompanied by high water table, often within 2 m of soil surface. Sub soil waters are generally salty and, therefore, their use for irrigation presents major constraints. In general, these soils have good physical properties but poor natural drainage. The formation of saline soils is generally associated with rise in watertable due to introduction of irrigation and inadequate drainage.

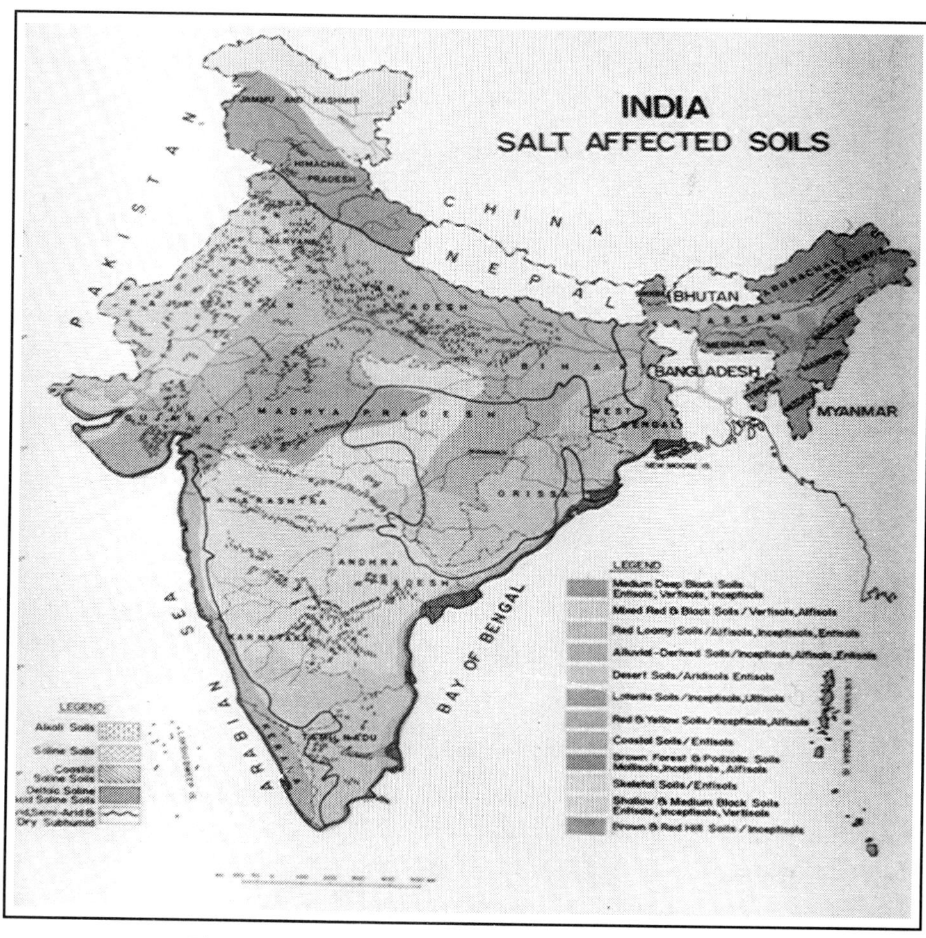

Fig. 6. Distribution of salt affected soils in India

Reclamation and Management Options

Relative Salt Tolerance of Plants

The plants and cultivated crops, in general, differ in their ability to tolerate higher salt concentration in the root zone soil. Various mechanisms involved in such tolerance includes ability of the plants to take up K inspite of high sodium concentration; exclusion of salts at the root, stem, branches and node/stem level; accumulation of salts in the vacuole; excretion of salts through the leaves and osmo-regulation through the accumulation of organic acids. Based upon these criteria several species have been screened and relative tolerance of different crops to exchangeable sodium in the soil solution is given in Table 3. Depending upon the climate, water availability and length of growing season crops can be chosen out of this list for growing in salt affected soils of Ladakh region.

Salt Tolerant Varieties

The choice of tolerant varieties within a crop is essential for better production and economic returns from salt-affected soils. Such varieties have been identified and also evolved through selection and hybridization in crops like rice, wheat, mustard, barley and sugarbeet. Some of the promising varieties recently released and in pipeline for general cultivation in salt-affected soils are listed in Table 4. Rice is a warm season crop and may not be suitable for Ladakh region. However, salt tolerant wheat, mustard and barley varieties can be tried to make productive use of salt lands.

Chemical Reclamation Technology

Chemical based technology involves application of chemical amendments like gypsum as a source of Ca to exchange with Na on the exchange complex. The dose of an amendment required for reclaiming sodic soil is governed by its initial ESP, tolerance level of crops to be grown, texture, mineralogy of soil and depth to be reclaimed. It also depends on the nature of soil, its stage to deterioration and crops to be grown. Field experiments further showed that for initiating the reclamation of sodic soils and for cultivation of shallow rooted crops like rice, wheat, barley and berseem, application of 50 percent of the gypsum requirement, which amounts to 10-15 t/ha in 0-15 cm soil, is sufficient. Field studies have shown that mixing gypsum in shallow depths was more beneficial than mixing in deeper layers. Mixing limited quantities of gypsum in deeper depths results in its dilution and therefore, lesser soil improvement. The organic matter content of sodic soils is often very low. Organic manures, including farmyard manure, composts and green manures, have long been known to facilitate the reclamation of alkali soils. Decomposition of organic matter results in the evolution of carbon dioxide and organic acids; lowering of soil pH and release of cations by solubilization of $CaCO_3$ and other soil minerals, thereby increasing the electrical conductivity and replacement of exchangeable sodium by cations like calcium and magnesium and thus lowering the exchangeable sodium percentage. Organic materials together with inorganic amendments quicken the reclamation of sodic soils and therefore have an important place in reclamation of sodic soils.

Table 1: Analysis of Soil/Water samples from Leh

Location/ place	pH	ECe (dS/m)	Na Meq/l	K Meq/l	Ca+Mg Meq/l	CO3 Meq/l	HCO_3 Meq/l	Cl Meq/l	SO_4 Meq/l
Soil									
Bank of Pangonglake	8.06	102.0	2519.6	88.5	1600.0	0.0	125.0	1400.0	2576.0
Bank of Pangonglake	7.25	186.7	3817.4	329.6	2325.0	0.0	150.0	2550.0	3645.5
Tangtsay	9.21	21.6	276.0	24.43	75.0	0.0	30.0	200.0	142.4
Choglamsor	9.35	19.2	208.4	21.24	10.0	30.0	50.0	150.0	7.8
Water									
Pangonglake 1	9.32	14.90	121.30	4.55	54.0	40.0	11.0	92.0	34.65
Pangonglake 2	9.25	14.85	117.8	4.69	50.0	38.0	11.0	90.0	30.52
Muglack 1	8.20	1.30	3.03	0.13	10.6	0.0	2.0	1.2	8.56
Muglack 2	7.52	1.15	1.06	0.18	10.4	0.0	5.8	1.4	3.85
Choglamsor (River Sindhu)	7.82	0.34	0.48	0.48	3.4	0.0	3.0	1.8	0.10

Table 2: Extent and distribution of salt-affected soils in India

States	Area (in m ha.)
Gujarat	2.23
Uttar Pradesh	1.37
Maharashtra	0.61
West Bengal	0.44
Rajasthan	0.38
Tamil Nadu	0.37
Andhra Pradesh	0.27
Haryana	0.23
Bihar	0.15
Punjab	0.15
Karnataka	0.15
Orissa	0.15
Madhya Pradesh	0.14
Andaman & Nicobar Islands	0.08
Kerala	0.00
Total	6.74

Source: CSSRI, NBSSLUP, NRSA (Reconciled figures)

Table 3: Relative tolerance of crops and grasses to soil exchangeable sodium (ESP)

Tolerant ESP, 35-50	Moderately tolerant ESP, 15-35	Sensitive ESP <15
Karnal grass (*Leptochloa fusca*)	Wheat (*Triticum aestivum*)	Gram (*Cicer arietinum*)
Rhodes grass (*Chloris gayana*)	Barley (*Hordeum vulgare*)	Mash (*Phaseolus mungo*)
Para grass (*Brachiaria mutica*)	Oat (*Avena sativa*)	Chickpea (*Cajanus cajan*)
Bermuda grass (*Cynodon dactylon*)	Shaftal (*Trifolium resupinatum*)	Lentil (*Lens esculenta*)
Rice (*Oryza sativa*)	Lucerne (*Medicago sativa*)	Soybean (*Glycine max*)
Dhaincha (*Sesbania aculeata*)	Turnip (*Brassica rapa*)	Groundnut (*Arachis hypogea*)
Sugarbeet (*Beta vulgaris*)	Sunflower (*Helianthus annus*)	Sesamum (*Sesamum oriental*)
Teosinte (*Euchlaena maxicana*)	Safflower (*Carthamus tinctorus*)	Mung (*Phaseolus aureus*)
	Berseem (*Trifolium alexandrinum*)	Pea (*Pisum saccharatum*)
	Linseed (*Linum siuqatissimum*)	Cowpea (*Vigna unguiculata*)
	Onion (*Allium cepa*)	Maize (*Zea mays*)
	Garlic (*Allium sativum*)	Cotton (*Gossypium hirsutum*)
	Pearl millet (*Penisetum typhoidies*)	
	Cotton (*Gossypium hirsutum*)	

Table 4: Important varieties for sodic and salinity stress (based on the stress level with less than 50% reduction in grain yield)

Crop	Level of stress	Varieties	Remarks
Rice	pH2 <10.2 and ECe <10 dSm^{-1} pH2 <9.9 and ECe <7 dSm^{-1} Medium tolerant pH2 < 9.6	CSR-10, CSR11, Try1, IR4630-22-2-5-1-3, CSR-13 and CSR19CSR-21, CSR23, CSR27, CSR31 and CSR 42 CSR 30	Dwarf, early maturing (120 days) and high yielding salt tolerant varieties CSR 21 is long slender variety Basmati type variety
Wheat	Tolerance pH2 <9.6 ECe <8.5 dSm^{-1} Medium tolerance pH2 <9.3 ECe <6.5 dSm^{-1}	Kharchia 65, KRL 2-10, KRL-19 and WH 157 KRL 1-4, KRL-19, PBW 65 and Raj 3077	Wheat varieties cannot grow economically beyond pH2 9.3
Barley	Up to pH2 9.3 ECe 11.0 dSm^{-1}	CSB1, CSB2, CSB3, DL200, Ratna, BH97, DL 348	Hulled barley varieties. Economic yield can be obtained up to pH2 9.6
Mustard	pH2 9.2 ECe <8 dSm^{-1}	CS-52 and CS614-4-1-4, Pusa Bold, Varuna, and Kranti	Economic yield can be obtained up to pH2 9.2
Sugarbeet	pH2 9.5 to 10.0 ECe 10 dSm^{-1}	Ramonskaya-06, Polyrava-E, Tribal, and Maribo-Resistapoly	Sugar beet is highly tolerant to alkalinity up to pH2 10.0

High pH, excess exchangeable sodium, higher amounts of $CaCO_3$, negligible to low organic matter content and inhospitable soil physical environment of alkali soils influence the transformation and availability of native and applied fertilizer nutrients. Black alkali soils are known to be deficient in both the available nitrogen as well as phosphorus whereas alluvial alkali soils are deficient in nitrogen but medium to high in available phosphorus. The efficiency of nitrogen fertilizer is generally low in alkali soils partly due to volatilization losses as ammonia and partly due to sub-optimal plant growth. To compensate the losses of nitrogen in sodic soils, it is recommended that crops should be fertilized with 20% extra nitrogen than recommended for normal soils. In order to have higher efficiency, split application of nitrogen in rice and wheat is recommended.

P application in sodic soils under reclamation should be based upon soil test for available P. The clay fraction of the Indo-Gangetic alluvium is dominated by illite which is a good reserve of potassium in these soils. The rate of potassium release and its uptake are not ordinarily affected in sodic soils. As the water soluble and ammonium acetate soluble K^+ fraction is at a high to very high rate in these soils, no responses to applied potassium have been reported. In order to maintain Na : K balance in sodic soils, it is always necessary to maintain a high K^+ level in these soils. Singh *et al.* (1987) and Singh, 1998 recommended application of 20-40 kg Zn ha^{-1} during reclamation phase of sodic soils. Integrated use of organic and inorganic sources found more beneficial than sole application of inorganics in terms of sustainability of crops in sodic environment.

Diversification Options

The choice of crops to be grown in alkali soils is restricted due to adverse physical and chemical properties. As the reclamation proceeds, there is continuous improvement in the soil properties both in surface and subsurface layers. The reclamation is faster in surface than in the subsurface layers. There is a distinct possibility to diversify cropping systems in the reclaimed alkali lands after 3 years of amendment application. Depending upon the soil pH in the 0-15 cm layer, crops that can be successfully grown in these soils are listed in Table 5.

Alternate Land Use

Most of the sodic lands in possession with individual farmers have been reclaimed through conventional techniques and are producing good crops of rice and wheat. However, a sizeable area is constituted by village community lands and govt. and private lands reserved for specific purposes. Special techniques for reclamation and management of such sites have been developed. The most promising forest species identified for planting in highly alkali soils are pahari kikar (*Prosopis juliflora*), *Acacia nilotica, Casuarina equisetifolia* and pharans (*Tamarix articulata*) (Table 6). *Eucalyptus tereticornis* can also be grown in alkaline soils having $pH_2 < 10.0$. A package of cultural practices such as irrigation, fertilization, spacing and pruning schedules for the promising species such as *Prosopis juliflora* and *Acacia nilotica* has been developed and standardized. Long term studies at Karnal further suggested that prolonged occupation of sodic soils particularly by trees such as *Prosopis* and *Acacia* may restore the productivity of such abandoned soils much above the present agricultural production levels.

Table 5: Crop diversification as related to soil pH and number of years after reclamation

Soil pH (0-15 cm layer)	Expected years after reclamation	Crops	
		Kharif	Rabi
9.2-9.3	3-4	Cotton, sorghum and pearl-millet	Mustard, rapeseed Persian clover and berseem
8.9-9.0	5-8	Sugarcane, groundnut, pigeonhea and soybean	Sunflower, safflower gram, pea and linseed
8.5-8.8	after 8-10 years	Green gram and all other crops	Potato, onion, garlic and tomato

Table 6: Biomass accumulation of selected trees in 10 years in highly alkaline soils

Trees species	Biomass production (kg/tree)		
	Bole weight	Branches + leaves	Total
Prosopis juliflora	113	43	156
Acacia nilotica	85	44	129
Casuarina equisetifolia	84	28	112
Eucalyptus tereticornis	66	24	90

In addition to biomass production, growing of trees on sodic soils results in their amelioration by improving physical, chemical and biological properties of the soil. Long-term field studies at several locations indicated that growing leguminous trees like *Prosopis, Acacia* and *Casuarina* ameliorate these soils at a much faster rate than non-leguminous trees, because of the build-up of organic matter and the recycling of important nutrients (Table 7). It has also been proved that highly sodic soils reclaimed through tree plantations can be successfully used for growing all kinds of agricultural crops including wheat, oat are grown after removing the trees. Several salt-tolerant grasses have been identified which can be grown as sole or as understorey with trees to improve the productivity of salt-affected pasture lands. The promising grasses are Karnal grass (*Leptochloa fusca*), para grass (*Brachiaria mutica*), Rhodes grass (*Chloris gayana*) and *Panicum leavifolium*.

Table 7: Ameliorating effects of tree plantation on alkali soils

Species	Original		After 20 years	
	pH_2	Organic C (%)	pH_2	Organic C (%)
Eucalyptus tereticornis	10.3	0.12	9.18	0.33
Acacia nilotica	10.3	0.12	9.03	0.55
Albizia lebbeck	10.3	0.12	8.67	0.47
Terminalia arjuna	10.3	0.12	8.15	0.58
Prosopis juliflora	10.3	0.12	8.03	0.58

Earlier it was considered highly difficult to raise fruit trees in sodic soils of $pH_2 > 10.0$. But recent experiments at CSSRI, Karnal and Lucknow with some salt-tolerant species of fruits have shown that there are some fruit plants that can be grown in highly sodic soils by adopting proper site-preparation techniques and by using higher dose of organic and inorganic amendments. A special planting technique, called pit-augerhole technique has been developed and standardized for raising fruits in sodic soils. This planting technique consists of making pits of 60cm x 60cm x 60cm dimensions manually, and in the center of each pit an auger bore of 20-25 cm diameter and 140-160 cm depth (to pierce through the *kankar* pan) is made using tractor-mounted augers. These pits-cum-augerbores are filled back with mixture of original soil + 8 to 10 kg gypsum +10-20 kg FYM + 20 kg river sand. A field trial on the identification of intercrops for fruit plantations showed that if the pH of the soil is less than 9.5, crops of babuna (*Matricaria camomila*) and isabgol (*Plantago ovata*), aromatic and medicinal crops can be grown in association with fruit trees successfully.

Reclamation and Management of Waterlogged Saline Soils

Sub-surface drainage is provided to control waterlogging and associated soil salinity by maintaining water table below a specific depth. The system consists of a network of concrete or PVC pipes(with filter) installed manually or mechanically at a design, spacing and depth below soil surface to control watertable depth by draining excess water and disposing it out of the area by gravity or by pumping from an open well called sump. The technology is called as sub-surface drainage technology. This technology results in significant increase in crop productivity and an option for diversification of crops in the area. So far more than 50,000 ha area has been reclaimed by adapting this technology in different states of India. Disposal of saline drainage effluents, however, is a matter of serious environmental concern. Disposal strategies depend upon the volume and quality of effluents and the availability of an outfall. The effluents volume can be reduced by recycling it as irrigation to semi-tolerant and tolerant crops. The residual effluents, after concentration, can be disposed through storage in ponds for evaporation or discharge to a outfall. This technology results in generation of additional 128 man days of employment per ha per year. Further, it results in three fold increase in income of the farmers and reducing flood hazards. The cost benefit ratio of the sub-surface drainage technology has been worked out between 1.26 to1.49. However, high cost involved in installing drainage system, difficulty in maintenance, highly skill oriented, need of community participation, environmental problems of drainage effluent disposal and other socio-economic problems are limiting its wide scale adaptation. Several alternate technologies as substitute/complimentary to sub-surface drainage such as biodrainage, saline, aquaculture and bio-remediation etc. are being tried for the reclamation and management of waterlogged saline lands.

Biodrainage

Biodrainage removes excess soil water through transpiration by trees using solar radiation energy of the sun. It is an option to prevent the development of waterlogged and saline soils especially in areas having no possibility of disposing saline drainage effluent. Also there are large areas with water stagnation due to seepage from higher elevation, surface disposal of urban and industrial effluents and floods during rains where conventional system of dewatering through pumping and surface drainage is not possible. Biodrainage simply lowers the water table. It does not remove salts but minimizes their accumulation in the root zone. It has several advantages; does not require skilled labour and can be easily adopted by the farmers, low cost; no problem of disposal of drainage effluent; eco-friendly, generates revenue for farmer from the sale of wood and can be adopted alongwith the conventional subsurface drainage to reduce the drainage effluents. Haryana state has undertaken a massive programme to reclaim 3000 ha of waterlogged and saline soils in the districts of Rohtak, Sonepat, Jind, Hisar and Sirsa. Punjab Government is taking this programme in Malout area and the technology has been adopted to control waterlogging along the Indira Gandhi Nahar Pariyojna (IGNP) in Rajasthan.

Bio-remediation of Waste Water

The use of waste water in agriculture has increased many folds in the recent past due to scarcity of good quality water. The waste water particularly from industries contains high concentration of heavy metals which enter into human beings and animals through food chain. It calls for development of low cost technologies to remove heavy metals from the waste water. Many microbes have been reported to act as adjuvant to remove heavy metals/toxic substances from waste waters. Several microbes including bacteria and fungi have been isolated which have the capacity to detoxify waste waters loaded with heavy metals such as Lead, Cadmium etc. Similarly several plants have the capacity to remove heavy metals from the waste waters. Such plants have also been identified.

Aromatic and Medicinal Crops

Crops like Isabgol (*Plantago ovata*) and matricaria (*Matricaria chemomila*) which are cultivated during winter season in northern plains can be successfully grown in both sodic and saline soils having pH upto 9.5 and EC 6-8 dS/m. These crops can also be cultivated using saline and brackish ground water. Vetivar grass and palmarosa are also identified as moderately tolerant. Other medicinal plants which may have scope for cultivation in moderate salty environments include:

sena, mulathi, Lepidium and *ashvaganda*. Flowers like *chrysanthemum* can also be grown using saline water irrigation.

Keeping in view the climate, soil and water resources of Ladakh cold desert some of the above options can be tried for reclamation and management of salt-affected soils.

Future Research and Development Needs

- Generation of database about extent, distribution, nature and characteristics of salt affected soils in cold desert
- Pedological studies to know the primary sources of salts (geogenic or anthropogenic) in the Leh and Ladakh cold desert
- Processes involved in salt transport and deposition in valley lands and water bodies
- Processes responsible for waterlogging and salinity development in pasture lands. There is a strong case to study waterlogging and salinity related changes in natural flora/succession in native pasture lands
- Socio-economic and environmental impact assessment of salinity in Ladakh region
- Development of location specific preventive, reclamatory and management strategies
- Collaborative research efforts by CSSRI, Karnal; SKUAST, Srinagar and Defense Research and Development Organisation

References

Chhabra, R. and Abrol, I.P. 1977. Reclaiming effect of rice grown in sodic soils. *Soil Sci.* 124, 4955.

Dargan, K.S. and Chillar, R.K. 1980. Studies on green manuring and nitrogen requirement of rice and residual effect on wheat on a sodic soil. Proceeding of International Symposium on salt-affected soils. CSSRI, Karnal, pp. 440-446.

El-Mowellhey, N. 1998. Proc. Intern. Symp. "Sustainable Management of Salt Affected Soils in the Arid Ecosystmem". Cairo, Egypt. 21-26 September, 1997, Pages 1-2.

Khosla, B.K., Dargan, K.S., Abrol, I.P. and Bhumbla, D.R. 1973. Effect of depth of mixing gypsum on soil properties and yield of barley, rice and wheat grown on saline sodic soils. Indian J. Agric. Sci. 43, 1024-1031.

Setter, T.L. and Waters, I. 2003. Review of prospectus for germplasm improvement for water logging telerance in wheat, beraly and oats. Plant and Soil. 253, 1-33.

Setter, T.L., Ram, P.C., Singh, B.N. and Singh, B.B. 2004. Germplasm improvement for soil sodicity and water logging tolerance of wheat. Proc. International Conference on sustainable management of sodic land, Lucknow, pp. 27-47.

Sharma, D.P., Singh Gurbachan and Mehta, K.K. 1984. Effect of cropping patterns on organic carbon and available nitrogen status of sodic soils. Current Agriculture. 8, 164-168.

Singh, Gurbachan and Gill, H.S. 1992. Ameliorative effect of tree species on characteristics of sodic soil at Karnal. Indian J. Agric. Sci. 62, 144-146.

Singh, Gurbachan. 1986. Response of wheat to sources, doses and methods of nitrogen application in a sodic soil. Indian Journal of Agronomy. 32, 352-357.

Singh, Gurbachan. 1987. Nitrogen management for rice-wheat sequence in sodic soils. Indian Journal of Agronomy. 32, 387-391.

Singh, Gurbachan., Dagar, J.C. and Singh, N.T. 1996. Growing fruit trees in a highly alkali soils.a case study. Land Degradation and Rehabilitation. 8, 257-268.

Singh, Gurbachan., Singh, Harinder and Bhojvaid, P.P. 1998. Amelioration of sodic soil by trees for wheat and oat production. Land Degradation and Rehabilitation. 10, 214-253

Singh, Gurbachan., Singh, N.T. and Abrol, I.P. 1994. Agro-forestry techniques for the rehabilitation of salt lands. Land Degradation and Rehabilitation. 5, 223-242.

Singh, K.N. 1994. Crop and Agronomic Management In: Salinity Management in Sustainable Agriculture, pp.124-144. Rao, D.L.N., Singh, N.T. Gupta, R.K. and Tyagi, N.K. (Eds.) CSSRI, Karnal.

Singh, N.T. 1998. Reclamation and management of alkali soils. In agricultural salinity management. N.K. Tyagi and P.S. Minhas (Eds.), pp. 261-278, CSSRI, Karnal.

Szabolcs, I. 1979. Review of research on salt affected soils. Natural Resource Research, IS, UNESCO, Paris.

❑❑❑

12

Application of Plastics for Intensive Agricultural Production in Hilly Areas

PRATAP RAY BHATNAGAR
Central Institute of Post Harvest Engineering & Technology
PAU, Ludhiana, Punjab, India

INTRODUCTION

India has been endowed with variety of agro-climatic conditions and each condition has its unique nature in which agricultural production occurs. Such conditions are coupled with adverse conditions like low temperatures and shortage of irrigation water in hilly areas, hot climate in arid and semi arid areas, high intensity rainfall, accumulation of water in root-zone and creating conditions of water logging, submergence, etc which affects the agricultural production adversely, resulting in poor productivity, poor quality, making production of some of the products difficult during various seasons, and most important is forcing the land to remain unutilised for a considerable period. It all causes low return from the agricultural system in many parts of the country and reason for low livelihood earning of the farmers. Plastics have played an important role in solving most of the problems facing the agriculture production all over the world. The scope of adoption of plastics in agriculture is very wide, to conserve the natural resources, enhancement of production, productivity and quality of produce. In the developed countries, plastics have become an inevitable part of agricultural production system. But, the use of plastics in agriculture is very small in India as only 0.7% of the plastic produced in the country is being used in agriculture sector.

Plastics are gaining importance in all spheres of life including agriculture due to the useful properties as high strength to weight ratio, flexibility to molding to any shape and size with varying strength, colour and transparency, corrosion resistance, etc. It can be created with any strength and resistant to any type of deterioration by adding some additives. This makes it suitable for almost all types of uses; however, economics may be one of the constraints.

Plastics in agriculture are used in such forms as fibres, films, sheets, pipes and tubes, woven fabrics, nettings, moulding, bags, etc. for a variety of uses. While plastic industry continues to make rapid progress by way of making available newer and better materials, it is for the professionals to make plans for its use in the agriculture system (from seed to seed / market) based on the research experiences conducted under varying farming condition which changes from location to location. Good efforts have already been made to generate information on applicability of plastics in surface covered cultivation, pressurized irrigation systems, lining of ponds and conveyance system, mulching and soil solarisation, fish hatching devices, fish culture mechanism, and related aspects. However, testing of the technologies in the farmer's field condition, refining them to suit the real location specific farming conditions under varying socio-economic conditions and development of package of practices to achieve efficient utilization of the available resources are some of the challenges targeted under the aegis of All India Coordinated Research Project on Application of Plastics in Agriculture, funded by Indian Council of Agriculture Research. The AICRP has four centres located in the Himalayan Region i.e. VPKAS, Almora, Uttarakhand; SKUAST-K, Srinagar, Jammu & Kashmir; CSKHPKV, Palampur, Himachal Pradesh; and ICAR Research Complex for NEH, Barapani, Meghalaya. The paper discussed the achievements made by the centres relevant to plasticulture development in the hills.

Protected Cultivation

Protected cultivation means modifying the environment of plants to realise the better production by avoiding the harmful effect of climatic factors. Protected cultivation in the forms of greenhouses, net houses, low tunnels, mulches etc. offers several advantages to grow crops of high quality and yields, thus using the land and other resources more efficiently, about 7, 23,000 ha was occupied by polyhouses and large plastic tunnels in the world in 2002. On the other hand, only 600-700 ha could be brought under greenhouse cultivation till now with varying levels of climate control in India. In 2005, the estimated consumption of greenhouse plastic was 1.6 to 2.1 KT in India. Floriculture and cut flowers for export as well as domestic market, is the main use of polyhouses and estimated to occupy about 95% of the area under greenhouse cultivation (also called as Controlled Environment Agriculture, CEA) in the country, while coloured capsicum and some other exotic vegetables occupy rest.

Different types of polyhouses are in use with varying cost of construction (Rs/m^2 of floor area) and degree of climate control. The high cost greenhouses are designed to control the greenhouse temperature and humidity in a narrow range using automated systems and cost involvement of Rs 30,000/m^2. The medium cost greenhouses are constructed with relatively less control on the climatic parameters and have relatively wider range of temperature and humidity variations and construction cost varying from Rs 600-1000/m^2. Both the structures are made strong enough to sustain wind pressure and other anticipated forces and have longer life i.e. 15 years or more (excluding cladding material). The low cost structures have bamboo frame, which has the life of 3-5 years and constructed at a cost of Rs 200/m^2.

Present status of polyhouse use was found out by the AICRP team in the Kashmir and Ladakh valley, Uttarankhand and Himachal Pradesh. On the basis of the surveys carried in Kashmir Valley and Ladakh region, the present status of greenhouses installed by different development departments, NGO's, DRDO, private firms, individuals in Kashmir Valley and Ladakh region has been presented in Fig.1. Survey data reveal that both the regions have 10558 different types of greenhouses covering 58.37 ha area out of which 3327 are in Kashmir valley and 7231 in Ladakh region covering an area of 19.44 hectare and 38.93 ha respectively. The greenhouses are being used in Kashmir valley for the growth of vegetable seedling, production of off-season vegetables and propagation of fruit plants especially walnut because it requires specific temperature under greenhouse for maximum graft success. However, the Ladakh region has its own type of greenhouse, known as Ladakhi greenhouse, made from the materials locally available. The said greenhouse increases inside temperature up to

20^0 C higher than the ambient temperature and as a result the people of Ladakh could grow leafy vegetables even during severe cold.

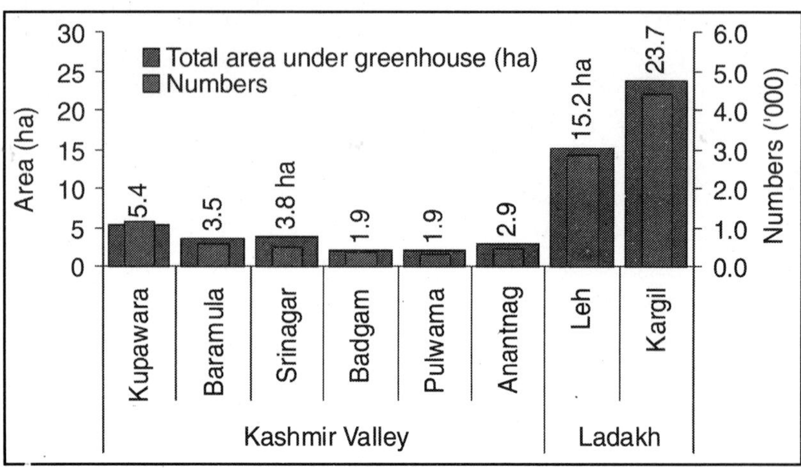

Fig. 1: District Wise Numbers of Green Houses and Area Under Greenhouse Cultivation in Kashmir and Ladakh Region of J&K.

In Uttarakhand, a total of 1016 greenhouses were constructed till 2004, out of which 969 are in operational condition in nine hill districts (Chamoli, Pauri-Garhwal, Uttarkashi, Rudraprayag, Almora, Champawat, Bageshwar, Nainital and Pithoragarh) out of total 13 districts. The structures were constructed mostly with the State Governmental support. The Champawat district was very poor in adoption of the plasticulture techniques due to constraints of lack of technical know-how, availability of materials and skilled manpower.

In Himachal Pradesh, a survey conducted by the team indicated that the sizes of polyhouses were very small in Kinnaur (24-48 m^2) and Lahaul & Spiti as compared to Shimla (100-400 m^2) which is mostly being used for flower cultivation such as Begonia, Geranium, Lilium and Carnation, while few polyhouses are being used for nursery of apples and grapes. Vegetables were also being grown in the polyhouses.

Designs of low cost bamboo/wooden greenhouse have been developed for different regions in the Himalayan regions and constructed in Himachal Pradesh, Uttarakhand and Meghalaya by the AICRP team. Such designs are being accepted by the farmers and constructed on a large scale by the farmers and earning about Rs 4000-5000 or more per structure of about 80-100 m^2 floor area. Evaluation of ten such polyhouses in the farmer's field conditions was conducted in Almora district, revealed that minimum 80 m^2 of floor area is minimum required to realize the economic viability with benefit-cost (B-C) ratio of 1.67 to 2.2 with growing off-season vegetables.

The agro-techniques for growing different vegetables and fruits have been worked out for efficient utilization of the floor space in the greenhouses. It is obvious that more benefits would be accrued only when off-season vegetables are grown in the polyhouses, hence, different cropping systems have been evaluated at different locations and suggested economical systems based on the climate and market opportunities of the location (Table 1).

Table 1: Cropping System Identified for Round the Year Cultivation of Off-Season Vegetables at Different Locations

Location	Cropping system	Economics
Almora, Uttakhand	Capsicum+Tomato+Spinach	B-C: 3.94
	Squash+Frenchbean+Tomato+Spinach	B-C: 2.57
Palampur, HP	Capsicum+Capsicum	B-C: 5.20
	Capsicum + Cucumber+Frenchbean	B-C: 4.95
Barapani, Meghalaya	Capsicum+Tomato+Lettuce	Profit: Rs 109.1/ m^2
	Tomato+Frenchbean+Cabbage	Profit: Rs 76.0/ m^2

Agro-techniques for strawberry cultivation in the polyhouse at Srinagar (J&K) were worked out with drip irrigation and plastic mulching. Results indicated a 2.5 times yield of strawberry with 45 days advancement can be achieved with walk-in-tunnel type polyhouse as compared to open field condition (yield 5.22 t/ha). In low tunnels, the earliness was 20 days with 1.6 times of yield. At Barapani (near Shillong, Meghalaya), strawberry inside rainshelter (polythene covered with bamboo frame) produced 682 g/plant (1.1 kg/m^2) good quality fruits, while outside only 362 g/plant produced which was infested by diseases, bird bites, spots due to frost, etc. that reduced its price substantially.

The fish production during winter goes down due to low temperature and fish remains in hibernation. Such conditions are very crucial as disease infestation is higher during this period. Polythene housing was tried to raise the temperature of the pond water in Megahalaya with bamboo frame. It increased the water temperature by 3-4°C with increase in fish yield by 12-15%. However, further evaluation in this direction is required to work out economics.

Efforts are being made to popularize the polyhouse production which has the potential to enhance the income of the small and marginal farmers of hilly terrains who get very less from their land through traditional cultivation. For this entrepreneurship development programs, training to ITI trained personnel and exposures to farmers are being undertaken so that trained personnel can become a means for further upscaling. Participatory evaluation is also being undertaken in the farmer's field with input sharing (mostly cost construction of frame is born by the farmer and only UV stabilized polythene is provided from project side) and more than 50 polyhouses has already been constructed in Jageshwar block of Almora and Darim in Nainital.

Polythene Lined Ponds and Irrigation Systems

Irrigation sector is the largest consumer of plastics in agriculture. About 180 KT of PVC and about 40 KT of HDPE pipes were used in 2005 for irrigation purposes. More than 40 KT of polythene was used for canal lining by 2005. In addition, it has potential use in micro-irrigation systems, which saves 30%-50% water along with improved quality and productivity of agricultural produce. In 2002, a 3.4 million ha was under micro-irrigation in the world, which amounted to about 1.5% total irrigated area. In India, the area under micro-irrigation is increasing rapidly in the recent year and approached nearly 12.0 lakh ha has been brought under the micro-irrigation including drip and sprinklers which is about 1.6% of total irrigation potential created in the country under major and medium irrigation Projects. However, it has vast potential and opportunity to increase the area under micro-irrigation considering the limited water availability for crop production and ever-shrinking of available water resources in the country.

Water is the most important input for agricultural production. However, in uplands areas which comprises of about 90% of total geographical area of hills (it may vary a little from region to region in Himalayan region), water scarcity exists for irrigation purposes. Mostly the agricultural production is rainfed or based on some diverted spring water, which contribute almost negligible in overall production. However, there is very high opportunity for rainwater harvesting exists in upland humid area, which can provide good amount of water for irrigation, though on limited availability based on the catchment-command area ratio, and opportunity for water harvesting through construction of ponds. Secondly, numerous low discharge springs are found in uplands, which cannot be used directly for irrigation. Some of them are developed for drinking purposes. Storing the water from such low discharge springs, which may be perennial in nature, can provide irrigation for horticultural crops, as the region is more congenial for such production. Lining of such ponds is very important to retain the water for longer periods. Polythene lining is one of the cheap, easy to construct and simple technology for the purpose. However, designs and needs for safety mechanism are very important for satisfactory performance of the ponds for longer periods.

Various designs of low-density polyethylene (LDPE) lined ponds have been designed at different centres. Previously, a design of LDPE film line tank on hilly terrace was prepared in which the polythene was covered by river boulders or stone plates. But, as the river boulders/ stones are not easily available in uplands and also mining of stone is discouraged in some of the districts of Uttarakand hills, a new design have been prepared at Almora, Uttarakhand. Coal tar sheet covered LDPE tanks were developed, which cost of Rs 115/m^3 for the tank capacity higher than 100 m^3. Coal tar sheet cover to LDPE is effective to protect LDPE from strength deterioration in sunlight and is a cheaper (Rs 60/m^2) alternative as compared to earlier design with stone pitching.

The technology of coal tar sheet covered LDPE tanks was used by the farmers in Almora and Nainital districts of Uttarakhand. A 24 and 55 LDPE film lined tanks have been constructed by farmers in Jageswar (Almora) and Mukteshwar (Dist. Nainital) blocks with technical support provided by the APA team, which are being used for life saving irrigation for off-season vegetable cultivation and fisheries to some extent. The B-C ratio of various tanks was found to vary between 2.11 to 2.73 and the average income has increased by approximately Rs 5000/ year. The tank was found high in demand at the farmer's field since it helps in realizing the successful cultivation of high value off-season vegetables that boosted the farm income. In Meghalaya, a design for LDPE lining of pond has been prepared and then dug out as per design. Pond bed and sides was made weed and stone free. Steps at 50 cm vertical interval were made on sides of the pond. On top of the sides, continuous trench was dug for anchoring the LDPE film. A 10cm thick layer of sieved sand was spread uniformly on bed and sides to provide cushion to the LDPE film. The LDPE film (250 μ LDPE) was joined to suit the shape and size of the pond using bitumen. The LDPE film was laid in the pond with enough stretching or tightness. It was covered with soil cover of 30 cm at the bottom. Stone pitching was done on only sides to safeguard the sides of the pond against erosion and any other external forces (Fig. 2). The LDPE film lining reduced the seepage loss from 55 to 2.9 l/m²/day i.e. by 94.7%.

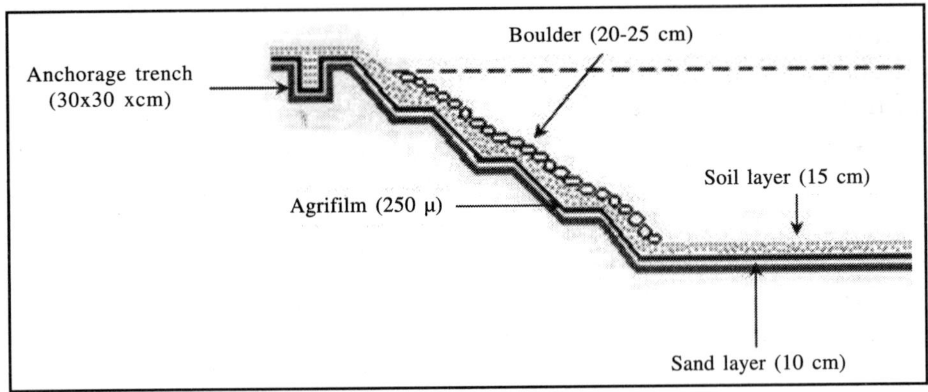

Fig. 2: Plastic Lined Pond and Layout of Laying Plastic in Pond

In the rainy season seepage water from adjacent hill seeps below the polythene and created uplift hydraulic pressure, causing bottom lining of pond lifted and tends to rupture the plastic film. A safety mechanism has been developed to protect the lined ponds using perforated plastic pipes laid in trenches below polythene lining. Pipes were wrapped with coir rope to prevent choking of the perforations, which were enveloped in sand cover. Due to this arrangement seepage water got disposed out and hydraulic pressure was released. This protected the lined pond and the water remains through out the dry period.

The water stored in the polythene lined tanks needs to be utilised very efficiently as it is very preciously and available in very limited quantity. Efforts have been made to develop efficient methods of water applications and irrigation scheduling with such a system.

Use of LDPE lined tank to supplement the water for vegetable production in hilly terraces resulted in enhanced productivity per unit of area. The increase in productivity was 14.7% for French bean, 27.3% for tomato, 21.1% for garden pea, 27.3% for cabbage, 21.4% for potato and 16.7% for capsicum in Uttarakhand. Gravity fed micro-tube based drip irrigation systems was also designed at Almora for using the gravity head available between the hilly terraces for system operation without use of pump. Experiment conducted to find out variation in discharge for changing length of micro tube for gravity fed micro tube irrigation system. The set up had 7 laterals with length between 12.3 m to 27.0 m with gravity head of 10.4 m at the first lateral. The average discharge from one micro tube was 3.20 lph, which varied between 1.20 to 10.0 lph. Gravity fed micro sprinkler irrigation was also evaluated with a gravity head of 2.5 m, which gave an average discharge of 17 lph for single micro-sprinkler with CV - 35% and radius of throw of about 50 cm (20% overlap against desired 50%).

Rainwater harvested in the LDPE film lined pond was recycled through sprinkler irrigation system for irrigating the experimental crops. The system was operated as per the daily water requirement of the crops, estimated for three different stages using standard crop and canopy factors. Full circle sprinkler head with multiple nozzles of size 5.5 x 3.2 mm, average discharge 0.48 lps at 2.0 kg/cm^2 (equivalent to precipitation rate of 2.5 mm/h), maximum throw was 12.4m covering an area 483.2 m^2, were used for irrigation. A production of 426, 280, 44 and 120 q/ha of cabbage, cauliflower, pea, and lettuce was obtained using the system in Meghalaya. Considering the cost of pond and irrigation system with its maintenance requirement, the B-C ratio was worked out to be 2.7 over a period of 15 years.

Variation in uniformity, discharge and area covered by sprinklers on hill slope was studied for terraced cultivation on 41% slope. The sprinkler irrigation system was laid in our research farm having two laterals having three sprinklers in each lateral, and linked with rainwater harvesting LDPE lined pond. The elevation difference between two laterals was 3.5 and water source to first lateral was 4.5 m. A significant decrease of 55% in area covered by first sprinkler (nearest to the water source, 254m^2) and last 6th sprinkler (farthest from the water source, 113 m^2) was observed. The is discharge also reduced from 1.94 to 1.25 l/m^2/h between 1st sprinkler and 6th sprinkler respectively and, respectively recording a reduction of about 35%. Uniformity coefficient of 0.71 and 0.67 for 1st and 6th sprinkler was obtained. However, no appreciable variation was found in uniformity coefficient between the sprinklers.

Packages for drip irrigation to different vegetables have also been developed for optimised use of harvested water in the ponds.

Plastic Mulching

Plastics are very effective material for mulching, which is gaining importance in Indian agriculture to reduce soil evaporation, weed control, solarisation and soil temperature control. Black plastic film as well as clear plastic film is being used for the specific purpose as per need. Infrared transmitting (IRT) mulches, which transmit most of the solar heat portion of light radiation but absorb most of the visible portion, were also introduced to the market, which provide weed control along with increase in the soil temperature. Unfortunately, labour requirements to remove plastic mulch from the field after the growing season can be high. New bio- and photo-degradable polyethylene and combinations of polyethylene-paper and polyethylene-starch show promise in eliminating the need for mulch removal. However, their techno-economic feasibility needs to be thoroughly studied.

In India plastic mulching is yet to gain momentum. Black or two-colour film is more popular in India. Some also use clear film. It is estimated that 500 t of PE film is used in mulch application in India in 2005. Globally, 12.86 million ha was under plastic mulch in 2002, out of which 9.6 million ha was only in China.

Horticultural production with plastic mulch needs drip irrigation since other means of water application cannot meet out the crop water demand. Plastic mulch with drip irrigation enhanced the strawberry productivity by 36% in open field and 22% in walk in tunnel type greenhouse as compared to without mulch i.e. respectively 7.0t/ha and 18.4 t/ha, respectively in two conditions in Srinagar, J&K. On the other hand, if the plastic mulched crop is irrigated with other methods e.g. sprinkler irrigation, the yield have been found to reduce. In Meghalaya, a sprinkler irrigated cauliflower yield was reduced by 38% with plastic mulching, while the reduction was only 13% for Broccoli. Less water penetration in the soil was the reason for reduced yield under sprinkler irrigation.

Farm Machineries

A lot of scope exists to utilize the plastics as a material to reduce weight of the farm implements and machinery used in terraced cultivation in hills. The reduction in the weight enables easy and comfortable transportation of the equipment on the hilly paths where it can be taken by labourers only. Use plastics also reduced the drudgery in operation and make the equipment cheaper. Efforts in this direction are being made under the project.

A pedal operated paddy thresher with plastics (polycarbonate sheet) body has been developed to reduce the weight (35 kg) to nearly

50% of similar metallic body thresher (70 kg) with a throughput capacity of 100 kg/h and operated by one person. Present cost of the pedal thresher is Rs 3500 with nominal saving in cost, but it enhanced transportability on hilly terrains, higher threshing efficiency, ergonomically effective (operated in sitting position with power transmission through pedal as compared to standing operation and power transmission through cam drive in conventional thresher), straw is recovered for further use without any damage. This enabled continuous working up to 4 hours as compared to only one hour for similar design. Hence, the operational cost is less as compared to traditional threshers of similar kind. The machine was well accepted by the farmers as 350 such machines have already been supplied to the farmers in Uttarakhand.

A plastics thermocole Langstroth beehive have been developed at Palampur using HDPE sheet and thermocole. The components of the beehive i.e. brood chamber, bottom board, top cover and inner were fabricated using plastics sheet of HDPE. The 10 mm thick thermocole was used as an insulator material in between the plastic sheets for maintaining the temperature inside the box. The performance of the beehive was more or less similar to wooden beehive but had more area of honey, more life and better handling.

The maize sheller with iron body was modified by replacing the iron body with plastics (PVC) body. The change of material resulted in the reduction of the weight by 50.5%. The modified plastic covered maize sheller has weight of only 89 g against the 180 g of conventional tubular maize sheller. The shelling efficiency was not affected by changing the material of construction.

Packaging and Post Harvest Processing

Packaging industry is the largest consumer of the plastics, as it accounts for 42% of the total consumption. However, actual use in packaging of the agricultural commodity is not available; however, it must be a significant proportion. Plastics packaging of different kind have been investigated and are in use for enhancing the shelf life of the perishable or semi-perishable agricultural commodities. In addition to plastic packaging is being done for agricultural inputs e.g. fertilizers, pesticides, etc. for their safe transport. About 1000 t of PP leno bags were consumed in 2005 for packaging of fruits and vegetables. For storage of food grains with Cover & Plinth technique, about 5000 t of PE was used in 2005. Plastic crates are steadily becoming popular for on-farm handling and transfer of horticultural produce.

Efforts have been initiated to develop techniques for packaging and transportation of horticultural products of Himalayan region. Efforts have been made to design low cost structures for drying of the vegetables and fruits to ensure availability of dried fruits and vegetables during non-availability periods in the Kashmir valley. Poly tunnel drier has been designed and constructed which is portable and movable and could be placed in a sunny location. The drying in poly tunnel drier was quicker and requires 1.5-3.0 days depending on nature of item, than in the open condition 5-10 days. The inside average temperature of upper and lower tray of loaded drier was higher than the ambient by a difference of about 20 & 17.2°C respectively, whereas it was 22 & 19.3°C respectively for unloaded drier. The maximum temperature inside was 59.5°C (for upper tray) & 47.6°C (for lower tray) in loaded drier and 61.8°C (for upper tray) & 58.9°C (for lower tray) in unloaded drier. This is because the drier is based on the polyhouse effect and it traps the solar radiation resulting in a subsequent increase in temperature. The chimneys drag the moist air within the chamber and remove excess moisture from the vicinity of the product placed on perforated trays. Further evaluation and standardisation of the design and dehydration process is in progress.

CONCLUSIONS

There are good opportunities to enhance to agricultural productivity through application of plastics in different activities which has been realized at many places in the country as well as abroad. Enough technologies have also been developed in this direction which needs to be further investigated for their relevance in different agro-climatic and topographic conditions especially in Himalayan region which is very suitable for horticultural production with high level of production as well as good quality of produce, but it is coupled with several problems like manpower availability, transportation, etc. Polyhouses, low tunnels, mulching, and polythene lined ponds; low cost high efficient irrigation systems, etc are some of the technologies which need popularization to realize its actual benefits. AICRP on Application of Plastics has undertaken this challenge and undertaking needful steps to achieve this target in collaboration with all related organizations operative in different locations.

□□□

13

Medicinal Plant Biodiversity of Trans Himalaya : Sustainable Utilization and Conservation

O.P. CHAURASIA, BASANT BALLABH, X JANIFER RAJ, S.K. SAURAV, DIVAKAR SHARMA, AMOL TAYADE AND SHASHI BALA SINGH
Defence Institute of High Altitude Research, DRDO, Leh-Ladakh, J&K, India

INTRODUCTION

Cold deserts are found in the interior of Asia and the inter mountain zone of North America. Sixteen percent area of total landmass is under cold deserts. In India cold desert comes under the trans-Himalayan zone. They are confined to Ladakh in Jammu & Kashmir (J&K) and Lahaul & Spiti in Himachal Pradesh (HP). In J&K cold desert lies between 32° 15'-36° N and 75° 15' -80° 15' E. It covers approximately 68,321 sq. km besides 27, 555 sq. km. area which is under illegal occupation of Pakistan and China. In H.P. the area lies between 31° 44' 57" -32° 59' 57" N and 76° 46' 29" -78° 41' 34" E and covers approximately 6,488 sq. km.

The climate is characterized by great extremes of heat and cold, coupled with excessive dryness. The summer months are brief while the winter months are long with the temperature going as low as -59° C (lowest recorded from Dras) -23° C at Kargil and -30° C at Leh. Rainfall is very little in these areas (Less than 100 mm in a year). Snow falls more often but the quantity is not much ever being more than six to eight inches deep. Similarly average relative humidity remains around 40–50 percent which makes the area totally arid. Physiographically the trans Himalayan zones may be divided into six valleys viz. Leh, Nubra, Changthang, Suru, Zanskar and Lahaul - Spiti.

Botanical Curiosities

The flora of trans Himalayan zones comes under alpine and high alpine zones and is dominated by wild annual and perennial herbs followed by dwarf bushes or shrubs. The vegetative growth starts at the commencement of summer when melting snow provides abundant moisture to the herbaceous vegetation to run its full course with great rapidity. The flora is in full bloom in the month of August but starts degenerating by the end of September or beginning of October.

The most significant feature of high altitude plants is adaptations on their habitats to withstand the adverse eco-climatic conditions. The commonest adaptation observed is the cushion habit. Species of *Astragalus, Caragana* form spinescent cushions while the species of *Androsace, Arenaria, Draba, and Rhodiola* form soft cushions with rossette arrangements of their leaves. The rigid mat forming plants are also frequent and represented by species of *Arenaria, Astragalus* and *Oxytropis*.

Another remarkable feature of this zone is vegetative propagation. Most of the taxa are perennial in nature and perennate by means of roots, root-stock, runners, bulbs, bulbils, rhizomes, tubers etc. These herbs also exhibit a very interesting cycle of growth due to scanty rainfall, low temperature, snow storms, blizzards and extremely frigid conditions at this zone.

Medicinal Wealth

Since the dawn of history man has been in search of ways to find cure and relief from mental and physical illnesses. The history of medical botany dates back to the origin of civilization. In India, the history of medicinal plants can be traced back to the Vedic period (4500 B.C. to 1500 B.C.). The identity of several plants like Semal, Pithuan and Pipal referred to in the suktas of Rig Veda can be fixed with reasonable certainty. Atharva veda contains detailed information on approximately 2000 medicinal plants and their uses. After the Vedic era, the works of Charak and Susruta namely Charak Samhita and Susruta Samhita deal with 700 drugs of daily and specific uses.

It has been estimated that out of about 2000 drugs that have been used extensively in India, only 200 each are of animal and mineral origins while the rest are of plant origin. According to WHO report, over $3/4^{th}$ of the five billion world population cannot afford the products of modern medicines and have to rely upon the use of traditional medicines of plant origin. WHO currently encourages, recommends and promotes the inclusion of plant derived drugs in national health care programme because of low cost and within the reach of a common man and as such are time-tested and considered to be much safer.

The people of Ladakh and Lahaul-Spiti have its own medical system called "Amchi system of medicine" (Tibetan Medicine) and the practitioners are called Amchis (Superior of all). Amchi medicine is principally based on a skillful use of plants, minerals and animal products. The origin of Amchi medicine is traced back to India, where more than 2500 years ago Lord Buddha delivered a medical practice called Bzi (*Chatus tantre*) which was translated in Tibetan language during 8th century AD by Acharya Chandrananda and Varochana. This system was introduced in cold arid zones during $10^{th}/11^{th}$ century AD approximately.

The Amchi system of medicine has been in vogue in Ladakh for past many centuries. Amchis are looking after 60 percent of public health. Amchis have not only theoretical texts but also practical experience handed over from generation to generation. They have also a reputation of having high ethical standard in social system of Ladakh. Generally combinations of two or more herbs are prescribed and administered.

Ethnobotanical Studies

Keeping in view the rich ethno-medicinal wealth of trans Himalaya, Defence Institute of High Altitude Research (DIHAR), Leh - Ladakh, has carried out detailed ethnobotanical explorations of all the six valleys covering far-flung areas and higher passes and collected approximately 1000 plant specimens alongwith ethno-medicinal information with the help of tribals, shepherds, senior citizens and Amchis. The plant specimens were identified with the help of floras and Botanical Survey of India / Forest Research Institute, Dehradun and the High Altitude Ethnobotanical Herbarium had been established. An inventory of widely used ethno-medicinal plant species has been compiled and tabulated in (Table 1) along with family, local name, parts used and their respective therapeutic uses.

Sustainable Utilization

Based on ethnobotanical survey of Ladakh and Lahaul-Spiti it was observed that medicinal plants were quite useful against high altitude maladies faced by low landers viz. stress, fatigue, loss of appetite, loss of memory, UV radiation, sleep disturbance, frostbite, etc. DIHAR has developed certain herbal health products using high altitude medicinal plants to improve the working efficiency of low landers deployed in the high altitude cold desert region of India. The briefs of herbal products developed are as under:

i. Multivitamin Herbal Beverage from Seabuckthorn

Seabuckthorn (*Hippophae sp.*) is wild thorny bush of cold desert of India, growing in abundance under wasteland conditions. Fruits of seabuckthorn are storehouse of vitamins and other bioactive compounds. DIHAR has standardized process for preparation of herbal beverage from fruits of seabuckthorn and got it patented.

Qualities: The beverage is a rich source of various vitamins (C, A, B, K & E) with anti-oxidant, anti-aging and anti-stress properties. It does not freeze up to - 22^{0}C.

Current status: Introduced in Army ration and technology transferred to 07 vendors : Ladakh Food Ltd, New Delhi, Kohinoor International Agro-Products, Srinagar, Arctic Bio Pro Ltd, New Delhi, Mamta Agro Food Ltd, New Delhi, Victory Corporate Services Ltd, New Delhi, Leh Nutrition Project, Leh.

ii. DRDO Herbal Tea

Seabuckthorn and many other medicinal plants of high altitude are rich source of flavonoids, carotenoids etc. The main ingredients of DRDO herbal tea are *Achillea sp, Carum sp, Hippophae sp, Origanum sp, Mentha sp., Rhodiola sp.*, etc.

Qualities : The tea is rich in flavonoids, vitamins, etc. and has medicinal value viz. soothing, energizer, anti-stress and stimulant. It has its own natural colour, flavour and taste.

Current status : Technology transferred to two vendors' viz. Arctic Bio Pro Ltd, New Delhi & Victory Corporate Services Ltd and introduced in Army special rations.

Table 1 : List of Medicinal Plants of Trans Himalaya

S. No	Botanical Name With Family	Common Name	Parts Used	Medicinal Uses
1.	*Achillea millefolium* (Asteraceae)	Milfoil/Yarrow/ Chuang	Leaves	Tonic, cough, cold & toothache
2.	*Aconitum heterophyllum* (Ranunculaceae)	Aconite, Attees, Bona-karpo	Roots	Antipyretic, astringent & stomach complaints
3.	*Aconitum violaceum* (Ranunculaceae)	Aconite, Pattees, Bona-nagpo	Roots	Antipyretic & fever
4.	*Allium carolinianum* (Amaryllidaceae)	Skotche	Leaves	Indigestion
5.	*Arctium lappa* (Asteraceae)	Burdock, Jangli kuth, Pizums	Roots	Diuretic, depurgative & lung diseases
6.	*Arnebia benthamii* (Boraginaceae)	Ratanjot, Balchar	Roots	Hair tonic & blood purification
7.	*Arnebia euchroma* (Boraginaceae)	Ratanjot, Demok	Roots	Hair tonic & blood purification
8.	*Artemisia brevifolia* (Asteraceae)	Wormseed	Leaves	Antiworm & stomach complaints
9.	*Artemisia dracunculus* (Asteraceae)	Tarragon / Sanche	Whole plant	Diuretic & toothache
10.	*Berberis ulicina* (Berberidaceae)	Berberry, Khizar	Roots & bark	Tonic
11.	*Bergenia stracheyi* (Saxifragaceae)	Pashanbhed, Tiang kidney problems	Roots	Astringent, tonic &
12.	*Biebersteinia odora* (Geraniaceae)	Khardung / Drakpose	Leaves	Antiseptic & skin sores
13.	*Caltha palustris* (Ranunculaceae)	Marsh marigold	Leaves & roots	Gonorrhoea
14.	*Capparis spinosa* (Capparidaceae)	Caper bush, kabra	Whole plant	Astringent, liver complaints & rheumatism
15.	*Capsella bursa-pastoris* (Brassicaceae)	Shepherds' purse	Whole plant	Lung complaints & stomach problems

Contd...

16.	*Carum carvi* var. *gracile* (Apiaceae)	Caraway, Koniyot	Seeds	Carminative, digestive & febrifuge
17.	*Centaurea depressa* (Asteraceae)	Vashakha	Whole plant	Fever
18.	*Clematis tibetana* (Ranunculaceae)	Clematis, Zakgic	Roots	Indigestion
19.	*Codonopsis clematidea* (Campanulaceae)	Burrkutang	Roots	Rheumatism & stimulant
20.	*Colchicum luteum* (Liliaceae)	Hirantutiya, Tukapa	Corm & seeds	Colchicine & gout
21.	*Corydalis govaniana* (Papaveraceae)	Stong-zil	Leaves, roots	Antipyretic, diuretic & febrifuge
22.	*Cremanthodium ellisii* (Asteraceae)	Ming-chen-ser-po / Lukumentok	Whole plant	Contagious disease & fever
23.	*Cuscuta reflexa* (Cuscutaceae)	Large dodder, Amarlata	Whole plant	Purgative & warts
24.	*Dactylorhiza hatagirea* (Orchidaceae)	Spotted heart orchid, Salam panja, Ambolakpa	Roots	Energetic, health tonic & nervine tonic
25.	*Delphinium brunonianum* (Ranunculaceae)	Himalayan larkspur, Lunde-kaown	Whole plant	Colic & insecticidal
26.	*Dioscorea deltoidea* (Dioscoreaceae)	Dioscorea, Singli-mingli	Tubers	Oral contraceptive pills & source of disogenin
27.	*Dracocephalum heterophyllum* (Lamiaceae)	Zinkzer	Whole plant	Cold & cough
28.	*Euphorbia stracheyi* (Euphorbiaceae)	Khron-bu	Latex	Boils
29.	*Euphrasia officinalis* (Scrophulariaceae)	Eyebright, Kangchuk	Whole plant	Eye complaints
30.	*Ferula jaeschkeana* (Apiaceae)	Wild heeng, Sampharu	Roots	Rheumatism & septic wounds
31.	*Gaultheria trichophylla* (Ericaceae)	Winter green	Leaves	Antiseptic & source of gautherine
32.	*Gentiana algida* (Gentianaceae)	Titkas	Whole plant	Analgesic & bronchitis
33.	*Heracleum pinnatum* (Apiaceae)	sPru-ma	Roots	Leucoderma, menses regulation & source of xanthotoxin

Contd...

34.	*Hippophae rhamnoides* ssp. *turkestaica* (Elaeagnaceae)	Seabuckthorn, Sastalulu	Fruits, seeds, leaves	Source of multi-vitamins, carotenoids & anti-oxidants
35.	*Hippophae salicifolia* (Elaeagnaceae)	Seabuckthorn, Sastalulu	Fruits, seeds, leaves	Source of multi-vitamins, carotenoids & anti-oxidants
36.	*Hippophae tibetana* (Elaeagnaceae)	Seabuckthorn, Sastalulu	Fruits, seeds, leaves	Source of multi-vitamins, carotenoids & anti-oxidants
37.	*Hippuris vulgaris* (Hippuridaceae)	Dam-bu-kara -machong	Leaves & stems	Antiseptic & febrifuge
38.	*Hyoscyamus niger* (Solanaceae)	Henbane, Khurshani Ajawain, Gay-langthang	Whole plant	Asthma, Toothache & source of hyoscyamine drug
39.	*Hypecoum leptocarpum* (Papaveraceae)	Meerang / Parpata	Leaves	Septic wounds
40.	*Hypericum perforatum* (Hypericaceae)	St Jonh's Wort	Whole herb	Neuralgia & sciatica
41.	*Hyssopus officinalis* (Lamiaceae)	Hyssop (E), Juff / Tengu	Whole herb	Sprains & stimulant
42.	*Inula racemosa* (Asteraceae)	Puskarmool, Manu	Roots	Anthelmintic, bronchial asthma & expectorant
43.	*Juniperus communis* (Cupressaceae)	Juniper, Sukpa	Fruits	Purgative & anti-inflammatory
44.	*Lagotis cashmeriana* (Scrophulariaceae)	Hong-len	Whole plant	Febrifuge & anti-inflammatory
45.	*Lancea tibetica* (Scrophulariaceae)	Spa-yang-rtsa-ba	Whole plant	Tonic & wound healing
46.	*Meconopsis aculeate* (Papaveraceae)	Himalayan Blue Poppy, Akshatsermum	Whole plant	Ulcer, liver & lung problems
47.	*Mentha longifolia* (Lamiaceae)	Horse mint, Phuloling	Leaves	Headache & stomachache
48.	*Nepeta longibracteata* (Lamiaceae)	Prainku / Chhagnamgo	Whole plant	Stomachache
49.	*Onosma hispidum* (Boraginaceae)	Ratanjot, Gajowan	Roots	Stimulant & cardiac tonic
50.	*Origanum vulgare* (Lamiaceae)	Wild oregano, Marjoram	Whole plant	Carminative, stimulant & mild tonic

Contd...

51.	*Pedicularis longiflora ssp. tubiformis* (Scrophulariaceae)	Lug-ru-serpo	Whole plant	Diuretic
52.	*Peganum harmala* (Zygophyllaceae)	Wild rue, Sepan	Roots, seeds	Asthma, rheumatism & diuretic
53.	*Physalis alkekengi* (Solanaceae)	Winter cherry, Shoklo	Fruits	Gout, rheumatism & vermifuge
54.	*Physochlaina praealta* (Solanaceae)	Langthang	Leaves & seeds	Narcotic, vermifuge & source of atropine drug
55.	*Picrorhiza kurrooa* (Scrophulariaceae)	Kutki, Kaor	Roots	Appetizer, bitter tonic & liver protactant
56.	*Podophyllum hexandrum* (Berberidaceae)	Mayapple, Ban kakri, Ol-mose	Roots & fruits	Anti-cancer, radioprotection & Gynaecological diseases
57.	*Polygonatum multiflorium* (Liliaceae)	Soloman's seal (E), Ramnge	Rhizome	Kidney problems & health tonic
58.	*Prangos pabularia* (Apiaceae)	Prangos, plans	Roots, seeds & leaves	Carminative & source of coumarin
59.	*Rheum webbianum* (Polygonaceae)	Rhubarb, Revalchini	Roots	Laxative, purgative & source of anthraquinone compounds
60.	*Rhodiola imbricate* (Crassulaceae)	Rose-root, golden arctic root, shrolo	Root-stocks	Anti-stress, restores memory & health tonic
61.	*Rubia cordifolia* (Rubiaceae)	Maddar, Manjith, b-Tsod	Roots & stem	Anthelmintic & leucoderma
62.	*Saussurea lappa* (Asteraceae)	Costus, Kuth, Rusta	Roots	Asthma, anthelmintic & carminative
63.	*Saussurea obvallata* (Asteraceae)	Brahma Kamal, Span-rtsa-bo	Flowers-buds	Cold & cough
64.	*Swertia petiolata* (Gentianaceae)	Wild Chirayta, Tikta	Leaves	Fever & tonic
65.	*Taraxacum officinale* (Asteraceae)	Dandalion, Khur-mang	Roots	Analgesic, diuretic, liver disorders & tonic
66.	*Thalictrum foetidum* (Ranunculaceae)	Lesser meadow rue, sNgo-sprin	Roots & leaves	Diuretic & fever
67.	*Thlaspi alpestre* (Brassicaceae)	Bre-ga	Whole herb	Digestive & fever

Contd...

68.	*Thymus serpyllum* (Lamiaceae)	Wild thyme	Whole herb	Carminative, carminative, expectorant & stimulant
69.	*Tribulus terrestris* (Zygophyllaceae)	Puncture vine, Gokhroo, gZe-ma	Fruits	Diuretic, sexual weakness & tonic
70.	*Ulmus wallichiana* (Ulmaceae)	Elm, Umbok	Bark	Wounds & hair-tonic
71.	*Urtica hyperborea* (Urticaceae)	Stinging hairs, Zazot	Roots	Rheumatism
72.	*Verbascum thapsus* (Scrophulariaceae)	Flannel Mullein, Dandashal	Leaves & seeds	Asthma & frost-bite

iii. Herbal Appetizer

Loss of appetite is the major problem of the troops deployed in high altitude conditions. Based on traditional knowledge DIHAR has formulated Herbal appetizer using high altitude medicinal plants. The major ingredients are *Achillea sp, Carum sp, Hippophae sp, Origanum sp, Rheum sp, Saussurea sp.*

Qualities : The herbal product is having bioactive compounds like Aloin, t- anethole, Carvone, Thymol and is having medicinal properties like appetizer, carminative and laxative.

Current status: Under toxicological evaluation.

iv. Herbal UV Protective Oil

There is a heavy influx of UV radiations in high altitude areas which result in various skin diseases especially sun burn and skin dryness. A herbal UV Protective oil has been formulated using extracts of indigenous high altitude medicinal and aromatic plants.

Qualities: The Herbal oil possesses carotenoids, vitamin 'E', saturated and unsaturated fatty acids, etc. It is useful for enhancement of microcirculation, antioxidant protection, epidermal regeneration, natural UV blocking and skin dryness.

Current status: Under toxicological evaluation.

Rare, Endangered & Threatened Medicinal Plants

The high altitude medicinal plants play major role in traditional systems of Indian medicines especially Amchi system of medicine since a long time. Their unscientific exploitation, over grazing, is uprooting for fuel, increasing tourism, road construction and natural calamities have made some important plant species endangered and threatened. The important endangered medicinal plants are *Aconitum heterophyllum, Colchicum luteum, Dactylorhiza hatagirea, Ephedra gerardiana, Hyoscyamus niger, Meconopsis aculeata, Picrorhiza kurrooa, Podophyllum hexandrum, Rheum webbianum and Saussurea lappa.*

Collection of Herbal Materials

It is believed that herbal materials collected from natural localities were preferred to be used in view of their 'purity' as they grew in untouched woods and forests free from disturbances of man himself and quality of these crude drugs is superior in comparison to cultivated medicinal forms. However, collection of herbal plants from wild is unscientific, labour-oriented and undependable. Many of the crude drugs are poisonous and dangers of collection by unskilled collectors are obvious because even the experts have difficulty in correct identification of some species unless they are in flowers.

The local Amichis collect the useful parts of herbal plants during their various stages of growth as per the accumulation of crude drugs. Usually underground parts are storage organs for the plant and accumulate active principles during summer months. Roots, rootstocks and rhizomes of perennials are generally collected after two to four years of growth. This applies to majority of drugs collected from wild. Tubers, corms and bulbs are usually collected at the end of flowering/ fruiting when all the aerial parts show signs of senescence or at the time of sprouting of plants. These parts are properly cleaned and dried in shade.

Leaves are collected throughout the growing season. Usually tender leaves are considered to be of highest quality so far as active principles is concerned. The whole herb is collected just before or at the beginning of flowering. Flowers or complete inflorescence is gathered at the beginning of flowering period. Seeds or fruits are collected at the time of maturity. Bark is collected either in spring when the trees / shrubs begin to bud or in autumn after they have shed their leaves.

The crude drugs collected from natural resources are properly cleaned and dried and they are packed in cloth/leather and are stored properly in dry and cool places to keep their potency.

Conservations

Despite of its long history of contribution for public health and socio-economic development, there is not much sound data regarding numbers and volume of Trans Himalayan plants, collected from wild sources for domestic and commercial uses. Since 1980 the trans-Himalayan plants were only used by the local people for their domestic needs and local Amchis for treating patients of their locality in sustainable manner. The inaccessible, remoteness and harsh climatic conditions have provided natural safety for these plants. With the expansion of roads and increased population have exposed this herbal wealth and brought new commercial interest. Therefore, it is very important to develop a mechanism for sustainable and coordinated exploitation of these wild plants which will provide ecologically sustainable, socially acceptable and economically equitable production and utilization system. Following measures have been suggested to Govt. agencies, various NGO's and farmers for conservation of medicinal plants:-

i. *In-situ* conservation by the establishment of nature reserves or biosphere reserves.

ii. *Ex-situ* conservation through medicinal & aromatic plant reserve / medicinal plant garden.

iii. Creation of medicinal & aromatic gene bank for cold arid zones.

iv. Propagation and cultivation study of medicinal plants.

v. Sustainable development and creation of ecological awareness.

vi. Calculated and planned exploitation of medicinal plant resources.

vii. Active role of Govt. agencies, NGO's and Army to preserve the medicinal wealth.

viii. To popularise the potential of medicinal wealth among the local farmers' for their commercial cultivation.

ix. To conduct regular training among the farmers and troops for cultivation, harvesting and sustainable utilization.

Field Gene Bank

Cultivation of high altitude medicinal plants is one of the important aspects for its conservation and sustainable utilization. This Institute has established an alpine herbal garden (Field gene bank) within the premises of DIHAR at an altitude of 3,500 m above msl with the view to standardize the propagation and cultivation practices of medicinal plants growing at different altitudes and localities. In this direction efforts have been made to collect the germplasm from natural localities and at present germplasm of 55 medicinal plants are being maintained. The important medicinal plants being maintained at Field Gene Bank are *Aconitum heterophyllum, Dactylorhiza hatagirea, Ephedra gerardiana, Hyoscyamus niger, Inula racemosa, Podophyllum hexandrum, Rheum webbianum, Saussurea lappa*, etc.

CONCLUSIONS

The trans-Himalayan cold desert of Ladakh and Lahaul-Spiti is a trove of medicinal plant biodiversity which plays major role in traditional systems of Indian medicines. Based on traditional knowledge, certain herbal health products have been developed to ameliorate high altitude maladies using high altitude medicinal plants faced by low landers under high altitude environment. In addition *ex-situ* cultivation will play important role in conservation and sustainable utilization of medicinal biodiversity. The role of Govt agencies, NGO's and industries is very important to achieve a balanced conservation, cultivation and sustainable utilization practices in trans-Himalayan cold desert which will decrease the threats of extinction of certain rare and endangered species, increase availability of plant raw materials and finally improve the economy of the tribal population.

References

Annonymous. 1948. Wealth of India: Raw Materials (CSIR). New Delhi.

Ballabh, B. and Chaurasia, O.P. 2007. Traditional medicinal plants of Ladakh used against cold, cough and fever. J. Ethnopharmacology. 112, 341-349.

Ballabh, B., Chaurasia, O.P. and Ahmed, Z. 2007. Herbal products from high altitude plants of Ladakh Himalaya. Current Science. 92(12), 1664-1665.

Chauhan, N. S. 1999. Medicinal & Aromatic plants of Himachal Pradesh, Indus Pub. Company, N. Delhi.

Chaurasia, O.P. and Singh, B. 1996-2001. Cold Deserts Plants Vol., I-V. FRL C/O 56 APO.

Chaurasia, O.P. and Singh, B. 2000. An ethnobotanical profile of cold desert flora of Nubra valley (Ladakh). Bull. Bot. Surv. Ind. 42. (1-4), 81-90.

Chaurasia, O.P. and Gurmet, P. 2004. A checklist on medicinal and aromatic plants of trans-Himalayan cold desert. FRL (DRDO), Leh-Ladakh.

Chaurasia, O.P. and Ahmed, Z. 2005. Challenging Nutrition at Heights. Food and Nutrition. 3(1), 22-24.

Chaurasia, O.P., Parimelazhagan, T. and Ahmed, Z. 2006. High Altitude trans Himalayan Medicinal Plants of Defence Importance. In Herbal Drug: A twenty first century Perspective. Jaypee Brothers Medical Publishers, New Delhi, pp. 182-189.

Chaurasia, O.P., Ahmed, Z. and Ballabh, B. 2007. Ethnobotany and Plants of Trans Himalaya. Satish Serial Publication, New Delhi. ISBN: 81-89304-33-X.

Jain, SK. 1993. "A manual of Ethnobotany". Scientific Publishers, Jodhpur.

Kachroo, P. Sapru, B.L. and Dhar, U. 1977. "Flora of Ladakh". Bishen Singh Mahendra Pal Singh, Dehra Dun.

Mani, M.S. 1990. Fundamentals of High Altitude Biology. Oxford & IBH Pub. Company, Delhi.

Polunin, O. and Stainton, A. 1989. Flowers of the Himalaya. Oxford Univ. Press, New Delhi.

Singh, B. and Chaurasia, O.P. 2000. Medicinal flora of Indian cold desert: Proceedings of XXV International Horticultural Congress Part 13, Acta Horticulture. 523, 65-72.

Dwivedi, S.K., Attrey, D.P. and Chaurasia, O.P. 2005. Studies on preparation and preservation of seabuckthorn (*Hippophae rhamnoides* L.) squash beverage. Ind. J. Horticulture. 66(1), 78-80.

Parimelazhagan, T., Chaurasia, O.P. and Ahmed, Z. 2005. Seabuckthorn: Oil with promising medicinal value. Current Science. 88(1), 8-9.

Ved, D.K. et al. 1998. CAMP Workshop Report, Kullu- Himachal Pradesh, organized by FRLHT, Banglore.

❑❑❑

14

Conservation Status and Cultivation of Selected Medicinal Plants in the Indian Himalayan Region

S.S. SAMANT[1], S.K. NANDI[2] AND J.S. BUTOLA[1]
[1]G.B. Pant Institute of Himalayan Environment & Development, Himachal Unit, Mohal-Kullu- 175 126, Himachal Pradesh, INDIA
[2]G.B. Pant Institute of Himalayan Environment & Development, Kosi-Katarmal, Almora-263 643, Uttarakhand, INDIA

INTRODUCTION

The conservation and cultivation of medicinal plants have become global concern due to increasing demand of the commercially viable species in the pharmaceutical industries. Of the estimated 30,000 known medicinal plant species of the world, about 8,000 are found in India. In India, >70% population rely on plant based medicine. Such high dependence on herbal medicine is due to the fact that remedies based on these plants often show minimal side effects, and the relatively high cost of synthetic medicines. In recent years, the demand for Indian medicinal plants has increased considerably at local, regional, national and global markets and the India is second largest volume exporter of raw herbal drugs to the global market. Major part of the exported raw material includes medicinal plants of the Himalayan Region. It has been realized that medicinal plants of this region offer an advantage in having much greater possibilities of providing novel biomolecules in view of the environmental stress they are subjected.

The unique topography, climatic conditions and large altitudinal range of the Indian Himalayan Region (IHR) support diverse habitats, species, communities and ecosystems. The region, one of the identified Mega biodiversity Hot Spots and recognized as one of the best habitats for medicinal plants, is known to support representative, natural, unique and commercially viable medicinal plant species. Medicinal and aromatic plants form an integral part of the daily activity of most of the hill communities and inhabitants are known to collect these plants from natural habitats mainly for their own use or for trade. Majority of these are known constituents in Ayurvedic, Unani and other Traditional Systems of Medicines and have been used since long. The Tibetan system has been reported to exclusively use medicinal species of the IHR. Besides above use, many species are exploited for essential oils and aromatic purposes. The extract of these plants are also constituents of various modern medicines and pharmaceutical preparations, and are currently in high demand worldwide. The increasing demand has led to much dependency by the pharmaceutical companies on collection of medicinal and aromatic plants from the wild; only a minor fraction of the requirement is met from cultivation. Economic assessment of some high altitude species indicates that their cultivation can lead to major sources of income and help in socio-economic upliftment of the rural communities. In spite of efforts made through agro-technological innovations and progress in the field of biotechnology, only selected plants of the IHR are under cultivation. The major reasons may be non-availability of proper agro-technologies, lack of proper marketing facilities for the produce, etc. The rapidly declining natural stock for cultivation or non-availability of quality planting material of most of the commercially viable medicinal plants and increasing demand of the raw material have necessitated developing agro-technologies and promoting their cultivation in the respective altitudinal zones of occurrence. In view of the importance of medicinal plants in the regional, national and global markets and potential for the socio-economic development of the rural communities across the IHR, the present paper provides information on the agro-technologies developed on some of the commercially viable high altitude medicinal plants.

MATERIALS AND METHODS

Information on diversity and conservation status of medicinal plants is based on Samant *et al.* (1998). The agrotechnology is based on the experiments conducted in the herbal gardens and medicinal plants nurseries of the Institute, information collected from the farmers and also, from the available information (Chauhan, 1999; Maikhuri *et al.*, 2001; Samant *et al.*, 2001; Kumar and Samant, 2001; Nautiyal and Nautiyal, 2004; Anonymous, 2006; Butola and Badola, 2008).

RESULTS

Diversity and Conservation Status

The IHR harbours 1748 medicinal plant species, of which 25.8% are native to the Himalayan region; 5.66% species are native to the Himalayan and neighbouring biogeographic domains; 3.55% species are endemic and 11.9% near endemic to the IHR. The maximum diversity of medicinal plants has been reported in the sub-tropical zone, i.e. < 1800 m, the diversity decreasing with increasing altitude. Across the IHR, Sikkim and North Bengal (Darjeeling district) support the maximum medicinal plant diversity (707 species) followed by Uttarakhand (701 species) and Himachal Pradesh (643 species). In view of the large collection from the wild, about 120 species of medicinal plants of IHR have been categorized as critically endangered, endangered, vulnerable and/or near threatened as per the International Union for Conservation o Nature and Natural Resources (IUCN) (Ved *et al.*, 2003, 2005) and 17 species of the medicinal plants have been listed in the Red Data Book of Indian Plants (Nayar and Sastry 1987, 1988, 1990).

Species Description, Agrotechnology and Post Harvest Processes

The agrotechnology and post harvest processes of the commercially viable high value medicinal plants namely *Aconitum heterophyllum* Linn., *Angelica glauca* Edgew., *Dactylorhiza hatagirea* (D.Don) Soo, *Hypericum perforatum* Wall., *Jurinella macrocephala* (Royle) Cl., *Nardostachys grandiflora* DC., *Picrorhiza kurrooa* Royle ex Benth., *Podophyllum hexandrum* Royle, *Saussurea costus* (Falc.) Lipsch., *Inula racemosa* Hk. f. and *Bunium persicum* (Boiss.) Fedt. have been investigated and compiled. The details are as follows.

1. *Aconitum heterophyllum* Linn.; Family: Ranunculaceae

Local/common name: Patish, Atish, Karwi; *Trade name:* Atees, Atis, Ativisha, Bonga

Status and distribution: This species is Critically Endangered for the globe and distributed in Pakistan, Nepal and Indian Himalayan region between 3000-42,00m in the sub-alpine forests, alpine meadows/thatches and grassy slopes.

Taxonomic description: Erect herb, 30-100cm tall, with tuberous fusiform or ovoid roots. Lower leaves orbicular or broadly ovate, cordate or more or less 5-lobed, teeth obtuse or acute; upper leaves lanceolate, sharply toothed. Flowers in racemes, greenish blue with purple veins. Roots are tuberous.

Flowering: July-August; *Fruiting:* August-September

Active ingredients: Tuberous roots contain a noncrystalline, non-toxic alkaloid Atisine (0.4%). Nearly 0.79% of the total alkaloid in the plant is found in the root. Other alkaloids in this species are heteratisine (0.3%), histisine, heterophyllisine, heterophylline, heterophyllidine, atidine, and hitidine. Aconitic acid, tannic acid, a mixture of oleic, palmitic, stearic glycerides and vegetable mucilage are also present in addition to starch and sugars.

Uses: The roots are used in several Ayurvedic and Unani preparations as anti-pyretic, analgesic, anti-periodic, aphrodisiac, astringent, anti-venom, anti-inflammatory, anti-rheumatic, vermifuge, etc. It is also used to treat stomach disorders (diarrhoea, dyspepsia, dysentery, anorexia, etc.), to reduce obesity and to treat cough and headache caused from eating excessive amounts of greasy foods, thirst associated with fever, yellowish sclera, nausea, vomiting, throat pain and piles.

Market value (Rs/kg; dry): 2000-2400 (Kullu); 3900-4300 (Solan); 3700-4400 (Dehradun); 4600-5000 (Amritsar) and 3300-4000 (Delhi).

Agrotechnology and Post Harvest Processes

Seed collection, drying and storage: Seeds are collected after maturation during late September and early October. Initially, seed viability is very high but, when seeds stored at room temperature, they germinate during the first six months of storage and subsequently lose viability rapidly due to increase in temperature. Seeds stored at low temperature (2-4^{0}C) retain good viability upto one year.

Climate and soil: This species can be cultivated between 2200 to 4200 m amsl. It requires cool and temperate climate. Sandy loam, porous and acidic soils are the best for seed germination, plant survival, growth and yield. The addition of forest humus to the soil increases survival and growth of plants as it keeps the soil damp which is prerequisite for growth and survival.

Field preparation and manuring: The land is dug or ploughed twice or thrice to increase porosity of soil. Manuring during the rainy season is done by digging trenches that prevent the rain splashing the manure. Approximately 75-80 qt of manure is required for initiating cultivation in one hectare land. Use of vermicompost or forest humus is found suitable for high production.

Propagation and cultivation: It is propagated through seeds or tuberous roots. In the field, germination takes place after 15 to 25 days and completed within 40-45 days of sowing. However, the germination time is relatively short in laboratory condition. In open beds, germination and growth are generally very slow and low as compared to polyhouse and laboratory conditions. Germination as well as true leaf initiation is early in the sandy soil. Seedlings raised during the winter in polyhouse and over 90 days olds are planted in the field (at high altitudes) during April or May at a distance of 30×30 cm, and 35×35 cm distance should be maintained for planting root segments. Seedlings transplanted during October need care as heavy frost and occasional snow fall damages seedlings. For vegetative propagation the top root segment with the innovation bud is found more successful. Top root segments produce a single shoot, which is found best for multiplication as compared to middle and basal segments. Topmost segments, when transplanted in a 10 to 15^0 slant, show high survival rate, i.e., 80 to 90 %. At lower altitudes, transplantation during May or June gives better results. About 90, 000 plants or 800 gm seeds are needed for cultivation in one hectare land.

Irrigation and weeding: Initially, regular watering is required to decrease the mortality rate of seedlings. During rainy season, watering is very rarely required. Frequent watering is required every 24 hours for 6 month old seedlings particularly at lower altitudes (2000-2200 m amsl). During the rainy season, water-logging causes bolting of plants and black spots appear on leaves. To avoid this, raised beds with slight slope are developed. Frequent (weekly) weeding during the summer and rainy seasons is required. During the winter months, irrigation is needed once a week to retain moisture and weeding every 15 to 20 days.

Harvesting: The plants raised through seeds take at least three years for maturation and harvested after third year. Through tubers, plants flower in the first year of plantation. Flowering takes place during late August or September. After completing the reproductive phase plants become mature for harvesting and yield high percentage of active contents. The reproductive phase of the species varies from lower to higher altitudes. The maximum quantity of active ingredients is found during July and August in the beginning of the flowering period. This period is suitable for harvesting. However, to get the maximum yield and seeds for multiplication, plants must be collected during the October-November.

Yield, post-harvesting techniques and income generation: After two years of cultivation using roots and vermicompost, the estimated yield is 1.22 qt/ha. This value may vary according to mortality rate of plants In this case, actual profit of Rs. 2-2.10 lakhs can be obtained. After harvesting, the roots are washed properly and old and new roots are separated and air dried. Oven or direct sun drying reduces the active contents of the roots. After complete drying, roots are packed in gunny bags for marketing, or can be stored in dry cold chambers.

2. *Angelica glauca*; Family: Apiaceae

Local/common name: Chora, Choraka (Sanskrit), Chamchora Chura (Himachal Pradesh), Chora and Chura (Punjab), Chohore (Kashmir), Choru, Gandrayan or Chipi (Uttarakhand) and Chur (Pakistan); *Trade name:* Chora, Gandrayan and Himalayan Angelica

Status and distribution: This species is endangered for the globe and distributed between 2400-3800 m from temperate to alpine zones of Kashmir, Himachal Pradesh and Uttarakhand. It is endemic to the Indian Himalayan Region.

Taxonomic description: Perennial, glabrous herb, with thick aromatic rootstock. Roots generally 20-50 mm thick, externally grayish brown with deeply, longitudinally, wrinkled and furrowed surface Stem 1.3m high, stout, fistular. Leaves large, 1-3 pinnatipartite segments 2.5-8cm long, irregularly serrate or coarsely toothed; leaf base sheathing, broad, inflated. Flowers white, in large, 15-30cm across, compound umbels bearing hermaphrodite flowers. Peduncles stout, 20 cm long. Involucres 5, long, linear bracts. Rays 20 to more Involucels 6-10, linear bractlets. Calyx teeth 0. Petals 5, obovate emarginate, sometimes tinged purple and stamens 5. Fruit mericarp oblong, flattened, 1.25cm by 0.6cm, lateral ridges expanded into wings and bearing single vittae.

Flowering: July-August; *Fruiting*: September-October

Active ingredients: Roots contain valeric acid, angelic acid and angelisine resin. Steam distillation of roots yields a pale to brownish yellow essential oil (0.4-1.3% dry basis), which contains lactones (dimeric lingusticum), sesquiterpenes, α-phellandrene, selinene, α-cadinene, umbelliprenin and terpene alcohols. Petroleum ether extract of the fruit contains iso-imperatorin, prangolarin and coumarin called tert-o-methyloxypeucedanin, furocumarin, angelicin, bergapten, xanthotoxin and umbelliferene.

Uses: Roots are used as stimulant, cardioactive, carminative, digestive, sudorific, expectorant and stomachic properties. It is useful in cough, ulcer of palate, menorrhoea, menopausal, anemia, infantile atrophy, debility, bronchitis, vomiting, typhoid, fever, spasms, increasing appetite and in various stomach disorders in the human and animal (diarrhoea, rinderpest, dysentery, constipation, anorexia and flatulent colic, etc.) and to treat loss of kidney heat, accumulation of excess lymph in the joints. The species is used in various diseases in women, and known as 'female ginseng'. The root powder is mixed with warm water and given to children in stomach ailments and vomiting. Due to strong fragrant odour, it is important for industrial uses, i.e., fixative of perfume. Coumarin compounds present in various plant parts are likely to be related to natural defense of plants against fungal attack, which is believed to play important role for having anti-inflammatory and anti-microbial properties. Coumarin is the parent molecule of Warfarin which acts as a Vitamin Kantagonist.

Market value (Rs/kg; dry): 80-90/- (Kullu)

Agrotechnology and Post Harvest Processes

Seed collection, drying and storage: Seeds are collected during late September or early October. Viability of the seeds is generally found low. Seeds lose viability rapidly when stored at room temperature and therefore, storage of seeds at low temperature (4^0C) is recommended for retaining viability for a long time, even upto four years.

Climate and soil: It can be cultivated between 2200 - 3400 m amsl in the temperate, sub-alpine and alpine zones. Rich porous soil, moist and shady places with rich forest humus are highly suitable for the growth and productivity of the species.

Field preparation and manuring: The land is dug or ploughed twice or thrice upto 2 feet for better growth and yield. Stones and weeds are removed from the soil. Approximately 20 t of vermicompost or forest humus is required for initiating cultivation in one hectare land. Vermicompost and forest humus are suitable manures for the growth, survival and yield of the plants. Generally, manuring is done during the field preparation, however, manuring may also be done after the completion of the vegetative growth phase during October or in the winter after 2 or 3 years of cultivation. The areas where forest humus is easily available, may be used in place of manure.

Propagation and cultivation: It is propagated through seeds or root cuttings. Germination takes place after 30 to 40 days and sometime after 50 days of sowing. Seed germination and germination time can be improved through pre-sowing treatments of Potassium Nitrate (150 mM) for 24 hrs or Sodium hypochlorite for 30 minutes. Four months old seedlings, raised inside polyhouse or in the nursery are be transplanted during the commencement of rainy season. If the site is moist or has irrigation facility, transplantation can be done during April and May. The distance between the seed raised plants should be 35 cm and 45 cm between the plants rose through vegetative propagation. Vegetative propagation through root cutting is most successful for the establishment of seedlings in cultivation. Apical parts of rhizome (2-3 cm) are planted during April-May. Approximately 80,000 plants or 2-3 kg seeds are required for cultivation in one hectare land. Raised beds are suitable for growth of the plants.

Irrigation and weeding: In the initial stages of cultivation, regular irrigation is required. However, in the dry season, irrigation is required twice in a week. During summer and rainy season weeding is usually done at least twice in a month. During winter season one time weeding in a month is suitable. Water-logging should be strictly avoided which causes root decay and consequently, high mortality in plants.

Insect and pest management: At high altitudes (above 2500 m), cultivation of this species is generally free from invasion of insect/ paste but at lower altitudes, leaves, flowers and fruits are invaded by aphids and other insects/pests. To overcome this problem, 5 ml Neemarin is mixed with one liter water and sprayed on the affected parts. Spray of Neemarin is done in the initial stage of invasion.

Harvesting: The fruits usually mature during late September or early October. The roots are harvested after the senescence of seeds.

The plants developed from seeds are harvested after 2-3 years and those developed from the roots can be harvested every year.

Yield, post-harvesting techniques and income generation: Using rhizome cuttings and vermicompost, approximately 18 qt/acre at 1800 m and 47 qt/acre above 2200 m can be obtained every year under cultivation. However, through seeds, the yield is approximately 28 qt after two years using vermicompost. The yield varies place to place depending upon the climatic conditions.

After harvesting, the apical portion of rhizome is separated and transplanted in the field for developing new plants. The remaining portion is washed with water to remove soil. The roots are cut into small pieces and kept in the partial shade for drying. After complete drying, roots are packed in cloth/jute bags and stored in dry places to avoid insect infestation.

3. *Bunium persicum* (Boiss.) Fedt.; Family: Apiaceae

Local/common name: Kala-zira, Kala-jirah, Umbu, Siahzira, Krishnajeerak; *Trade name:* Kala-zira

Status and distribution: Vulnerable in Himachal Pradesh and distributed in South-eastern Europe, the Caucasus, Siberia and Western Asia, and gentle exposed mountain slopes in C. Asia between 2000-3500 m in dry scrubby slopes. It is native of Baluchistan, Afghanistan, Kashmir and Himachal Pradesh.

Taxonomic description: A glabrous herb with profuse branches attains a height of 50-60 cm and spreads of 40-50 cm. Roots globular, tuberous type. Leaves dissected with narrow segments. Flowers small, white, in compound umbels, hermaphrodite. Fruits darkish-brown, ribbed with pointed ends with deep aroma.

Flowering: May-June; *Fruiting*: June-July

Active ingredients: Seeds contain 2.5% essential oil having 45-65% carvone, a mixture of ketone, carvone, terpene and traces of carvacrol and cuminaldehyde (40.6 %).

Uses: Seeds are used as spice and for flavouring dishes. In the indigenous systems of medicines, the fruits are regarded as stimulants, carminative and are useful in curing diarrhoea, dyspepsia, fever, flatulence, stomach-ache, haemorrhoids and obstinate hiccup.

Market value (Rs./kg, dry): 425 (Shimla)

Agrotechnology and Post Harvest Processes

Seed collection, drying and storage: Seeds are collected in the last week of September to November first week. After collection, seeds are kept in shade and after 9-10 days, these are sun dried. The dried seeds are packed in bags and stored in cold and dry rooms.

Climate and soil: This species requires cold and humid climatic conditions. Sandy loam, moist and well-drained soil is best suited for its cultivation.

Field preparation and manuring: The hard ground is ploughed till soil becomes smooth and pliant. The manure is spread in the field and again field is ploughed to ensure the proper mixing of the manure with soil. 20- 40 Kg manure is added in 3.5 bighas.

Propagation and cultivation: This species propagates through seeds and bulbs. About 2-3kg seeds are mixed with 3 kg of sand and are sown by broad casting in 3-5 Bigha land. After broadcasting of seeds the soil is made plain using Pickaxe (Kudal).

Irrigation and weeding: This crop requires well irrigation facilities. Irrigation is done 2-3 times in a week but the frequency is increased in summer. Weeding is done 3-4 times in May or June with the help of small pickaxe (Kudal). After weeding, the land is irrigated.

Insect and pest management: The protracted rainfall allows insects to attack the flowers. Spray of Metasat, an insecticide weighing 4kg per 400 liter of water is done in 3.5 Bigha of land.

Harvesting: Kala zira is harvested after the 3rd year of sowing when a bulb is formed in its root. Usually the seeds are harvested in the last week of September to the first week of November.

Yield, post-harvesting techniques and income generation: In favorable climatic conditions, approximately 1 bigha of land provides 15kg of seeds. Seeds are collected between September last and November first week. Only the upper portion of the plant bearing seeds is plucked. After collection, seeds are kept in shade and after 9-10 days, these are dried in sunlight. Further, threshing and winnowing are done to segregate seeds. The separated seeds are then collected in bags and stored in cold and dry rooms.

4. *Dactylorhiza hatagirea* (D.Don) Soo: Family: Orchidaceae

Local/common name: Panja, Hathajari, Salep, Hathpanja, Aralu; *Trade name:* Salap, Salampanja, Panja.

Status and Distribution: This species is Critically Endangered in Indian Himalayan Region and distributed between 2500-4000m in Pakistan, Nepal, China and India (Uttarakhand, Sikkim, J&K, Himachal Pradesh).

Taxonomic description: A terrestrial herb. Roots tuberous, palmately divided in to 2-3 finger-like lobes. Leaves oblong-lanceolate, erect, obtuse, base sheathing. Flowering stems robust, erect, hollow, leafy. Flowers spotted rosy-purple in many flowered dense cylindrical terminal spike, bracts green, narrowly lanceolate. Sepals and petals equal, three together in a hood, the lateral sepals spreading, lip rounded, shallowly 3-lobed, spotted dark purple, spur straight, cylindric.

Flowering: June-July; *Fruiting:* August-September

Active ingredients: The tuber contains mucilage (45 %), phosphate (2.7%), chloride and glucoside-loroglossin and a trace of volatile oil and ash consisting chiefly of potassium and lime.

Uses: Tubers are used as a farinaceous food, nervine tonic and aphrodisiac. <2.5 gm powder of tuber is considered a full day's diet in adverse conditions by local people. Mucilage jelly is nutritious and useful in diarrhoea, dysentery, chronic fever and in leucorrhoea. Decoction of salep with sugar and flavoured with spices is a useful drink for the patients. In Unani System of Medicine, it is used in seminal debility, chronic diarrhoea and general weakness in debilitated women after delivery. It is also used in paralytic affections.

Market value (Rs/kg; dry): 1000-1200 (Kullu); 1000-1100 (Dehradun); 850-1000 (Amritsar) and 1500-1700 (Delhi).

Agrotechnology and Post Harvest Processes

Seed collection, drying and storage: Fruits are collected during September. After air drying for 2-3 weeks, seeds are separated from the fruits and stored at low temperature.

Climate and soil: This species can be cultivated between 2000 to 4200 m, amsl. It requires acidic and sandy loam soil with sufficient moisture, rich organic manure and open slopes.

Field preparation and manuring: Slight slope is suitable for cultivation of this species. The land is ploughed twice or thrice to increase porosity of soil. The waste material is separated from the soil. Approximately 20 t of FYM or forest humus is used for cultivation in one hectare land. Forest humus is most favorable for higher survival, growth and yield.

Propagation and cultivation: Seeds are very minute and therefore, it is mixed with sand before sowing. Seeds have very low germinability. Vegetative propagation through tuber cuttings is quite successful. So far only 20-30% germination is achieved in this species when sown in natural habitat. Furthermore, seeds germinate best in soil collected from its natural habitat. Collection of immature and fresh seeds from nature showed optimum germination. Seeds require symbiosis with fungal species in nature for germination. Small slices (4 mm) of tuber containing meristematic tissues are reported to develop plantlets when transplanted at 5-7 cm depth and 30×30 cm apart. About 1,10,000 plants are required for cultivation in one hectare land.

Irrigation and weeding: Initially, irrigation is required every day. 80-90% humidity is to be maintained for rooting and development of plants. In later stage, watering within a gap of 10 days is sufficient. During summer and rainy seasons weeding is required at 10 days interval. In other seasons, monthly weeding is sufficient.

Harvesting: Tubers are harvested after seed maturation during mid September to October. The plants developed from the tuber get ready for harvesting in 2 years. The yield can be taken every year if the whole tuber is used as propagule.

Yield, post-harvesting techniques and income generation: About 3.70 qt/ha (dry tubers) can be obtained if the species is cultivated nearby villages of the natural habitats. Total Rs. 3,69,600 can be earned if sold in the market. After harvesting, old tubers are separated from young tubers and steeped in hot water for one hour. Outer membranes of tubers are removed through this process, and tubers turn light yellow in color. These are sun dried and can be preserved for a long time. Drying is done in partial shade. Dried tubers are packed in small gunny bags and stored in dry rooms.

5. *Hypericum perforatum* Wall.: Family: Hypericaceae

Local/common name: Basanti, Basant, St. John's Wort; *Trade name:* Basant, Khoontir

Status and distribution: This species is Vulnerable to Indian Himalayan Region. It is native to Europe, Western Asia and North Africa, and is widely distributed throughout the temperate areas of the World. In India, it is distributed in Northwestern Himalaya including Himachal Pradesh. It grows in shady damp forests and open grassy slopes along with shrubberies between 2000-3000 m above MSL.

Taxonomic description: A perennial herb. Stem erect, angled, branched, glabrous, leafy. Leaves oblong, obtuse, lower surface pale, glabrous, entire, gland-dotted, sessile. Flowers yellow, bracteate, in terminal corymbs. Seeds numerous, brownish and hard.

Flowering: April-May; *Fruiting:* May-June

Active ingredients: Hypericin (0.05-0.3 % w/w on dry weight basis) is the main active content. Its aerial parts contain anthraquinone derivatives, hypericin, usigtoercin and protohypericin. It also contains kaempferol, quercetin, rutin, phenolic acid and tannins.

Uses: It is a multipurpose species used for fuel, fodder, dye and medicine. Paste of leaves is applied on cuts to check bleeding. Leaves' infusion is given in malarial fever. It has astringent, expectorant and diuretic properties, used in pulmonary and urinal troubles, diarrhoea and in the therapy of depression state. It is also used as anthelminitic. An oil is known as St. John's Wort Oil (Oleum Hypericin) is prepared by infusing the fresh flower in olive oil and used externally for the treatments of burns, wounds, sores, ulcers, swellings and sometimes against rheumatism, lumbago and intestinal worms. Hypericin has antidepressant and antiviral properties. It is suggested for having a potential in the treatment of Acquired Immuno Deficiency Syndrome (AIDS) and also reported to possess anti-tumor and anti-cancer properties. An infusion of the plant, taken as a tea, has been helpful in treating night-time incontinence in children.

Market value (Rs/kg; dry): 50-60 (Delhi)

Agrotechnology and Post Harvest Processes

Seed collection, drying and storage: Seeds are collected in late May. After drying for two weeks, seeds are stored in cool and dry places. Seeds are small in size and dark brown in color.

Climate and soil: This species can be cultivated between 1600-3600 m, amsl. Partial sunny sites and medium fertile moist soil are suitable for its cultivation.

Field preparation and manuring: The land is dug or ploughed twice and approximately 20 t/ha of forest humus or FYM (Cow dung) is mixed in the soil. Quantities of the manure may be reduced during second and third years according to soil condition.

Propagation and cultivation: It propagates mostly by seeds. However, it can also be propagated by slips. Seeds are sown during November-December in raised beds. Seeds start germinating after 15-20 days. Three-four months old seedlings should be transplanted at a spacing of 30×30 cm during March-April. The plants flower during the same season. Approximately, 100 g of seeds are required for one hectare land.

Irrigation and weeding: During the summer season, it requires regular irrigation and in winter, twice in a week is sufficient.

Harvesting: Harvesting is done after seed maturation during May-June.

Yield, post-harvesting techniques and income generation: Approximately 2.5-3.5 t/ha (fresh herb) yield can be obtained through harvesting of one year old crop. After drying, yield ranges between 1-1.5 t/ha. The whole plant is used for commercial purpose. Rs. 90,000 can be obtained from one year yield if sold dried.

6. *Inula racemosa Hook* f.; Family: Asteraceae

Local/common name: Manu, Poshkar, Poskarmul, Pushkarmool; *Trade name:* Mannu, Pushkarmool

Status and distribution: Endangered species (as per the Red Data Book of Indian Plants), also found under cultivation. It is distributed in East Asia to Western Himalaya between 2000-3200 m, above MSL.

Taxonomic description: Stout herb, upto 1.5 m tall. Leaves leathery, rough above, densely hairy beneath and toothed. Radical leaves, stalked, elliptic, lanceolate, cauline leaves ovate-oblong, semi-amplexicaule, often deeply lobed at the base. Flower heads many, 3.5-5.0 cm in diameter, yellow in colour and arising in terminal racemes, flowers hermaphrodite. Achenes 0.4 cm long, slender and with reddish pappus.

Flowering: July-August; *Fruiting:* September-October.

Active ingredients: The main ingredient is Inulin and an essential oil containing isoalantolactone and di-hydralactone. Alantolactone, β sitosterol and its glucosides have been found in alcoholic extract of roots.

Uses: The roots are used in Tibetan System of Medicine. It has a sweet, bitter and acrid taste with a neutral potency and antipyretic/ antiseptic, anti-inflammatory, antiasthmatic, anthelmintic, expectorant and diuretic properties. It is used in the treatment of contagious fevers that have not fully ripened and pain in the upper body, especially between the neck and the shoulders. The seeds are bitter and aphrodisiac.

Market value (Rs/kg; dry): 200-225 (Kullu)

Agrotechnology and Post Harvest Processes

Seed collection, drying and storage: Seeds are collected in September-October. After drying for two weeks in room temperature, seeds are stored at low temperature.

Climate and soil: Temperate, sub-alpine and alpine climates are suitable for the growth of the plant. The plant prefers moist soils and can also grow in heavy clay soil. It can be cultivated in acid, neutral and basic (alkaline) soils in exposed places.

Field preparation and manuring: The soil is ploughed 2-3 times. Deep ploughing is considered better for this crop owing to its deep penetrating roots, which may have a depth of 30cm in the soil. Usually the fields are ploughed after the rain or after irrigation. After ploughing, stone particles and weeds are removed and fields are leveled. 20 qt FYM is required in one acre of land.

Propagation and cultivation: It is propagated both, through seeds and root cuttings. Seeds are sown in spring or autumn in a cold frame. Root divisions are propagated in spring or autumn. Propagation through root cuttings is most suitable. Seeds of this species are very small and often difficult to sow in agricultural fields in large quantity. On the other hand, the root cuttings (upper cuts) are easy to plant and successful in sprouting. Root cuttings are raised in November. In case of drought, these are stored in pits and sowing done in May. The root cuttings are sown at a distance of 20 to 25cm and 7-8cm deep in soil.

Irrigation and weeding: It requires a large quantity of water for better growth. Irrigation is done 4-5 times in a month during summer season. Weeding is done after 20-30 days of germination. After germination 2-3 times weeding in a month is necessary because at this time the seeds are tender and very small. However, after 2-3 months of sowing, the leaves of plant grow very large and broad, hence one time weeding in a month is sufficient. The weeding is done with the help of small pickaxe (Kudal).

Harvesting: Harvesting of roots is done in October after 3rd year. However, in September the aerial parts are removed and the crop is left for one month. After this, the roots are dug with the help of pickaxe in October. After harvesting the roots are cut into small pieces. The roots to be used for cultivation are stored in pits beneath the soil. Whereas, the remaining roots are sun dried for 8-10 days. After complete drying, the material is graded for the trade, as in the case of *Saussurea costus*.

Yield, post-harvesting techniques and income generation: The material is stored in a pit in the field. To store roots in winter season, particularly to save them from snow, 5 feet long, 4 fits wide and 5 fit deep pit is dug. Wherein, the roots remain warm and viable to propogate. The grasses are laid above the lower surface of the pits so that stones must not come in the contact with the roots and damage of soft rooting material can be avoided. Such storage pits have the capacity to contain approximately four quintal of root stocks. Upper portion of the pits (after filling) is covered with the leaves of Rai (*Picea smithiana*) and Tosh (*Abies pindrow*) trees to keep material safe from winter damage. Finally the pit is covered with soil approximately upto 1.5 feet. The roots remain viable for 4-6 months. After the melting of snow, propagules are dug and sown in the month of April or May.

7. *Jurinella macrocephala* (Royle) Clarke: Family: Asteraceae

Local/common name: Dhooplakkar, Jari-dhoop; *Trade name:* Guggal Dhoop

Status and distribution: This species is endangered to the Indian Himalayan Region and distributed in Pakistan, Nepal, China and India. It usually grows in alpine grassy slopes and rarely in rock crevices and glacial moraines between 3000 - 4500 m.

Taxonomic description: A stemless perennial herb, with woody rootstocks. Leaves oblong lanceolate, spreading, all radical, deeply pinnatifid, white lanate above, densely white tomentose beneath. Heads 4.0 cm across, numerous, rosulate, short-peduncled; peduncles stout, tomentose, often cottony at the base. Outer involucral bracts lanceolate or ovate-lanceolate; inner oblong lanceolate. Pappus brownish barbellate, subequal and deciduous. Inflorescens head or capitulum. The heads range from 5-25. The plants with twenty flowering heads is considered best for dhoop.

Flowering: July-August; *Fruiting:* August-September

Active ingredients: Roots yield a caoutchouc-like material.

Uses: Roots are considered to be a stimulant and given in fever after childbirth. Decoction of the root is given for colic pain and bruised roots are applied to eruptions. The aromatic roots form the chief ingredient of Dhoop/Havan Samigri and are used as incense in homes and religious ceremonies.

Market value (Rs/kg; dry): 90-100/-

Agrotechnology and Post Harvest Processes

Seed collection, drying and storage: Seeds are collected during October. After drying in shade for two weeks, seeds are filled in airtight containers and stored in cool condition.

Climate and soil: This species prefers sub-alpine and alpine climates and can be cultivated between 2500-5000 m amsl in sandy and porous soil.

Field preparation and manuring: The land is dug or ploughed twice or thrice to make the soil porous. Approximately 60-80 qt manure is required for one acre land. Raised beds are preferred to prevent water-logging.

Propagation and cultivation: It is propagated through seeds and roots. Seeds are sown at a distance of 30 cm during May and June. The aerial parts dry up during winter. This cycle is repeated for 4 to 5 growing seasons after which flowering and fruiting takes place. At lower altitudes, seeds are sown during October and November inside greenhouse and by the end of March, seedlings are well developed and ready to transplant. In one acre land nearly 33,000 plants are required.

Irrigation and weeding: The crop is irrigated in every two days interval particularly during March, April and May. Weeding is required frequently during the early stages of plants and at the seedling stage. After the second year of growth, monthly weeding is sufficient.

Harvesting: The species requires at least two years for maturation. However, harvesting after 3 or 4 years is suggested for better yield and quality. Harvesting is done after maturation in late September or October. The reproductive phase commences earlier at lower altitudes.

Yield, post-harvesting techniques and income generation: Estimated yield per acre is approximately 5-6 qt after three to four years of growth. After harvesting, roots are cut into slices and dried in the shade and packed in cotton bags for trade.

8. *Nardostachys grandiflora* DC.; Family: Valerianaceae

Local/common name: Jatamansi, Balchharh, Masi, Jantamansi, Mansi, Machhi, Spikenard; *Trade name:* Balchharh

Status and distribution: This species is critically endangered for Himachal Pradesh and endangered for Indian Himalayan Region, and distributed from Himachal Pradesh to Sikkim and Bhutan, Myanmar and also extends into Tibet and Western China in the alpine regions between 3000-5000 m, amsl in alpine meadows/thatches and moist rocky slopes.

Taxonomic description: Aromatic, perennial herb, erect, 10-60 cm high with long, stout, woody rhizomatic rootstock covered with tail like brown fibres left over from the withered leaves towards the stem, while the root continues to penetrate deep in the soil. Leaves radical in nature, long, narrow, flowers creamy white, often rosy or pale pink in appearance arising in terminal corymbose cymes. Fruit small, 4 mm long, covered with minute hairs.

Flowering: July-August; *Fruiting:* September-October

Active ingredients: The rhizomes of the species contain essential oil having jatamansone, jatamansic acid, lupeol, jatamansinone, jatamansinol, jatamansin, dihydrojatamansin, oroselol, oroselone, seslin, valeranone, valeranal, nardostachone, nardosinone and an alkaloid-actinidin.

Uses: The rhizomes are used in perfumery, offered to the Lord in religious shrines as 'dhoop'. It is used as a tonic, stimulant, antispasmodic, deobstruent, emmenagogue, laxative, diuretic, for stomach ache and wound healing. An infusion of the rhizome is reported to be useful in epilepsy, hysteria, leprosy, respiratory disorders, nervous excitement, palpitation of the heart and a tincture is given in intestinal colic and flatulence. Rhizomes are used as a substitute for valerian, and extracts are used in over 26 Ayurvedic preparations. The roots have remarkable properties to tone up the brain and also useful in the treatment of heart disease, high blood pressure and insomnia. The rhizome oil is useful for growth of hairs and also yields a dye. Spikenard oil possesses antirhythmic activity with possible therapeutically usefulness in cases of auricular blutter.

Market value (Rs/kg; dry): 150-250 (Solan); 155-165 (Saharanpur); 200-230 (Dehradun); 220-240 (Amritsar) and 175-180 (Delhi)

Agrotechnology and Post Harvest Processes

Seed collection, drying and storage: Seeds are collected in late September and October and dried in shade for 2-3 weeks. Seeds are viable for a very short period. Seed viability can be maintained for five or six months by low temperature storage.

Climate and soil: This species can be cultivated between 2000-4200 m, amsl. Cool, moist and partially sunny areas are suitable for cultivation. It requires sandy loam, acidic and humus rich soil.

Field preparation and manuring: The land is ploughed twice to make soil porous and fine textured. 15 tones of FYM (cow dung) or forest humus are required for one hectare land.

Propagation and cultivation: It is propagated through seeds and roots. Seeds are sown in open beds during April or May. Germination takes place within 7-14 days. To reduce the vegetative growth phase by four to six months, seedlings are raised at lower altitudes immediately after maturation of the seeds or during the winter months in net house. These seedlings are well developed in March and April and after two-leaf stage, these can be transplanted to nursery beds at higher altitudes.

Propagation through roots is found better than the seed sowing with regard to higher production within a short period. Segments of the roots are planted by digging a 10×10×10 cm pit during May or October when the plant completes the reproductive growth phase of seed setting and underground fibrous roots become dormant. Seedlings or root segments should be planted 30×30 cm apart. Approximately 1,08,000 seedlings or 600 gm seeds are required for cultivation in one hectare land. The survival percentage is low in both seedlings and rhizome cuttings due to desiccation problem during the winter.

Irrigation and weeding: The plants require excessive watering to decrease mortality rate. However, no watering is required in monsoon. During the dry season, watering is done in every two days. Mulching through broad leaf litter on the soil surface during the rainy season decreases weed invasion. Generally, weeding is done weekly in the first year of seedling growth and during the second and third years, weeding is done twice in a month.

Harvesting: Harvesting is done after seed maturation during the months of September or October to achieve the high quantity of the active ingredients. Plants cultivated at lower altitudes complete reproductive phase during August or September, whereas, plants cultivated at higher elevations are harvested in October.

Yield, post-harvesting techniques and income generation: After three years of cultivation, approximately 4 qt/ha through seedlings and 4.5 qt/ha through rhizome cuttings is obtained. A net profit of Rs. 1,56,947/- can be earned. After harvesting roots are washed carefully to remove the soil particles and dried in partial shade or warm air. After complete drying, roots are packed in gunny bags for marketing or may be stored for short time in a dry cold chamber.

9. *Picrorhiza kurrooa Royle ex Benth.*: Family: Scrophulariaceae

Local/common name: Kadu, Kaur, Karru, Kutki, Tikta, Karvi, Picrorhiza; *Trade name:* Kutki

Status and distribution: This species is endangered to Indian Himalayan Region and distributed in alpine region of the Himalaya from Kashmir to Sikkim extending to West Nepal between 2700-4300 m, amsl in moist rocky slopes, near water springs and old land slide areas.

Taxonomic description: Erect, hairy, perennial herbs, with woody rootstock. Runners rooting at intervals produces new plants. Leaves coriaceous, tip rounded, base narrowed in to a winged sheathing petiole, margins serrate. Flowering scapes ascending, stout, hairy, longer than leaves, naked or with few, lanceolate bracts below the inflorescence, flowers bluish-violet, in terminal racemes. Capsule ovoid, acute, turgid.

Flowering: June-July; *Fruiting:* August-September

Active ingredients: Roots contain Picrorhizin. A glycosidal bitter principle kutkin, a non-bitter substance kurrin, vanillic acid, kutkiol occurring as acetate and kutkisterol from the drug. Kutkin, apocynin, alkanol, alkane, etc. have also been reported.

Uses: It is used in dropsy, considered to be anthelmintic, appetizer, blood purifier, blood pressure reducer, cardiac, expectorant, febrifuge and is useful in asthma, cold, cough, bile trouble, jaundice, leprosy, constipation and stomach troubles. Clinical trials on this drug have shown its therapeutic efficiency in a number of conditions of immunological disorders like asthma, arthritis urticaria, etc.

Market value (Rs/kg; dry): 180-200 (Kullu); 250-265 (Saharanpur); 270 (Dehradun); 250 (Amritsar) and 240-250 (Delhi)

Agrotechnology and Post Harvest Processes

Seed collection, drying and storage: Seeds are collected in September-October. After air drying for 2-3 weeks, seeds are packed in airtight container. Seeds are viable for a very short time in dry storage conditions. The viability of seeds can be retained for a long time (4-6 months) when stored at low temperature (4-6^{0}C) without moisture.

Climate and soil: This species usually prefers sub-alpine and alpine climates and can be cultivated between 2200-4200 m amsl. Partially shady moist slopes with a thick layer of humus and sandy textured loam soil rich in organic carbon and high moisture content is suitable for its cultivation.

Field preparation and manuring: Fields with gentler slopes are suitable. The ploughing is done at least two times. After ploughing, all the stones, gravels and pebbles are removed from the soil. Approximately 15 t/ha of well decomposed FYM or 20 t/ha of forest humus (preferably) is thoroughly mixed with the soil. Sand can be mixed with the clay soil to improve porosity. A higher concentration of livestock manure causes powdery mildew and needs excessive irrigation during the initial phase of manuring. Manuring is done during the winter or before transplanting.

Propagation and cultivation: This species is propagated through seeds and stolons. To increase germination, seeds are sown on the upper soil surface in Styrofoam seedling trays, and covered with a thin layer of moss. Seeds are sown during November-December in greenhouse or during March-April in open beds and in May in alpine areas. Seedlings developed in nursery are transplanted to nursery beds during March-April. Vegetative propagation through top segments of stolon is more successful. Rooting can be increased following water dip treatment for 48 hours. The use of trenches, approximately 2×1×1 ft size for 1000 cuttings, to retain high moisture in the soil is another technique for better rooting in stolon cuttings. Seedlings or stolon cuttings are planted at 30×30 cm distance. About 1,08,000 plants are required for one hectare land.

Irrigation and weeding: In the initial stage, regular irrigation (every 24 hrs) is required. Watering should be done every two days during the winter months. Weeding is done as and when required. Weekly weeding is required during the summer and rainy seasons.

Harvesting: Harvesting is done during September-October, the best period for bio-active ingredients. Thinning of stolons is done in the fields, with a sharp knife or scissors. Roots are dug. The remaining stolons are used to multiply plants for the next crop.

Yield, post-harvesting techniques and income generation: After one year of cultivation through stolon cuttings, about 3.8 qt/acre yield (dry roots) can be obtained. A net profit of Rs. 40,000/acre can be earned. After harvesting roots should be washed carefully to remove the soil particles and dried at room temperature (15-20^{0}C). Drying in this condition retains high picrotin (3 %) and picrotoxin (2 %) contents. Drying in direct sun or in an oven should be avoided in view of rapid loss of active contents. After complete drying, stolon or roots are packed in hermetic containers or jute sacks for marketing or may be stored for a short time in dry places.

10. *Podophyllum hexandrum Royle*; Family: Podophyllaceae

Local/common name: Bankakri, Bankakar, Kanda-ri-mokri, Giriparpat, Rikhpat, Papra, Papri, Indian Podophyllum; *Trade name:* Bankakri, Indian Podophyllum.

Status and distribution: This species in Endangered to the Indian Himalayan Region and distributed in temperate, sub-alpine and alpine regions of the Himalaya from Afghanistan to SW China and Myanmar between 2800-4000 m amsl. In India, it is found in Jammu & Kashmir,

Uttarakhand and Himachal Pradesh in shady moist slopes and bouldery habitats.

Taxonomic description: A perennial herb. Stem erect, glabrous, unbranched, leafy at the top. Leaves 2-4, peltate, orbicular, deeply cut in to 3 lobes; lobes ovate, sharply toothed, acute. Flowers cup-shaped, white or pink, solitary. Sepals 3, petal like, soon falling off. Petals 6. Berry oblong, pulpy, many seeded, red when ripe.

Flowering: May-June; *Fruiting:* June-July

Active ingredients: The rhizomes yield resin (7-18%) which is higher than the American plant, *P. peltatum* (4-8%). It contains podophyllin and podophyllotoxin as active principles. It also contains podophyllol, quercetin, kaempferol, astragalin, an essential oil, wax and mineral salts. The roots are richer in resin than the rhizome.

Uses: The rhizome and roots are known to be used as hepatic, stimulant, cholagogue, purgative, alternative, vermifuge and bitter tonic. Podophyllotoxin and peltatins have high potency in treating cancer. The resin or podophyllin is a drastic purgative and produces copious watery stools and used for treating tumor. It is thus, given with Belladona and Hyscyamus to avoid gripping.

Market value (Rs/kg; dry): 45-50 (Kullu); 80-100 (Amritsar) and 80-120 (Delhi)

Agrotechnology and Post Harvest Processes

Seed collection, drying and storage: Fruits are collected in July-August. After segregating from pulp, seeds are dried in partly sunlight and stored under low temperature (4^0C) to retain viability for longer period.

Climate and soil: This species usually prefers sub-alpine and alpine climates with cold winters and mild summer. It can be cultivated between 1800-5000 m, amsl in shady moist places and requires rich organic black sandy loam soil holding sufficient moisture.

Field preparation and manuring: The fields are ploughed 2-3 times and weeds, stones and gravels are removed from the soil. About 6 t/ha of forest humus or FYM (cow dung) is mixed with the soil. Requirement of manure is less during second and third years.

Propagation and cultivation: The species propagates through seeds and rhizomes. The seeds have dormancy and sometime take 1-2 years to germinate. Seeds are treated with hot water at 40-100^{0}C for 30-120 seconds or with Sodium Hypochloride for 30 minutes before sowing. These treatments give early and higher germination in seeds. Besides, seeds kept in the solution of cow dung for 48 hrs give better results in germination. Seeds are sown during October inside greenhouse in polybags at a distance of 5×5 cm and 1 cm depth. The seedlings are ready for transplantation during May-June. The multiplication of this species is done by rhizome cuttings (2-3 cm having bud) owing to delayed germination and retaining long juvenile phase. Rhizome cutting are treated with IBA (100 μM) for 24 hrs before planting and transplantation is done during May-June at a distance of 30×30 cm. About 1,10,000 seedlings or rhizome cuttings are required for one hectare land.

Irrigation and weeding: In the initial stage, regular (24 hrs) irrigation is required. During the winter months, 2-3 times irrigation in a week is required. Frequent weeding during summer and rainy seasons and once in winter season is required.

Harvesting: The prolonged juvenile and reproductive phases take more time in maturation of the species. Usually, takes 5-6 years through seeds and 3-4 years through rhizome cuttings to give marketable yield of rhizomes. Rhizomes colleted in May have a higher resin content than those collected in November. Rhizomes harvested in July-August i.e., after the formation of Berry and seed setting give maximum production of resin.

Yield, post-harvesting techniques and income generation: About 4 t/ha yield is obtained after three years cultivation in one hectare land which has monetary value of Rs. 1,15,000/-. After harvesting, roots are washed with running tap water and then cut into small pieces and kept in shade for drying. Drying in the sun also does not affect the resin content. After drying, rhizomes are sold in the market or kept in jute bags and stored at cool and dry places.

11. *Saussurea costus* (Falc.) Lipsch.; Family: Asteraceae

Local/common name: Kuth, Kushtha, Pachak, Kooth, Kuste-Hindi, Costus; *Trade name:* Kuth

Status and distribution: This species is critically endangered for Globe; native and endemic to the Indian Himalayan Region and distributed in Kashmir (Kishenganga, Chenab valley), Himachal Pradesh and Uttarakhand. In Kashmir, it is found in moist slopes and birch forests. Also, cultivated in the above stat

Taxonomic description: A perennial herb. Stem erect, robust, unbranched, glabrous below, pubescent above. Lower leaves pinnate, with triangular terminal lobe and irregularly winged leaf-stalk; stem leaves irregularly winged, clasping, scaberulous above glabrate beneath. Flower purple; in dense rounded terminal clusters; involucral bracts purple, numerous, ovate-lanceolate, slender pointed, rigid, with twisted and recurved tips, glabrous or pubescent. Fruit achene, curved, compressed and grayish in colour.

Flowering: June-July; *Fruiting:* August-September

Active ingredients: The root contains resionides, 1.5-2.5% of volatile oil, an alkaloid saussurine (0.05 %), insulin and other minor constituents such as tannin and sugar. Kushtin has also been isolated from the roots.

Uses: The roots are used as anodyne, anti-arthritic, antiseptic, aphrodisiac, astringent, carminative, digestive, diaphoretic, deodorant, disinfectant, diuretic, emmenagogue, expectorant, febrifuge, narcotic, spasmodic, stimulant, stomachic, tonic and vermifuge. It is also useful in curing dysentery, rheumatism, bronchitis, cholera, jaundice, ulcer, skin diseases, stomachache, gas, toothache, cough, cold, fever, hiccough, hysteria, phlegm, oedema, bruises, cuts, heart troubles, menstruation trouble and promotes urination. The oil extracted from roots is known as 'Costus oil', which is used in high grade perfumes and in the preparation of hair oil and also said to be effective in the remedy of Leprosy. In the Himalayan states of India, the roots are used as insecticide to protect shawls and woollen fabrics and incense. In district Lahaul and Spiti of Himachal Pradesh, dried leaves of *Kuth* are smoked as tobacco and upper parts of its plants are used as fuel and fodder.

Market value (Rs/kg; dry): 100-120 (Kullu); 140-150 (Dehradun); 135-140 (Amritsar); 115-125 (Delhi).

Agrotechnology and Post Harvest Processes

Seed collection, drying and storage: The fruits are harvested in September or October and sun dried for a week before threshing. Seeds are segregated from the fruits and air-dried for 1-2 weeks. Seeds retain viability for longer period when stored under low temperature (4^{0}C).

Climate and soil: This species requires a cool and humid climate and can be cultivated between 2000 - 5000 m,above MSL. Sandy and porous soils with good depth are suitable for vigorous development of roots.

Field preparation and manuring: The field is ploughed 2-3 times upto 2 feet deep. About 15 t of FYM (cow dung) or forest humus is required for one hectare of land.

Propagation and cultivation: The propagation and multiplication is done through seeds or root cuttings. Seeds are sown at a distance of 30 cm in the fields or during September and October inside greenhouse. Seedlings are transplanted after 6-9 months growth. At lower altitudes, seeds mature little earlier than that in high altitudes. After flowering, plants continue to grow, but the root quality deteriorates. Root cuttings (2.5 cm, collar zone) are used for vegetative propagation which gives early yield than seeds. About 50,000 seedlings or 3.5-4 kg seeds are required for one hectare land.

Irrigation and weeding: In the initial stage, watering is required everyday. After one year, relatively little irrigation is required. During summer months irrigation twice a week is required. Weeding is required 3-4 times in a month during summer and rainy seasons.

Harvesting: Harvesting is done in 2-3 years after seed maturation during late September and October. The plants raised through root cuttings become mature in 1-2 years. After seed maturation, the root becomes hollow and quality deteriorates. Hence, for higher yield and quality, roots are harvested immediate after seed maturation.

Yield, post-harvesting techniques and income generation: About 4.65 t is obtained from one hectare land. Net profit of Rs. 1,17,683/- after 3 years or Rs. 39,228/- can be obtained after one year of cultivation. After harvesting, roots are washed with running tap water, cut into small pieces and kept in partial shade for drying. After drying, roots are packed in gunny bags for storage in dry chambers or marketed.

DISCUSSIONS

It is estimated that approximately 90% medicinal plants in use are collected from the wild in which 70% collection involves destructive harvesting. Notwithstanding, prevalence of destructive mode of harvesting accompanied by the restricted distribution range reflects a high incidence of threat. The high nativity (>31%) of medicinal plants makes the situation further complicated, when we look for their *in-situ* conservation. Due to destructive harvesting, the populations of many commercially viable high value medicinal plants facing high risk of extinction. This is the cause that 120 species of medicinal plants have been listed under Critically Endangered, Endangered, Vulnerable, Near Threatened, Least Concern, etc. and 17 species have already been placed in the Red Data Book of Indian plants. It is estimated that between 4000 and 10000 medicinal plant species might now be endangered at global level due to over exploitation and habitat degradation. The Protected Area Network has helped in the in-situ conservation of medicinal plants to some extent. Further, increased regulation and the introduction of sustainable wild harvesting methods can protect some species to certain extent. But these procedures are not adequate for meeting bulk demand of raw materials of medicinal plants to the pharmaceutical industries. Initiating cultivation of medicinal plants is viewed as the most viable long term alternative to ensure sustainable supply of the raw material.

In spite of efforts made through agro-technological innovations and progress in the field of biotechnology, only selected plants of the IHR are under cultivation. The major reasons attributed are constant removal of stock plants from the natural habitats, non-availability of location specific agro-technologies, lack of incentives by the Government, limited interest of the communities, lack of proper marketing facilities for the produce, etc. The rapidly declining natural stock for cultivation or non-availability of quality planting material of most of the commercially viable medicinal plants and increasing demand of the raw material have necessitated developing agro-technologies and promoting their cultivation in the respective altitudinal zones of occurrence. Cultivation of commercially viable medicinal plants of the IHR may help in the socio-economic development of the rural communities.

CONCLUSIONS

The present study provides comprehensive information on the agrotechnology and post harvest processes of the 11 high value commercially viable high altitude medicinal plants. The information would help to the stakeholders particularly the progressive farmers in the promotion of medicinal plants in cultivation and uplifting their socio-economic status. There is a need to develop and document the agrotechnology and post harvest processes of other commercially viable high altitude medicinal plants. So that these species could also be established in cultivation and become one of the major sources of their income generation. However, the Government and Non Government Organizations are required to provide the technical know how and quality planting material, and facilitate the marketing of the produce.

ACKNOWLEDGEMENTS

The authors are thankful to Dr. L.M. S. Palni, Director In-charge of the Institute for facilities and encouragements. The farmers of the Himachal Pradesh are acknowledged for providing information on agrotechnology and post harvest processes of some of the medicinal plants.

References

Anonymous. 2006. Package of practices for Forestry plants. Dr. Y.S. Parmar University of Horticulture and Forestry, Nauni, Solan, pp. 103.

Butola, J.S., Badola, H.K. 2008. Himalayan threatened medicinal plants and their conservation in Himachal Pradesh. Journal of Tropical Medicinal Plants, Malaysia 9 (1).

Chauhan, N.S. 1999. Medicinal and Aromatic Plants of Himachal Pradesh. Indus Publishing Company, New Delhi, pp. 632.

Dhar, U., Rawal, R.S., Upreti, J. 2000. Setting priorities for conservation of medicinal plants: a case study in the Indian Himalaya. Biological Conservation 95, 57-65.

Dhawan, B.N. 1997. Biodiversity-A valuable resource for new biomolecules. In: U. Dhar (Ed.), Himalayan Biodiversity: Action plan. Gyanodaya Prakashan, Nainital, pp. 111-114.

Edwards, R. 2004. No remedy in sight for herbal ransack. New Science 181, 10-11.

Holley, J., Williams, J.T. 1996. India uniquely positioned to capitalize on abundant medicinal plant heritage in Hindustan Centre of Diversity. 12, 2-36.

IUCN. 1993. Guidelines on the conservation of Medicinal Plants. The International Union for Conservation of Nature and Natural Resources (IUCN), Gland, Switzerland, pp. 38.

Kumar, A.K., Katakam, A. 2002. Credit for conservation. Frontline 19, 9-22.

Kumar, K., Samant, S.S. 2001. Documentation on Agrotechniques and Post Harvest Processing of Medicinal Plants of the Indian Himalayan Region. IUCN-Himal project, Final Technical Report submitted to IUCN-Himal Cell, WII, Dehradun.

Lange, D. 1997. Trade figures for botanical drugs world-wide. Medicinal Plant Conservation 3, 16-17.

Maikhuri, R.K., Nautiyal, S., Rao, K.S. 2001. Medicinal plant cultivation practices of Bhotiyas in Nanda Devi Biosphere Reserve villages of Garhwal Himalaya. In: Samant, S.S., Dhar, U. and Palni, L.M.S. (eds.), Himalayan Medicinal Plants: Potential and Prospects. Gyanodaya Prakashan, Nainital, pp. 317-328.

Nautiyal, M.C., Nautiyal, B.P. 2004. Agro-techniques for High Altitude Medicinal and Aromatic Plants (Silver Jubilee publication of HAPPRC). Bishen Singh Mahendra Pal Singh, Dehra Dun, pp. 62-66.

Nayar, M.P., Sastry, A.R.K. 1987, 1988, 1990. Red Data Book of Indian Plants. Vol. I-III. Botanical Survey of India,Calcutta.

Rawat, R.B.S., Garg, G.P. 2005. Medicinal Plants: Trade and Commerce opportunities with India. Indian Forester 131 (3), 275-287.

Samant, S.S., Palni, L.M.S., 2000. Diversity, distribution and indigenous uses of essential oil yielding medicinal plants of Indian Himalayan region. J. Med. Arom. Plant Sci., 22, 671 684.

Samant, S.S., Dhar, U., Palni, L.M.S. (Eds.). 2001. Himalayan Medicinal Plants: Potential and Prospects. Gyanodaya Prakashan, Nainital, pp. 435.

Samant, S.S., Dhar, U., Palni, L.M.S. 1998. Medicinal Plants of Indian Himalaya: Diversity, Distribution Potential Values. HIMAVIKAS Publication. No. 13, Gyanodaya Prakashan, Nainital, pp. 163.

Samant, S.S., Pant, S., Singh, M., Lal, M., Singh, A., Sharma, A., Bhandari, S. 2007. Diversity, distribution pattern, indigenous uses and conservation prioritization of medicinal plants of Himachal Pradesh, India, International Journal of Biodiversity Science and Management, 3, 234-251.

Tewari, D.N. 2000. Report of the Task Force on Conservation and Sustainable use of Medicinal plants. Planning commission,Government of India.

Ved, D.K., Kinhal, G.A., Ravi Kumar,K., Prabhakaran, V., Ghate, U., Vijaya Sankar, R., Indresha, J.H. 2003. Conservation and Management Prioritization for the Medicinal Plants of Jammu & Kashmir, Himachal Pradesh & Uttaranchal. FRLHT, Banglore.

Ved, D.K., Kinhal, G.A., Ravikumar, K., Vijaya Sankar, R., Haridasan, K. 2005. Conservation Assessment and Management Prioritisation (CAMP) for wild medicinal plants of North-East India. Medicinal Plant Conservation 11, 40-44.

World Health Organization(WHO). 2003. WHO guidelines on good agricultural and collection practices(GACP) for medicinal plants. World Health Organization, Geneva, pp. 72, accessed on January 31, http://whqlibdoc.who.int/publications/2003/9241546271.pdf.

❑❑❑

15

Nutritional Components, Nitrogen Fixation and Abiotic Stress Tolerance for Food and Human Security in Seabuckthorn

YOSHINORI KANAYAMA
Graduate School of Agricultural Science, Tohoku University,
Sendai 981-8555, Japan

ABSTRACT

Seabuckthorn (*Hippophae rhamnoides*), which originated in Central Asia, is a small fruit tree in the family Elaeagnaceae. We examined its symbiotic nitrogen fixation and stress tolerance, features that could facilitate the extensive and sustainable cultivation of seabuckthorn under severe high-altitude environments. Compared to common deciduous fruit trees, seabuckthorn plants showed high tolerance to temperature and drought stress, and metabolites that may play a key role in stress tolerance accumulated during cold acclimation. Seabuckthorn root nodules showed high nitrogen fixation activity, similar to that in *Rhizobium*-legume root nodules. Even under nitrogen-free conditions, seabuckthorn plants with root nodules grew normally. The morphology of the root nodules and the 16S ribosomal DNA sequence indicated that the root nodules were formed by *Frankia* belonging to the *Elaeagnus* group. The nutritional components of the fruit were evaluated and compared in two subspecies: *H. rhamnoides* ssp. *mongolica* and ssp. *rhamnoides*. In both subspecies, fruit contained high levels of beta-carotene (provitamin A), alpha-tocopherol (vitamin E), and L-ascorbic acid (vitamin C). Sugars and acids that are important to fruit flavor were also measured in both subspecies. Together, the results suggest that seabuckthorn could be valuable for nutrition, environmental sustainability, and food security. We hope this study will promote the breeding and cultivation of seabuckthorn for the benefit of humans and the environment.

Keywords : Abiotic Stress, nutrition, nitrogen fixation, Seabuckthorn

INTRODUCTION

Seabuckthorn (*Hippophae rhamnoides*), which originated in Central Asia, is a small fruit tree in the family Elaeagnaceae. Fruit of seabuckthorn is rich in vitamins and could be valuable as a healthy food source. Further, seabuckthorn could also be valuable for food security because it is stress tolerant and is a symbiotic fixer of nitrogen. The later characteristics could facilitate extensive sustainable cultivation in severe environments such as high-altitude locations.

Previous reports have described the stress tolerance of seabuckthorn. For example, Li et al. (2005) and Yang et al. (2005) found tolerance to temperature and drought stress in Chinese ecotypes of seabuckthorn. They also reported a relationship between stress tolerance and abscisic acid (ABA) content. However, information is still lacking on abiotic stress tolerance and the physiological mechanisms of stress tolerance in seabuckthorn.

Seabuckthorn is an actinorhizal plant, i.e., it is a symbiotic plant infected with *Frankia*, an actinomycetic, nitrogen-fixing, nodule-forming endophyte of woody trees and shrubs. The numerous species of actinorhizal plants are classified into eight families that contain 25 genera comprising more than 220 species (Wall, 2000). While actinorhizal Myricaceae and Elaeagnaceae plants (including seabuckthorn) bear fruits, they have not yet become popular as horticultural crops. For cultivation of seabuckthorn, it is necessary to evaluate its nitrogen-fixing ability.

Rich in vitamins and unsaturated fatty acids, seabuckthorn fruit has potential as a food and medicinal crop. Gao et al. (2000) reported that seabuckthorn fruit contains high levels of antioxidants, such as phenolic compounds and ascorbic acid during ripening. On the other hand, Tang et al. (2001) reported that increasing the sugar/acid ratio is important for improving the flavor of seabuckthorn fruit. It is thus important to evaluate the accumulation of sugars in the fruit, as well as in functional components, and to investigate the biochemical mechanism of their accumulation.

Seabuckthorn includes several subspecies: ssp. *rhamnoides,* distributed in the North Sea and Baltic coast region, ssp. *mongolica* in Mongolia and Siberia, and ssp. *sinensis* in China. Here we used ssp. *mongolica,* which has promising fruit characteristics, and ssp. *rhamnoides* to study nitrogen fixation and abiotic stress tolerance. We also compared the nutritional components of fruits of ssp. *mongolica* and *rhamnoides* grown in Japan and discuss ways to improve fruit quality.

MATERIALS AND METHODS

Plant Materials

Seabuckthorn plants and seeds provided by the Siberian branch of the Russian Academy of Sciences were grown in a field of the Graduate School of Agricultural Science, Tohoku University, Japan. We used mature trees, seedlings and rooted cuttings from mature trees to study nitrogen fixation. Mature leaves and stems from mature trees of *H. rhamnoides* ssp. *mongolica* and spp. *rhamnoides* were used to examine stress tolerance. To determine sugars and other components, we sampled fruits and leaves, which were frozen in liquid nitrogen and stored at -80°C until use. Fruit ripening was judged by fruit softening and skin color.

Ion Leakage

Stems or leaves were harvested and subjected to low temperature or drought conditions. After treatments, the samples were incubated in water at room temperature, and the conductivity in the solution was determined. The samples were then heated and conductivity was determined again. Electrolyte leakage was expressed as a percentage of the total conductivity after heating.

Root Nodules and Nitrogen Fixation

Root nodules of rooted cuttings were used for the examination of histological sections. Root nodules of mature trees were sampled, sterilized, and crushed with nitrogen-free Evans solution (Huss-Danell 1978) containing 1% (v/v) polyvinylpyrrolidone; the roots of rooted cuttings were inoculated with the slurry, and the plants were supplied with nitrogen-free Evans solution. After 3 months, root nodules were sampled, fixed, and embedded in histoparaffin. Microtome sections were prepared and stained with toluidine blue.

DNA was extracted from the hyphal clusters of *Frankia* microsymbionts living in root nodules for the sequence analysis of 16S ribosomal DNA (rDNA), as described previously by Benson et al. (1996). *Frankia* 16S rDNA was amplified by polymerase chain reaction (PCR) using primers fD1 and rDB1, as outlined by Clawson et al. (1998), and sequenced.

Root nodules of mature trees were sampled to estimate seasonal changes in nitrogenase activity. Nitrogenase activity was measured by the acetylene reduction assay (Suganuma et al., 2003).

Determination of Nutritional Components and Enzyme Activity

Soluble carbohydrates were extracted in hot 80% (v/v) ethanol and determined using a high-performance liquid chromatograph (HPLC) equipped with a refractive index detector, as described by Suzuki *et al.* (2001). Titrable acidity was determined using 0.1 N sodium hydroxide (NaOH) and expressed as grams of malic acid. Ascorbic acid was extracted with 1.5% (w/v) metaphosphoric acid and filtered through filter paper. The filtrate was adjusted to pH 3.5–4.0 using potassium hydroxide (KOH) and used for the enzymatic determination of ascorbic acid by an F-kit (JK International, Tokyo, Japan). L-Galactose dehydrogenase (L-GDH) involved in the pathway of ascorbic acid biosynthesis was extracted, and its activity was analyzed. Protein content in the enzyme extracts was determined using the method of Bradford (1976).

RESULTS AND DISCUSSIONS

Abiotic Stress Tolerance in Seabuckthorn

High abiotic stress tolerance in crops could be important in extensive sustainable cultivation and food security under severe environments such as high-altitude areas. We therefore evaluated tolerance to temperature and drought stress using seabuckthorn stems and leaves. The estimated median lethal temperature (LT50) of seabuckthorn leaves was lower than those of other common fruit trees such as apple and pear. Electrolyte leakage in seabuckthorn stems treated with freezing conditions decreased from fall to winter with a decrease in temperature. Electrolyte leakage was almost zero in stems that were sampled at midwinter and subjected to -40°C. Both spp. *mongolica* and spp. *rhamnoides* showed similar levels of freezing tolerance. The results indicate that seabuckthorn plants have strong tolerance to very low temperatures.

Soluble carbohydrates extracted from the stems of seabuckthorn spp. *mongolica* and spp. *rhamnoides* were determined from fall to spring because soluble carbohydrates play a role in abiotic stress tolerance (Deguchi *et al.*, 2006; Kanayama *et al.*, 2006). Soluble carbohydrates in which sucrose was a major sugar increased during cold acclimation. Besides common sugars, an oligosaccharide increased during cold acclimation. Interestingly, this oligosaccharide was detected only in winter. The results suggest that soluble carbohydrates play a role in freezing tolerance, and accumulation of the oligosaccharide in particular is related to the high stress tolerance of seabuckthorn.

Seabuckthorn also showed higher tolerance to drought stress compared to other common fruit trees such as apple and pear. Seabuckthorn contained high levels of ascorbic acid in its vegetative tissue as well as in its fruit. The high content of ascorbic acid could also be related to stress tolerance, because ascorbic acid plays a role in abiotic stress tolerance as an antioxidant.

Nitrogen Fixation in Root Nodules

Symbiotic nitrogen fixation could contribute to extensive sustainable cultivation of seabuckthorn. Therefore, we also focused on seabuckthorn root nodules and their nitrogen fixation. Histological sections of a lobe of root nodules were first observed, and 16S rDNA from *Frankia* in the root nodules was then sequenced. The observed morphological properties were consistent with those of *Frankia*-actinorhizal root nodules (Bond, 1976; Wall, 2000). Phylogenetic analysis was carried out using the 16S rDNA sequence, as described previously by Clawson *et al.* (1998). *Frankia* 16S rDNA sequences cluster in four groups of host-plant specificity (Clawson *et al.*, 1998). The sequencing results showed that *Frankia* in the seabuckthorn root nodules belongs to the *Elaeagnus* group (Kato *et al.*, 2007). Presumably, continued cultivation of seabuckthorn would increase the density of symbiotic *Frankia* in the field and enhance the nodulation of seabuckthorn, as in legume crops.

To evaluate nitrogen-fixing ability in seabuckthorn root nodules, germinated seedlings were inoculated with the slurry as described in the Materials and Methods section and grown in vermiculite. The seedlings were supplied with nitrogen-free Evans solution. Apparent growth differences were observed between plants with and without root nodules 1 month after inoculation. When nodulation and plant growth were observed 3 months after inoculation, seedlings without inoculation grew poorly and had light leaf color. In contrast, seedlings with root nodules grew normally, with dark green leaves even under nitrogen-free conditions. A high positive correlation was found between nodule fresh weight and plant fresh weight (Kato et al., 2007). Seabuckthorn seedlings that were well nodulated showed normal growth and seemed not to lack nitrogen, as judged by the leaf color, even though they were grown in a nitrogen-free nutrient solution. This indicates a high ability to fix nitrogen in seabuckthorn–*Frankia* root nodules.

Nitrogenase activity in soybean root nodules increases until flowering and then decreases. On the other hand, nitrogenase activity in *Alnus incana* (Betulaceae) root nodules starts to increase when leaves open and then decreases with leaf fall. In the cultivation of seabuckthorn, it is important to understand seasonal changes in nitrogenase activity. Nitrogenase activity was low in November when leaf yellowing and defoliation started and was not detected in January when trees had no leaves. High nitrogenase activity in root nodules was observed between May and September, suggesting that active photosynthesis in leaves plays a key role in nitrogen fixation (Kato et al., 2007). The level of nitrogenase activity determined in seabuckthorn root nodules was similar to that in legume root nodules, indicating the high potential of *Frankia*–actinorhizal symbiosis.

Nutritional Components

Seabuckthorn fruit contains high levels of functional food components, which are important for human health as micronutrients. The content of ascorbic acid (vitamin C) in ssp. *mongolica* fruit (approximately 100 mg/100 gFW) is high compared to that in popular fruits rich in ascorbic acid such as strawberry and kiwifruit. Fruit of ssp. *rhamnoides* contained even more ascorbic acid, approximately 200 mg/100 gFW.

Understanding the biochemical mechanism of ascorbic acid accumulation in fruit is important for improving fruit quality. Thus, we analyzed the activity of enzymes in the biosynthetic pathways of ascorbic acid. Three pathways have been proposed in plants, namely the galacturonate pathway that uses galacturonate from pectin degradation during fruit ripening, the *myo*-inositol pathway, and the L-galactose pathway that has L-galactose as an intermediate metabolite (Valpuesta and Botella, 2004). The galacturonate pathway plays a role in ascorbic acid accumulation during the ripening of strawberry fruit (Agius et al., 2003). However, the galacturonate pathway may not be important in seabuckthorn fruit, because ascorbic acid was relatively decreased during ripening. L-GDH activity in seabuckthorn fruit was much higher than that in several other species, even though L-galactose is not common compared to D-galactose. This high activity of L-GDH and the results of a tracer experiment using ^{14}C-L-galactose suggested that the L-galactose pathway is functional and related to the high content of ascorbic acid in seabuckthorn fruit.

Beta-carotene (provitamin A) and alpha-tocopherol (vitamin E) are also important. The mature fruit of both subspecies of seabuckthorn contained high levels of beta-carotene and alpha-tocopherol, while beta-carotene and alpha-tocopherol contents were greater in ssp. *mongolica* fruit than in ssp. *rhamnoides* fruit. As most popular fruits generally do not contain high levels of tocopherols, seabuckthorn fruit would be very useful as a rich source of vitamin E.

Glucose, fructose, and sucrose were the major sugars found in seabuckthorn fruit. Glucose content was higher than fructose and sucrose contents in the fruits of both subspecies. Total sugar content was higher in ssp. *mongolica* fruit than in ssp. *rhamnoides* fruit. In ssp. *mongolica* fruit, glucose content increased during ripening. Sugar content differed between the subspecies because of their different glucose accumulation. The fruit of ssp. *rhamnoides* contained ethylglucose, as described by Tiitinen et al. (2006). Because ethylglucose content is negatively correlated to glucose content, the conversion of glucose to ethylglucose in spp. *rhamnoides* fruit could be related to the low level of glucose in ssp. *rhamnoides* fruit. Sugar content is important to improve fruit flavor, since many cultivars and lines of seabuckthorn have high acidity.

High fruit acidity could be a problem if seabuckthorn is to be used as a food. Titrable acid was transiently increased during fruit growth and then decreased during ripening. Titrable acid in the mature fruit of ssp. *mongolica* was lower than that in ssp. *rhamnoides*. The higher sugar/acid ratio in ssp. *mongolica* fruit indicated better taste in that fruit. In contrast, the sugar/acid ratio was very low in the fruit of ssp. *rhamnoides*. Collectively, spp. *mongolica* fruit has favorable characters, which should be introduced into spp. *rhamnoides* and other subspecies including spp. *sinensis*.

Concluding Remarks

Seabuckthorn showed high tolerance to temperature and drought stress compared to common deciduous fruit trees. Some metabolites that may play a key role in seabuckthorn stress tolerance accumulated during cold acclimation. Seabuckthorn root nodules showed high nitrogen fixation activity, similar to that found in *Rhizobium*-legume root nodules. Even under nitrogen-free conditions, seabuckthorn plants with root nodules grew normally. Seabuckthorn fruit contained high levels of beta-carotene (provitamin A), alpha-tocopherol (vitamin E), and L-ascorbic acid (vitamin C) in both subspecies. The two subspecies showed different patterns of accumulation of vitamins and sugars. Collectively, seabuckthorn could be valuable for nutrition, environmentally sustainable food cultivation, and food security. This study should help to promote the breeding and cultivation of seabuckthorn for the benefit of humans and the environment.

ACKNOWLEDGMENT

The author thanks Koki Kanahama, Wataru Ohkawa, Kazuhisa Kato, Emi Chiba, Kazuyoshi Sato in Tohoku University for their help in the experiments.

References

Agius, F., Gonzalez-Lamothe, R., Caballero, J.L., Munoz-Blanco, J., Botella, M.A. and Valpuesta, V. 2003. Engineering increased vitamin C levels in plants by overexpression of a D-galacturonic acid reductase. Nature Biotech. 21, 177-181.

Benson, D.R., Stephens, D.W., Clawson, M.L. and Silvester, W.B. 1996. Amplification of 16S rRNA genes from *Frankia* strains in root nodules of *Ceanothus griseus, Coriaria arborea, Coriaria plumose, Discaria toumatou*, and *Purshia tridentata*. Appl. Environ. Microbiol. 62, 2904-2909.

Bond, G. 1976. The results of the IBP survey of root-nodule formation in non-leguminous angiosperms, pp. 443-474. In: P. S. Nutman (Ed.). Symbiotic nitrogen fixation in plants. Cambridge University Press. Cambridge.

Bradford, M. M. 1976. A rapid and sensitive method for the quantification of microgram quantities of protein utilizing the principle of protein dye binding. Anal. Biochem. 72, 248-254.

Clawson, M.L., Caru, M. and Benson, D.R. 1998. Diversity of *Frankia* strains in root nodules of plants from the families Elaeagnaceae and Rhamnaceae. Appl. Environ. Microbiol. 64, 3539-3543.

Deguchi, M., Bennett, A.B., Yamaki, S., Yamada, K., Kanahama, K. and Kanayama, Y. 2006. An engineered sorbitol cycle alters sugar composition, not growth, in transformed tobacco. Plant, Cell and Environ. 29, 1980-1988.

Gao, X., Ohlander, M., Jeppsson, N., Bjork, L. and Trajkovski, V. 2000. Changes in antioxidant effects and their relationship to phytonutrients in fruits of sea buckthorn (*Hippophae rhamnoides* L.) during maturation. J. Agric. Food Chem. 48, 1485-1490.

Huss-Danell, K. 1978. Nitrogenase activity measurements in intact plants of *Alnus incana*. Physiol. Plant. 43, 372-376.

Kanayama, Y., Watanabe, M., Moriguchi, R., Deguchi, M., Kanahama, K. and Yamaki, S. 2006. Effects of low temperature and abscisic acid on the expression of the sorbitol-6-phosphate dehydrogenase gene in apple leaves. J. Japan. Soc. Hort. Sci. 75, 20-25.

Kato, K., Kanayama, Y., Ohkawa, W. and Kanahama, K. 2007. Nitrogen fixation in seabuckthorn (*Hippophae rhamnoides* L.) root nodules and effect of nitrate on nitrogenase activity. J. Japan. Soc. Hort. Sci. 76: 185-190.

Li, C., Yang, Y., Junttila, O. and Palva, E.T. 2005. Sexual differences in cold acclimation and freezing tolerance development in seabuckthorn (*Hippophae rhamnoides L.*) ecotypes. Plant Sci. 168, 1365-1370.

Suganuma, N., Nakamura, Y., Yamamoto, M., Ohta, T., Koiwa, H., Akao, S. and Kawaguchi, M. 2003. The *Lotus japonicus Sen1* gene controls rhizobial differentiation into nitrogen-fixing bacteroids in nodules. Mol. Gen. Genom. 269, 312-320.

Suzuki, Y., Odanaka, S. and Kanayama, Y. 2001. Fructose content and fructose-related enzyme activity during the fruit development of apple and Japanese pear. J. Japan. Soc. Hort. Sci. 70, 16-20.

Tang, X., Kalviainen, N. and Tuorila, H. 2001. Sensory and hedonic characteristics of juice of sea buckthorn (*Hippophae rhamnoides* L.) origins and hybrids. Lebensmittel-Wissenschaft und-Technologie. 34, 102-110.

Tiitinen, K.M., Yang, B., Haraldsson, G.G., Jonsdottir, S. and Kallio, H.P. 2006. Fast analysis of sugars, fruit acids, and vitamin C in seabuckthorn (*Hippophae rhamnoides* L.) varieties. J. Agric. Food Chem. 54, 2508-2513.

Valpuesta, V. and Botella, M.A. 2004. Biosynthesis of L-ascorbic acid in plants: new pathways for an old antioxidant. Trend. Plant Sci. 9, 573-577.

Wall, L.G. 2000. The actinorrhizal symbiosis. J. Plant Growth Reg. 19, 167-182.

Wheeler, G. L., Jones, M. A. and Smirnoff, N. 1998. The biosynthetic pathway of vitamin C in higher plants. Nature. 393, 365-369.

Yang, Y., Yao, Y., Xu, G. and Li, C. 2005. Growth and physiological responses to drought and elevated ultraviolet-B in two contrasting populations of *Hippophae rhamnoides*. Physiol. Plant. 124, 431–440.

□□□

16

Carbon Pool of Orchards in Nubra Valley : A Socio-Economic Carbon Sequestration

G. PHANI KUMAR, SUNIL GUPTA AND SHASHI PRAKASH
Defence Institute of High Altitude Research, DRDO, Leh-Ladakh, J&K, India

ABSTRACT

A major problem being faced by human society is the global temperature which is believed to be rising due to anthropogenic activity that releases carbon dioxide to the atmosphere, i.e., climate change and global warming. The problem of increasing atmospheric carbon dioxide can be addressed in a number of ways. Orchards have received much recent attention because of their potential carbon pools that reduce carbon emissions to the atmosphere. Nubra valley is covered with more than 60,000 fruit trees (Apple, Apricot, Mulberry and Walnut etc.). Every year orchards of Nubra valley produce approximately 4,650 tons of fruits belonging to Apricot, Apple, Walnut, Mulberry, Grapes and Plums. Major focus of the present paper is on Socio-Economic Carbon Sequestration management of fruit yielding tree species contributing to the sequestration of carbon and at the same time locals could benefit through the fruit yield.

INTRODUCTION

Global warming is amongst the most dreaded problems of the new millennium. Carbon emission is supposedly the strongest causal factor for global warming. So, increasing carbon emission is one of today's major concerns, which is well addressed in Kyoto Protocol. Through photosynthesis, plants convert CO_2 into organic forms of carbon, viz. sugars, starch and cellulose, also known as carbohydrates. According to estimates reported by the Intergovernmental Panel on Climate Change (IPCC), emissions of CO_2 and other greenhouse gases have begun to affect the world's climate (IPCC, 1996).

Anthropogenic activity increases greenhouse gases in the atmosphere which are likely to bring about a global temperature rise of 1.4–5.8°C during the next century; with an associated rise in sea level of 0.09–0.88 m. Each 1°C rise in temperature will displace the limits of tolerance of land species some 125 km towards the poles, or 150 m vertically on mountains. Many species will not be able to redistribute themselves fast enough to keep up with the projected changes, and considerable alterations in ecosystem structure and function are likely. Hence the scientific community is now focusing on issues related to identification of carbon sinks (Houghton et al., 2001).

Harvested fruit is full of carbohydrates and fats, the out product of photosynthesis. As the fruit production increases, more carbon is held captive and the owner (village) receives capital gain in the form of socio-economic development. Carbon sequestration study is burning topic after Kyoto Protocol. Most of the biomass studies have been conducted in the African countries. Keeping this in view, Defence Institute of High Altitude Research (DIHAR), an establishment of Defence Research and Development Organization (DRDO) has carried out extensive studies for socio-economic status and carbon stock estimations of fruit yielding tree species in Nubra valley for the first time.

Study Area

The Nubra valley in Ladakh comprises the valley of Shyok river from its acute-angled bend down to its confluence with Nubra and further towards Indus (Fig.1). It is the northern most valley of Ladakh. Khardung-La pass (5,455 m) is the gate way of the Nubra valley that includes Siachen glacier. The area lies between two great mountain ranges i.e., the Ladakh range (on the south) and the Karakoram range (on the north). There is a great variation in altitude and it varies approximately between 2700m to 6000m msl. The valley remains cut off from other parts of country during winter months due to extreme weather conditions.

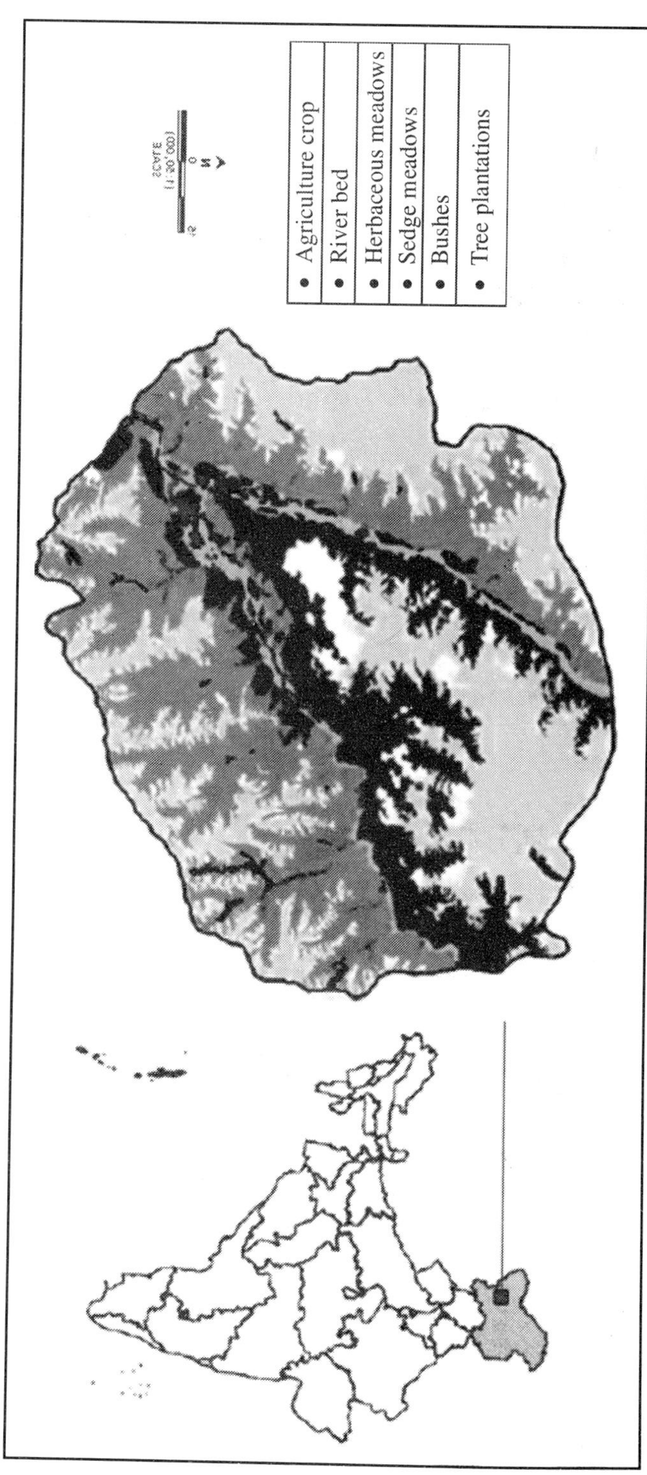

Fig. 1: Land Use and Land Cover Map of Nubra Valley

MATERIALS AND METHODS

International Forestry and Resources Institutions (IFRI) method was adopted for collecting botanical information (IFRI, 2002). Village-wise analysis was done for fruit yielding capacity of the orchards with the help of pre-requisite questioner marine. Apricot, Apple, Walnut and Mulberry plantations were analyzed for standing biomass through non-destructive, morphometric measurements (FAO, 1993). Estimates of carbon stock are based on algometric equations relating to diameter at breast height (DBH) and height of the tree (FAO, 2001). These equations are usually based on measurement of the fresh mass of each tree with sub samples taken to determine moisture content to convert to dry weight.

RESULTS AND DISCUSSIONS

Nubra valley is predominated by high altitude peaks of the mountains of Himalayas. The landscape is characterized by vast glaciers, boulders and sheer cliffs. The vegetation is largely xerophytic, although moist areas are found beneath glaciers, snowfields and along the river banks of "Nubra and Shyok". Fruit and timber produces are likely to play an important role in the material economy of sustainable society. Development and conservation of orchards will be fundamental to achieving sustainable socio-economic development in the Nubra valley.

The raising of fruit trees in kitchen gardens and agriculture fields is also a well-established tradition of this valley. Many different kinds of fruit trees are being grown, including Apricot, Apple, Walnut, Cherry, Mulberry, Plum, Grapes, Pears, Peaches and Pomegranates in well established orchards. It has been suggested that the demand for biomaterials may increase and it may also be possible to increase the production rates of forests and plantations through more intensive management (Sirkin and Houten, 1994; Haberl and Geissler, 2000).

The economy of most villages in Nubra valley is characterised by two elements: a high degree of self sufficiency and internal recycling of energy and nutrients. Practices and institutions in Nubra have evolved over centuries "by a process of constant refinement of response to the existing climatic and environmental conditions" (Rizvi, 1996). All of life seems to be organised according a single principle, "nothing is wasted and nothing thrown away; a use is found for everything" (Norberg-Hodge, 2000).

Apricot, Apple and Walnut are the common fruit yielding tree species of this valley. Apricot is one of the major cold resistant and highly sustained tree species in the Nubra geo-climate after willow and poplar species (Phani Kumar, 2008). In this valley, approximately 41,000 apricot plantations are being grown, of which 50 per cent plantations are under production and remainings are recently planted. Most of the recent plantations have been planted with the help of State Horticulture Department and DIHAR (DRDO). Every year more or less 4000 tons of apricot fruits are being harvested (Fig. 2). Some of the villages have apricot fruit processing industry to extract oil and to prepare jam and pickle etc.

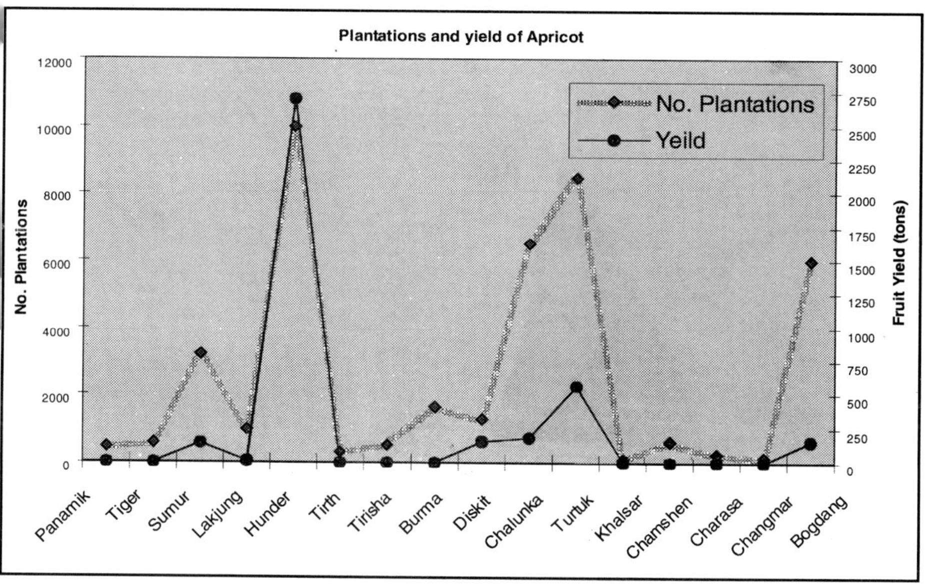

Fig. 2: Total Trees and Yield of Apricot Orchards in Nubra Valley

Many of the villages included these two species in their recent cultivation practices. Hence 70 per cent of these plantations are young and yet to start fruiting. Total fruit production of Apple and Walnut is 600 and 47 tons per year respectively (Fig.2 and 3). Apple and walnut orchards are established in most of the villages of this valley and approximately 14000 and 1200 trees are being cultivated respectively (Fig. 3 and 4). Only one of the villages of the valley i.e. 'Turtuk' is cultivating Mulberry, Grapes, Plums and Cherries due to its lower elevation and having sufficient irrigation resources. Every year approximately 7.5 quintal of mulberry fruit is being produced and the villagers are planning to launch Sericulture.

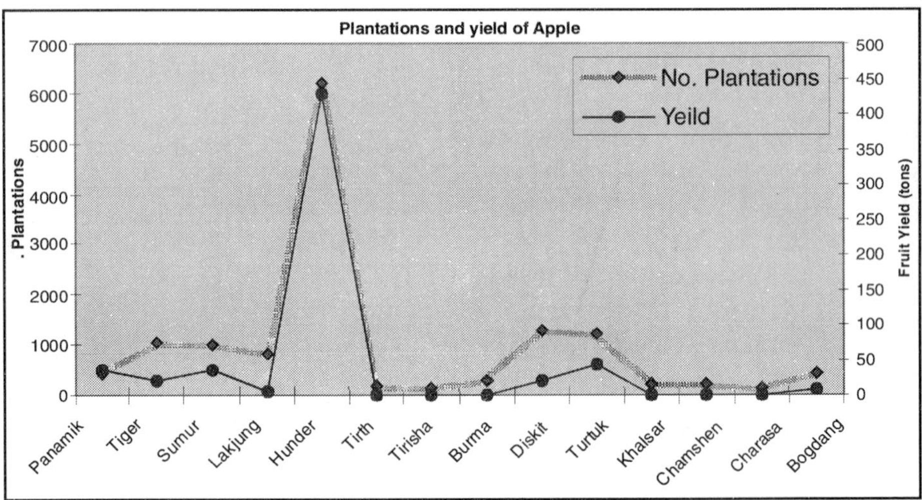

Fig. 3: Total Trees and Yield of Apple Orchards in Nubra Valley

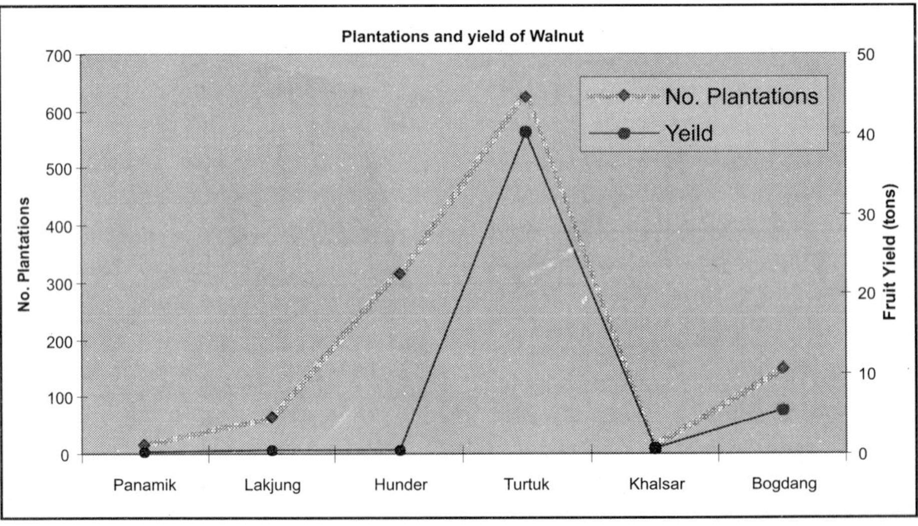

Fig. 4: Total Trees and Yield of Walnut Orchards in Nubra Valley

Standing biomass of present surveyed plantations is approximately 40,000 tons of dry weight of which Apricot, Apple and Walnut reported with 31,000 tons, 3,500 tons and 5,500 tons respectively (Table 1). 'Hunder and Turtuk' are two villages surveyed with highest number of plantations (16,500 and 10,350 respectively) and having oldest orchards i.e. more than 100 year old. Some of the fruit bearing crops like Apple, Peach, Plums, Almond, Walnut and Grapes have been estimated for standing biomass and their carbon sequestration capacity by several researchers (Palmer et al, 2002; Kroodsma and Field, 2006). The results stated that Almond and Walnuts having the highest capacity of standing biomass 100 and 75 t/ha, respectively and total CO_2 sequestrated 183 and 138 t/ha, respectively.

Table 1 : Standing Biomass of Mature Orchards of Apple, Apricot and Walnut (weight in tons)

S.No.	Village Name	Apple	Apricot	Walnut
1	Panamik	228.04(18.12)	237.54 (28.96)	113.04 (3.5)
2	Tiger	29.77(0.40)	329.19(6.29)	-
3	Sumur	77.53(4.45)	1116.51(56.08)	-
4	Lakjung	68.02(0.71)	61.96(0.64)	7.12 (0.005)
5	Hunder	1270.85(366.40)	18080.40(488.07)	54.97 (2.1)
6	Tirth	3.15(0.05)	40.97(0.21)	-
7	Tirisha	3.13(0.08)	78.73(1.10)	-
8	Burma	4.59 (0.31)	88.00(1.25)	-
9	Diskit	53.64(0.51)	404.59(12.06)	-
10	Chalunka	-	1562.43(31.73)	-
11	Turtuk	621.01(1.64)	4257.81(82.52)	3737.44 (74.1)
12	Khalsar	2.22(0.02)	7.96(26.72)	111.96 (4.5)
13	Chamshen	0.31(0.004)	65.74(8.37)	-
14	Charasa	3.08 (0.79)	15.03(0.08)	-
15	Changmar	-	24.06(1.06)	-
16	Bogdang	117.70 (3.21)	2251.47(144.73)	90.53 (3.7)

Note: Results parentheses are Standard Deviation

The Clean Development Mechanism (CDM), proposed in Article 12 of the Kyoto Protocol to the United Nations Framework Convention on Climate Change, will allow developing countries to sell or trade project based carbon credits, such as Carbon Emission Reduction Credits (CER credits), to or with industrial countries, if adopted. CER credits could provide an incentive for participation in climate change mitigation (Fearnside, 1999).

Salix sp. (willow) and *Populus* sp. (poplar) plantations of this valley sequestred 75,000 tons of carbon (Phani Kumar, 2008). Present study reports approximately 22,300 tons carbon has been sequestered by these three fruit yielding species i.e. Apricot, Apple and Walnut. Apricot orchards sequestered more carbon (18,000 tons) as compared with other two. Hunder village is full of green and covered with several orchards and sequestered more carbon than other villages (Fig.5). The potential for carbon credits based on standing biomass for Nubra orchards are limited when compared with forest stands. So, by comparison with forestry, fruit orchards seem to have a minor role in sequestering carbon, but these types of carbon sequestration programmes, encourage villagers to involve in Global carbon mitigation project and the villager could improve their socio-economy with the yield of orchards.

CONCLUSIONS

Apple, Apricot and Walnut are highly sustained cold resistant orchards of Nubra valley. Present analysis showed cultivation practices of these species have double benefit. Fast growth of these plantations will help in gaining of higher standing biomass through the photosynthesis and become rich carbon pools. In another point of view, socio-economy of the farmers will improve through the yield of orchards i.e. fruit, oils or other products exporting to the market. This practice appears to offer a relatively low-cost approach to sequestering carbon, furthermore the technology seems to be well–developed and effective in this valley. Only few of the villages have maintained the orchards and remaining are unaware of it due to lack of encouragement. So Horticulture department and other agriculture related organizations must come forward to start orchard plantation programmes in all the villages.

ACKNOWLEDGEMENTS

Authors are thankful to the Director, Defence Institute of High Altitude Research (DRDO) - Leh, for providing valuable suggestions Author gratefully acknowledge to Director, NEERI for the technical support. and also thankful to the workers, those helped for the extensive survey in Nubra Valley.

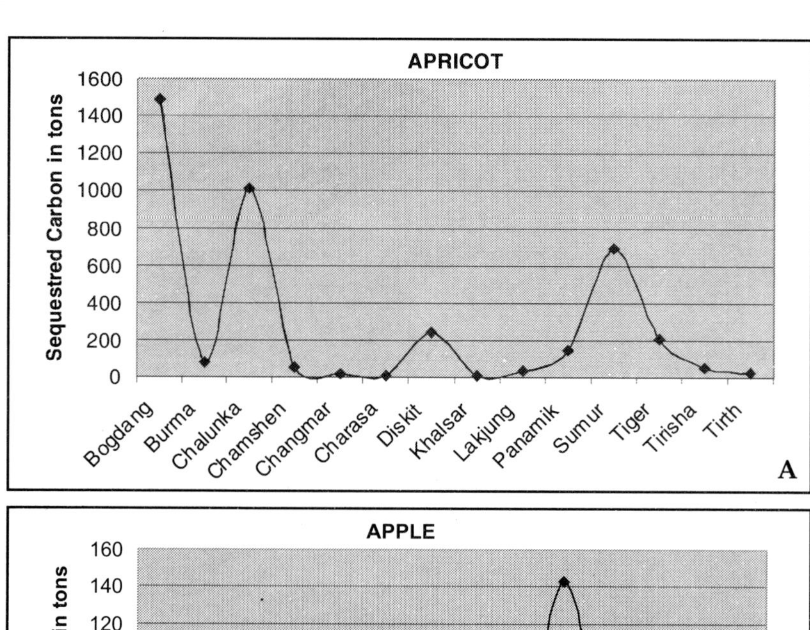

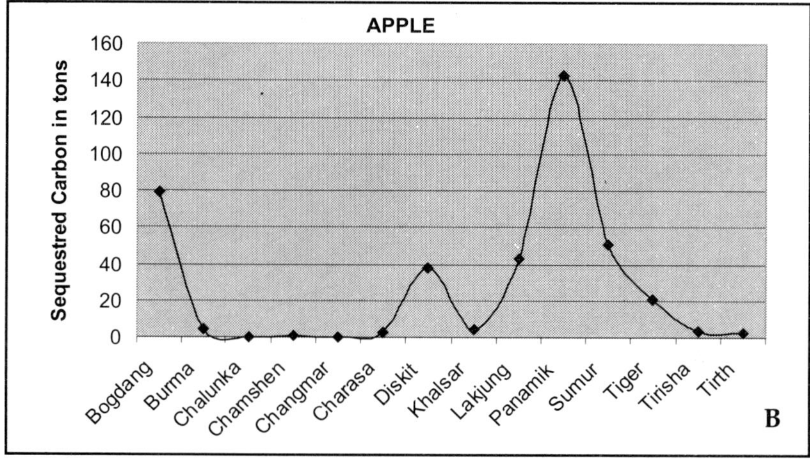

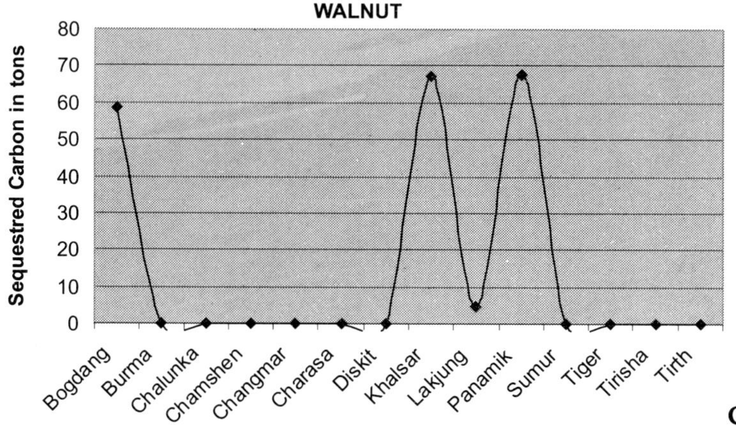

Fig. 5: Carbon sequestration capacity of different orchards of Nubra Valley

References

Bowen, G. D. and Rovira, A. D., 1999. The rhizosphere and its management to improve plant growth. Adv. Agron. 66, 1–102.

FAO. 1993. Forest Resource Assessment 1990: Tropical countries, FAO Forestry paper. 112, Rome.

FAO. 2001. Global forest Resources Assessment 2000-Main Report, FAO Forestry paper. 140, Rome.

Fearnside, P.M. 1999. Forests and global warming mitigation in Brazil: opportunities in the Brazilian forest sector for responses to global warming under the ''clean development mechanism''. Biomass Bioenergy. 16, 171–189.

Haberl, H., Geissler, S. 2000. Cascade utilization of biomass: strategies for a more efficient use of a scarce resource. Ecol. Eng. 16, S111–121.

Houghton, J. T., Ephramus, J. J. and G. J. (eds). 2001. Climate Change, The IPCC Scientific Assessment, Cambridge University Press, Cambridge.

IFRI. 2002. International Forest Research and Institution [IFRI] Field Manual [version 10.5]. Bloomington: Indiana University, Workshop in Political Theory and Policy Analysis.

IPCC. 1996. Climate Change 1995. The Science of Climate Change. Cambridge University Press.

Kroodsma, D.A. and Field, C.B. 2006. Carbon sequestration in California agriculture 1980-2000. Ecological Applications. 16, 1975-1985.

Norberg-Hodge, H. 2000. Ancient Futures: Learning from Ladakh, London

Palmer, J.W., Wünsche, J.N., Meland, M. and Hann, A. 2002. Annual dry-matter production by three apple cultivars at four within-row spacings in New Zealand .2002. Journal of Horticultural Science Biot. 77, 712-717.

Phani Kumar, G., 2008. Carbon Sequestration strategy of Nubra Valley with special reference to Agro-forestry. DRDO Tehcnology Spectrum, 187-192.

Rizvi, J. 1986. "Ladakh: Self-sufficiency or Dependence?" In: Ecology and Principles fro Sustainable Development, Proceedings of a Conference co-hosted by the Ladakh Project and the Ladakh Ecological Development Group in Leh, September 2-4, 1986.

Sirkin, T., Houten, M. 1994. The cascade chain: a theory and tool for achieving resource sustainability with applications for product design. Resour Conserv Recycl. 10(3), 213–76.

□□□

17

Stress Induced Changes in Burley Tobacco Seed Development and its Effect on Germination Characteristics

K. NAGESWARARAO AND V. KRISHNAMURTHY
Central Tobacco Research Institute, Rajahmundry, A.P, India

ABSTRACT

Studies on the influence of temperature and water stress on mother plant growth, seed development and seed characteristics in burley tobacco *N. tabacum* cv Banket A1, revealed that the day and night temperatures during mother plant growth and the seed development varied from 25° C to 40° C. Plants in a water stress treatment were subjected to field capacity to 40 per cent available water during vegetative growth and further seed development and maturation. Stage of seed development, ambient temperature and water stress affected the seed protein profile and free amino acid content.

High pre-anthesis and low post anthesis temperatures resulted in normal seed size, good seed germination, but transient dormant condition of the seed which lasted for 6-8 weeks after seed harvest. The combination of low pre-anthesis and high post anthesis temperatures during the growth of the mother plant was observed to be especially detrimental for good seed germination. Days to half germination of progeny were significantly delayed when the mother plant was grown under high post anthesis temperatures.

Common environmental history until anthesis, but seed development and maturation in different thermal periods, has resulted in delay of germination and relative vigor of the progeny. Water stress affected the vegetative growth, seed yield and germination characteristics. Harvesting the leaves by "priming" did not affect the seed yield and germination properties as compared to 'no leaf harvest'. It is concluded from the study that the growing environment of the mother plant significantly influenced the seed yield and variation in germination characteristics of the seed progeny.

Keywords : Burley tobacco, water stress, germination, mother plant, harvest by priming and seed development.

INTRODUCTION

Burley tobacco is grown in an area of 10,000 ha as a monsoon crop in the light soils of the agency area in the East Godavari, West Godavari and Visakhapatnam districts of Andhra Pradesh. Uniformity of the crop is important for achieving higher yields and superior leaf quality in tobacco, and it is largely dependent on the seedling quality and establishment. Germination characteristics of seed and the environmental conditions prevalent during the nursery period influences tobacco seedling production. Burley tobacco nurseries are raised during summer months, which coincide with high temperatures. Tobacco seed germination is sensitive to temperatures and any variation in the seed characteristics interferes with the seed germination. Higher temperatures interfere with the seed germination process of Banket A-1, an exotic burley tobacco variety adapted to cooler climates, resulting in production of uneven transplants (Nageswararao *et al.* 2002). Germination properties of tobacco seed are influenced by the environmental and cultural conditions prevalent during the mother plant growth and development (Koller, 1962, Gutterman, 1973 and Thomas and Raper, Jr. 1975b) and mother plant cultural history forms the basis for variation observed in progeny germination characteristics and further seedling production.

During the burley tobacco crop growth period, the plants were subjected to changing temperatures and soil moisture, depending on the pattern of rainfall during the crop season. These seasonal variations can cause variation in seed ontogeny resulting in the variation of progeny germination characteristics. Hence, an attempt has been made to find out the effect of temperature, water stress, and method of harvest on the seed development and maturation, yield and germination characteristics of burley tobacco seed.

MATERIALS AND METHODS

Pot culture studies were conducted with a recently introduced, exotic burley tobacco variety, Banket A-1, during the 1998-2000 season. Tests were completed on the effect of temperature, water stress and method of harvesting on seed development, maturation and germination characteristics of this variety.

Effect of Temperature on Seed Development

Low and high temperatures during the pre and post flowering periods were created using two different dates of planting. First planting was done in the month of August, and these plants were subjected to decreasing day/night temperatures from 60 days after transplanting (DAT) onwards. The second planting was done in the month of December and these plants were grown under increasing day/night temperatures from 60 DAT onwards. Uniform seedlings of variety Banket A-1 were planted in 50 kg porcelain pots filled with light soil (10 replications) on two sowing dates. The plants were given recommended doses of fertilizers and adequate watering. When the plants started flowering, from each plant, some flowers were tagged with black tags on the day of anthesis, tagged capsules were harvested and developing seed samples were collected on the 10, 15, 20 and 30th days after anthesis for estimation of amino acid content and protein profiles. Data on maximum and minimum temperature was recorded during the crop growth period. Inflorescences were covered with selfing bags to avoid cross-pollination and when the capsules turned brown the seed capsules were harvested, dried and stored in a desiccators and were tested for their germination characteristics in an incubator at 20°C.

Effect of Water Stress on Seed Development

Uniform seedlings of variety Banket A-1 were planted in 50 kg porcelain pots filled with light soil in two treatments with 10 replications each. The two treatments consisted of 1. Normal (plants were watered when available water (AW) in the pot decreased to 80% of field capacity (FC) 2. Water stressed (plants were watered to FC when AW decreased to 40% of field capacity in the pot). Plants were provided with adequate water upto 45 DAP in both normal and water stressed plant treatments. Stress treatments were imposed from 45th DAT and continued till the end of seed harvest. At the time of anthesis, flowers were tagged and developing seed samples were collected, with seed harvest and germination tests conducted.

Effect of Leaf Priming on Seed Development

Uniform seedlings of Burley cv. Banket A-1 were planted in two treatments with 10 replications each in 50 kg porcelain pots filled with light soil. The two treatments consisted of 1. Leaf harvest by priming and 2. No harvest (leaves left on the plant without harvesting). Plants were raised with recommended doses of fertilizer and cultural operations. Around 60 days after planting, when bottom leaves turned yellow, first priming was done in the first treatment and then the harvests continued in sequence. In the second treatment, the leaves were left on the plant without harvesting. Inflorescences were covered with selfing bags and the seed capsules were harvested when the capsules turned brown. Seed yield was recorded, the seed samples were preserved in desiccators and germination tests were carried out. Protein content was estimated using the method of Lowery *et al* (1951). Protein was precipitated by adding an equal volume of 20 per cent TCA solution. The precipitate was washed twice with acetone and then dissolved in 0.1 N NaOH. Protein content of the solution was determined using bovine serum albumin as the standard. Free amino acids in seeds at different stages of development were estimated (Nageswararao *et al.,* 2003).

RESULTS AND DISCUSSIONS

Uniform seedling establishment in the field is directly related to the quality of tobacco seedlings coming from nursery beds and this in turn depends on the seed characteristics, including, seed purity, seed size, seed viability, seed dormancy, seed vigor and seed germination. Germination, which is spread over several days, can produce variability in seedling size even if other environmental factors are common (Evans, 1972). In this study, tobacco seed of different size groups of 86 and 94 mg/1,000 seeds were tested for germination at 20° C (Table 1&2). The heavy seeds germinated earlier, in higher percentages, and over a shorter period of time than did the light weight seeds. Seeds from the two weight groups that germinated on the same date produced seedlings that did not differ significantly in size of transplants. However, seedlings from early and late germinating seeds within the same weight group differed in size of the transplants. Even though the late seeds germinated only a few days after the early ones under optimal conditions, they produced seedlings that continued to lag in seedling growth. Seed size and time of germination significantly influenced seedling growth and their performance under field conditions (Kasperbauer and Sutton, 1977).

Seed dormancy, seed vigor and seed germination capacity are determined during seed development on the mother plant. Germination can be influenced by external conditions to which the mother plant was subjected while the plant was still in the pre-floral stages of growth (Gutterman, 1973). In a field environment, cultural perturbation of the mother plant can produce measurable alterations in the germination characteristics of the tobacco seed and hence can be a source of plant to plant variability among the progeny. Hormone metabolism during seed development is critical in determining the germination characteristics of tobacco seed. Abscissic Acid (ABA) deficiency during seed development is associated with the absence of primary dormancy in the mature seed. However, dormancy, to a larger extent, may depend on an intrinsic balance of GA and ABA biosynthesis and catabolism, which will determine the dominance of either of the plant hormones (Ali-Rachedi *et al.*, 2004; Cadman *et al.*, 2006). ABA induces dormancy during maturation, and GA plays a key role in dormancy release and in the promotion of germination. Higher temperatures inhibit seed germination or induce a dormant condition in Banket A1 and this inhibition can be overcome by seed treatment with 50ppm GA_3 (Nageswararao *et al.*, 2002 and Krishnamurthy *et al.*, 2005). It is evident that events taking place during seed development and maturation on the mother plant the environmental and nutritional factors to which the plant is exposed during the seed formation has significant bearing on seed characteristics and subsequent progeny. Capsule and seed development patterns show that capsule and seed dry weight rapidly increased up to the 20th day after anthesis and then remained constant (Fig. 2). The increase in the weight of the capsule is mainly due to increased seed weight. Dry weight accumulation in the capsule as well as the seed was influenced by temperature and water stress.

The results revealed that seed produced from the treatments of differing temperatures and water stress differed in their seed characteristics, viz. percent germination, time taken for germination, seed weight, seed size and the occurrence of dormant phase.

Effect of Temperature on Seed Development and Seed Characteristics

Two planting dates exposed the plants to two different sequences of temperatures during the pre-floral and seed formation periods (Fig. 1). In the first date of planting, higher temperatures prevailed during the vegetative phase and cooler temperatures during the reproductive phase. For the second planting date, the vegetative phase was under cooler temperatures and the reproductive phase was during a period of higher temperatures. This condition has affected the days to flowering, time taken for seed formation (Table 1), and protein and amino acid content of the developing seeds (Fig. 3). Days to half germination of progeny were delayed when the mother plant was grown under high post anthesis temperatures (Table 1). A common environmental history until anthesis but seed development and maturation in different thermal periods has resulted in delay of germination and relatively poor vigor of the progeny. Changes in the levels of protein and free amino acids in the seeds of tobacco were studied. Large quantities of protein accumulate over short periods of time during the seed development in many plant species (Dure, 1975). The free amino acid and protein content of developing seeds revealed that the amount of free amino acids decreases and protein accumulation increases as seed development progresses in *Nicotiana tabacum* cv Banket A1 (Fig. 3). Seed maturation was complete in 30 days after anthesis in low post anthesis and 24 days in high post anthesis temperature subjected plants. The maturation period of tobacco seeds is defined as the period from anthesis to the time when the seeds reach a constant weight (Sano and Kawashima, 1982). Protein rapidly accumulated in the seeds between the 10th and 20th day. The amount of free amino acids in the seeds gradually decreased throughout maturation. The major free amino acid on the 10th day after anthesis was glutamine, which then drastically decreased between the 10th and 15th day. Higher post anthesis temperature resulted in smaller seed size and shorter seed filling duration, but the number of seeds per capsule did not change appreciably (Table 1).

Effect of Water Stress on Seed Development and Seed Characteristics

Water stress affected vegetative growth as well as the reproductive phase of burley tobacco. Days to flowering, time taken for seed formation, protein and amino acid content of the developing seed and germination properties of the seed were affected by water stress as compared to normal watering (FC-80% AW) (Table 2). Days to flowering and time period from anthesis to seed formation was lower in water stressed as compared to normally watered plant. Difference in amino acid and protein content was found due to water stress (Fig. 3) and also the seed size or 1000 seed weight was lower in plants subjected to water stress (Table 2). Water stress resulted in lower 1000 seed weight and the smaller or light seed germinated later than heavy seed produced from normal watering but the size of the seedling produced from the lighter seed were as good as from heavy seed.

Effect of Leaf Priming on Seed Development and Seed Characteristics

Leaf priming did not affect the seed yield or 1000 seed weight, days to flowering, time taken for seed formation (Table 3) or protein and amino acid content of developing seed (Fig 3). Harvesting or no harvesting did not affect the above characteristics.

Seed Germination Test

Results of germination tests showed that seed produced from the first planting exhibited a transient phase of dormancy of six weeks from the time of seed harvest, whereas, for the second planting, the seed germinated in the second week after harvest. Time taken for germination at 22.5° C was lower for seed havested from the second planting in which the vegetative phase of the plant and the time taken for flowering was longer in cooler temperature (Table 1). Seed from the first planting took more time for germination at 22.5° C and the mother plant vegetative phase was shorter. Time Taken for Germination was not Influenced by Harvest or no Leaf Harvest.

CONCLUSIONS

It is concluded that variation in the temperature during seed development of burley tobacco affected days to flowering, seed maturation period, capsule dry weight, seed size, germination count in the nursery beds and days to 50 per cent germination. Similarly, water stress during seed development also affected these characters, but priming or no priming did not influence these characters.

References

Ali-Rachedi, S., Bouinot, D., Wagner, M.H., Bonnet, M., Sotta, B., Grappin, P., Jullien M. 2004. Changes in endogenous abscisic acid levels during dormancy release and maintenance of mature seeds: studies with the Cape Verde Islands ecotype, the dormant model of *Arabidopsis thaliana*. Planta. 219, 479–488.

Cadman, C.S.C., Toorop, P.E., Hilhorst, H.W.M., Finch-Savage, W.E. 2006. Gene expression profiles of Arabidopsis Cvi seed during cycling through dormant and non-dormant states indicate a common underlying dormancy control mechanism. The Plant Journal. 46, 805 –822.

Dure, L.S.III. 1975. Seed formation. Ann. Rev. Plant Physio. 26, 259-278

Evans, G.C. 1972. The quantitative analysis of plant growth. Univ. of Calif., Berkeley.

Gutterman, Y. 1973. Differences in progeny due to day length and hormone treatment of the mother plant. pp. 59-80. In W. Heydecker (ed.) Seed Ecology. The Penn. State University Park. Pennsylvania.

Kasperbauer, M. J. and Sutton, T. G. 1977. Influence of seed weight on germination, growth, and development of Tobacco. Agron. J. 69, 1000-1002.

Koller, D. 1962. Preconditioning of germination in lettuce at time of fruit ripening. Amercan J. Bot. 49, 841-844.

Krishnamurthy, et al. 2005. Indian burley Tobacco – Its production and quality for export. Central Tobacco Research Institute, Rajahmundry. Andhra Pradesh, pp- 43.

Lowry, O.H., Rosebrough, N.J., Farrand, A.L.R., Randall, J. 1951. Protein measurement with the folin phenol reagent. J. Biol. Chem. 193, 265.

Minoru, Sano and Nobumaro, Kawashima. 1982. Changes in the contents of free and protein amino acids in tobacco seeds and placentae during maturation. Plant and Cell Physiology. 23(7),1245-1250.

Rao, Nageswara., Shivaraju, K. and Prabhakara babu, D. 2002. Effect of temperature and role of gibberellic acid on seed germination in burley tobacco. Tobacco Research. 28(2), 83-89.

Rao, Nageswara., Shivaraju, K. and Prabhakara babu, D. 2003. Tobacco Seed Storage: 3. Biochemical changes in stored seed and its relation to viability and seedling vigor. Indian J. of Agricultural Research. 37(4), 235-244.

Thomas, J.F. and Raper, C.D. Jr. 1975a. Differences in progeny of tobacco due to temperature treatment of the mother plant. Tobacco Science. 19, 37-41.

Thomas, J.F. and Raper, C.D. Jr. 1975b. Seed germinability as affected by the environmental temperature of the mother plant. Tobacco Science. 19, 104-106.

Thomas, J.F. and Raper, C.D. Jr. 1979. Germinability of tobacco seed as affected by culture of the mother plant. Agron. J. 71, 694-695.

contd...

Table 1: Effect of temperature during seed development on seed characteristics

Treatment water stress	Days to flowering	Maturation period	Capsule dry wt (mg/100caps)	Seed size (1000 seed wt)	Germination count		Percent Germination	Days to 50% germination
					10th	15th		
High pre anthesis & Low post anthesis	85	34	365	94	48	62	92	5
Low pre anthesis & High post anthesis	90	25	312	88	39	60	93	7

Table 2: Effect of Water Stress During Seed Development on Seed Characteristics

Treatment water stress	Days to flowering	Maturation period	Capsule dry wt (mg/100caps)	Seed size (1000 seed wt)	Germination count		Percent Germination	Days to 50% germination
					10th	15th		
FC-80%AW	89	33	358	95	50	64	93	5
FC-40%AW	82	24	284	86	42	62	94	7

Table 3: Effect of Harvesting During Seed Development on Seed Characteristics

Treatment Harvesting	Days to flowering	Maturation period	Capsule dry wt (mg/100caps)	Seed size (1000 seed wt)	Germination count		Percent Germination	Days to 50% germination
					10th	15th		
Priming	90	33	364	92	50	62	94	5
No harvest	91	34	374	94	48	64	92	5

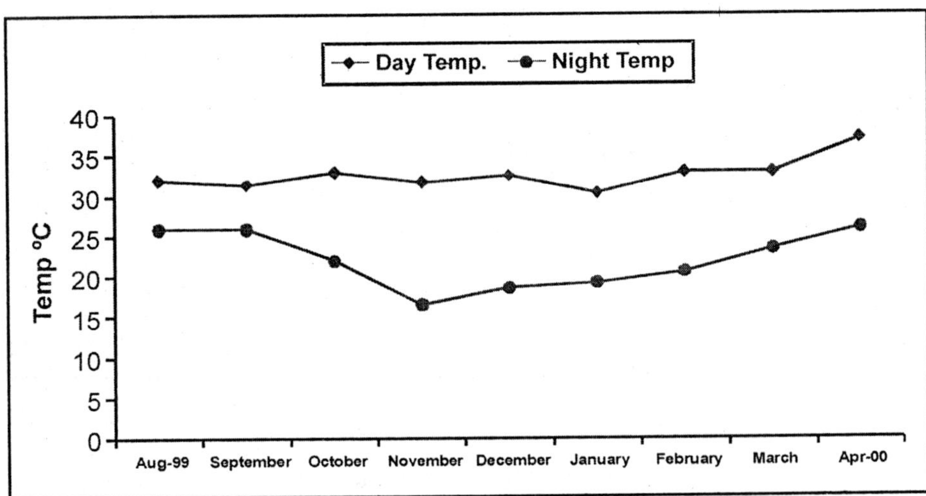

Fig. 1: Day and Night Temperatures During Plant Growth and Seed Development

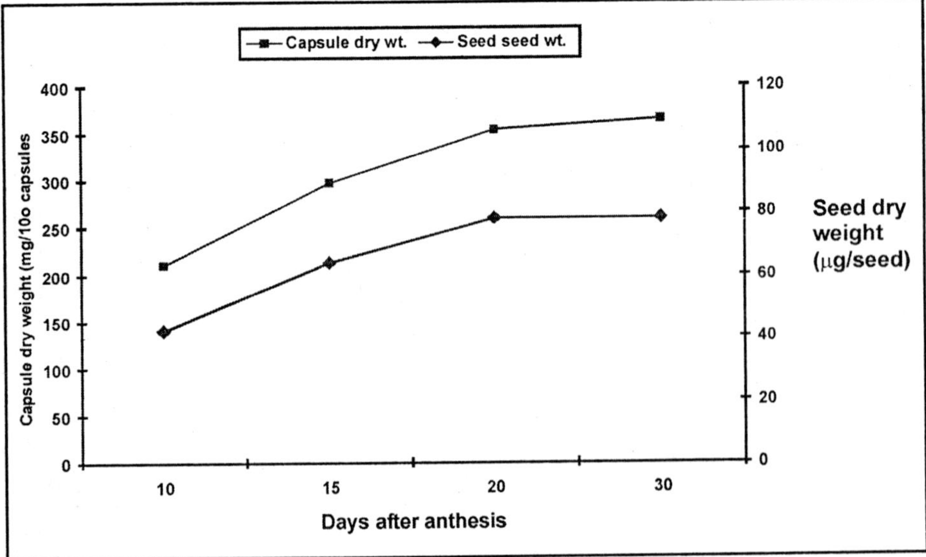

Fig. 2: Changes in Seed and Capsule Dry Weight During Seed Development

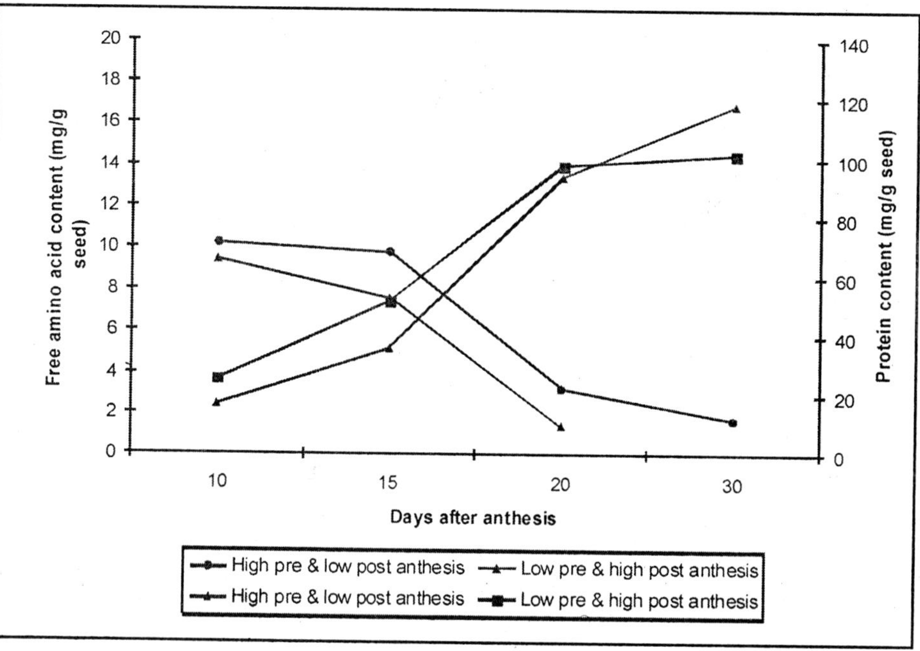
Free amino acid content (mg/g seed)
Protein content (mg/g seed)
20
18
16
14
12
10
8
6
4
2
0
140
120
100
80
60
40
20
0
10
15
20
30
Days after anthesis
High pre & low post anthesis
Low pre & high post anthesis
High pre & low post anthesis
Low pre & high post anthesis

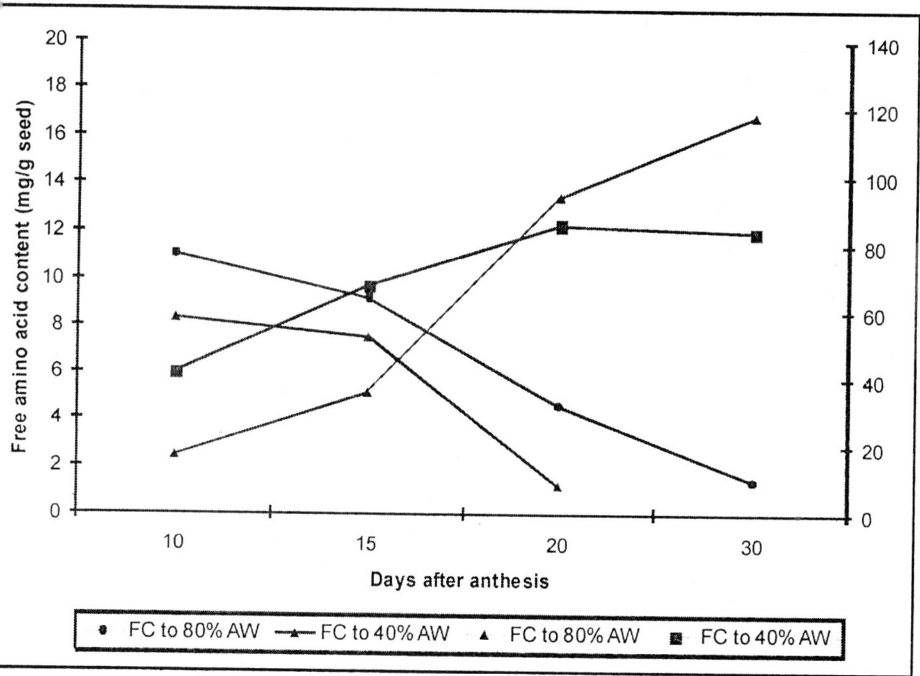
Free amino acid content (mg/g seed)
20
18
16
14
12
10
8
6
4
2
0
140
120
100
80
60
40
20
0
10
15
20
30
Days after anthesis
FC to 80% AW
FC to 40% AW
FC to 80% AW
FC to 40% AW

Fig. 3 Contd...

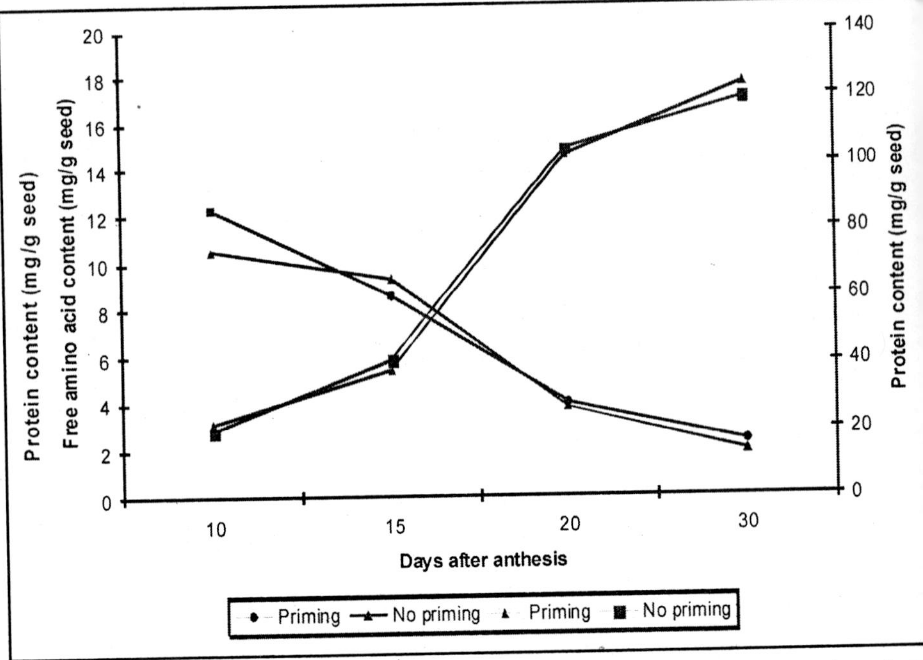

Fig. 3: Effect of Temperature, Water Stress and Harvesting on Free Amino Acid and Protein Content of Developing Burley Tobacco Seed

18

Photosynthetic Efficiency of Natural Flora and Introduced Species under Abiotic Stress in Cold Desert Areas of Himachal Pradesh

RAJAN BAWA AND C. L. THAKUR
Dr. Y.S. Parmar University of Horticulture and Forestry, Nauni (Solan) – 173 230, Himachal Pradesh, India

INTRODUCTION

Desertification is one of the most serious problems being faced by the world today. Large parts of the dry areas that cover more than one third of the earth's surface are being degraded. These areas include the more severe category of deserts called the Cold Deserts in the mountainous regions of the world - Spiti is one such area in the state of Himachal Pradesh in India. This area can be characterized under high altitude type, and is principally formed from physiographic causes and as a result of its location in the rain shadow.

The local people in the cold desert of Spiti have acquired rich knowledge concerning their crops, livestock and environment. This indigenous knowledge developed and adopted as a result of years of practical experience continues to be in the process of evolution for the urge of better and sustainable living. Only such adaptation practices and mechanisms that were inherent in the natural system and modified by the upcoming generations led to their coexistence till date even in such hostile conditions.

The cold deserts are characterized by weak market infrastructure, high level of emigration, predominance of female managed household, money order economy, poor linkages between formal and informal sections and low levels of social and political articulations. Besides these, the abiotic and edaphic factors are the basic factors which hinder the plant growth and are responsible for poor biomass production.

The abiotic and edaphic factors responsible for extreme xeric conditions include

- Short growing season
- Long freezing winters (5 to 6 months)
- Extreme variation in seasonal temperatures (-40°C to +35°C)
- Large daily and seasonal diurnal temperature variations
- Most of the precipitation in winter months in the form of show
- Hardly any precipitation during active growth period
- Soil moisture usually frozen during early spring
- Low relative humidity during growing season
- Fast winds (40 to 60 km/hr)
- Thin air - reduced O_2 and CO_2 levels
- Heavy influx of IV and UV radiations
- Coarse, highly porous and immature soils with poor nutrient contents
- Steep slopes
- Loose strata
- Soils prone to wind and water erosion

Physiological parameters responsible for poor plant growth include

- Poor germination of dispersed seeds - lack of moisture
- Reduced seed production
- Deformed canopy - high velocity winds
- Leaf injury - frost and desiccation
- Reduced leaf area
- Reduced photosynthetic activity - Low levels of O_2 and CO_2
- High transpiration rates
- Increased respiration losses
- Less of radial growth

- Incomplete lignification
- Late flushing of terminal buds
- Inadequate shoot primordial formation in terminal buds

Major biotic interferences responsible for sparse vegetation are

- Excessive grazing by domestic and migratory animals during active growth period
- Fuelwood and firewood shortage forces people to harvest shrubs along with the roots
- Heavy demand for dry fodder to stall feed the domestic animals during winters
- Over exploitation of certain medicinal and aromatic plant species by pharmaceuticals
- Seeds consumed by migratory birds and rodents

The Desert Development Project

The local people should not be deprived of the economic benefits for conservation of genetic diversity and the ecological balance. Thus the desert development programme was undertaken for the overall development of the area.

One of the main objectives of the desert development programme (DDP) was to make the area sustainable in the production of tree leaf fodder (during growing period), fuel and fire wood and small timber.

According to Negi (1996), before implementing the DDP, the old forest wealth of Spiti included the below mentioned naturally growing old trees

- At Poh (3,220 m) only 10 Juniper trees
- At Sumling (3,870 m) one old Poplar
- At Siling (3,700 m) 3 old Poplars
- At Tangti Yogma (3,750 m) one old Poplar
- At Tabo and Poh (3,280 m) 5 wild Apricots
- At Mane (3,675 m) some Poplars, Seabuckthorn forest and nearly 50 Junipers across the nala
- Besides, people had planted few trees of Poplars and Willows in Mane, Tabo, Poh, Kaza

During the period from 1978 to 1993, total plantations carried out under the programme was of 16, 36,735 plants in 830 ha of area at different altitudinal sites of Spiti. The species selected were, *Populus alba, Populus euphretica, Salix alba, Salix fragilis, Robinia pseudoacacia, Prunus armeniaca, Juniperus macropoda, Pinus wallichiana, Melilotus officinalis, Alianthus altissima, etc.* After three years, in 1996, the survival was quite satisfactory as depicted in Table 1.

However, few plantations totally failed due to one or the other reason, as listed in Table 2.

The basic reason for the failure of the plantations was that the following concepts were not taken care of

- Is the species capable of surviving at that altitudinal range
- Wrong choice of species for that particular area
- Can the species withstand the sub zero winter cold
- Soil type
- Availability of irrigation facilities (quantity and time)
- Photosynthetic efficiency of the species
- Transpiration rate of the species
- Water use efficiency of the plant species.

But presently, if we visit these sites, we find plantations surviving at few places only especially at higher altitudes. However, the plantations at lower elevations and with sufficient irrigation facilities, the plantations are growing and even other local flora (grasses, herbs, shrubs) has also established itself as under story.

DISCUSSIONS

The present study was undertaken to estimate the photosynthetic efficiency (PE) and the water use efficiency (WUE) of the traditionally grown crops and the introduced cash crops plus the natural vegetation and the introduced fuel and fodder tree species for comparing their PE and WUE so that in future, plantations of only those species should be carried out, which possess higher PR and WUE.

With the change in cropping pattern from subsistence based mixed cropping (wheat, barley, potatoes, local beans) to the growing of cash crop (peas), it is observed that the traditional crops were more suited to the abiotic conditions than peas, as peas have less of PE and WUE (Table 4,5). Hence we can conclude that peas can successfully be grown only in the lower areas, as also personally observed in upper areas of village Dhankar and Gete.

Similarly, we observe the PE, transpiration rate, stomatal conductance, and WUE, of the local flora and the introduced tree species, it is clearly evident from the data (Table 5, 6) that the PE and WUE of the local species is higher than that of the introduced tree species. While if we compare the efficiency with in the introduced tree species, we find that the highest values of WUE is of *Salix viminalis,* followed by *Populus euphretica, P. alba* and *Robinia pseudoacacia, P,* and least of *Alianthus altissima* (Table 6).

In the end, we can say that since higher transpiration potential allows greater carbon assimilation, but may also lead to development of water deficit when transpiration exceeds water availability, as in present case. The higher amount of stored carbohydrates (higher photosynthetic rates) in the stump can be one of the reasons for survival of the plants at higher elevations where the concentration of carbon dioxide is less. It was further observed that the plants with less of foliage (total leaf area) had higher rates of PE and WUE as in case of *Salix viminalis* and *Populus euphretica* (with less foliage) as compared to *Robinia pseudoacacia, P. ciliata* and *Alaianthus excelsa* with more of foliage (Table 6).

References

Negi, J.P. 1996. Greening the Mountain Deserts (Developmental perspectives in the Himalayan Mountain Desert) Editor R. Bawa., Bishan Singh., Mahindra Pal Singh, Dehradun.

Bawa, R. 2004. Structural, functional and Economic Linkages between Major Ecosystems of Cold Desert in Himachal Pradesh. Final Technical Report, UHF, Nauni, Solan.

contd...

Table 1: Survival rate of plantations carried out under DDP in Spiti area.

S No	Plantation Area	Area (ha)	Plants Planted	Survival (%)	Species Planted
1	Poh-II	9.25	24500	50	*Salix alba, Populus ciliata, P alba, Robinia pseudoacacia, Prunus armeniaca*
2	Kurith-I	7.00	23114	60	*Salix alba, Populus ciliata, P alba, Robinia pseudoacacia, Prunus armeniaca*
3	Kurith-I	2.50	3920	50	*Salix alba, Populus ciliata, P alba, Robinia pseudoacacia, Prunus armeniaca*
4	Poh-Jhula	1.25	3125	70	*Populus ciliata, P alba, Salix alba,*
5	Jarma	2.00	4000	70	*Populus ciliata, P alba, Salix alba,*
6	Sing chuck	4.00	7000	78	
7	Pomxang	3.00	6000	83	*Populus ciliata, Salix alba,*
8	Nydang-III	10.00	25000	72	*Populus ciliata, Salix alba,*
9	Nydang-I	10.00	21000	80	*Populus ciliata, Robinia pseudoacacia*
10	Nydang Thanka	5.00	6000	90	*Salix alba,*
11	Nydang-II	8.00	14000	85	*Salix alba, Populus ciliata, Prunus armeniaca*
12	Poh-I	2.50	5000	80	*Populus ciliata, Robinia pseudoacacia,*
13	Tulrang	5.00	10000	85	*Salix alba,*
14	Poh-III	10.00		65	*Juniper macropoda*
15		1.00	1800	60	*Salix alba, Populus ciliata, Robinia pseudoacacia,*
16	Mane	3.00	6000	60	*Salix alba, Populus ciliata*
17	Shitak	5.00	16100	60	*Salix alba, Populus ciliata*
18	Siluk	2.50	5000	20	*Salix alba, Populus ciliata*
19	Siluk-new	10.00	25000	50	*Salix alba, Populus ciliata*
20	Landupdin	6.00	18000	50	*Salix alba, Populus ciliata*
21	Manne-Dhany	5.00	10000	50	*Salix alba, Populus ciliata*
22	Schling	2.50		60	*Salix alba*
23	Subling	2.00	4000	60	*Salix alba, Populus ciliata, P alba*
24	Shego-III	7.00	28000	50	*Salix alba, Populus ciliata, P alba*
25	Shego-V	5.00	11500	75	*Salix alba, Populus ciliata, P alba*
26	Shego-II	15.00	40000	84	*Salix alba, Populus ciliata, P alba*
27	Shego-I	3.50	12000	75	*Salix alba, Populus ciliata, P alba*
28	Shego-IV	26.50	65000	50	*Salix alba, Populus ciliata, P alba*
29	Lara	5.00	12000	50	*Salix alba, Populus ciliata, P alba*
30	Lidang	5.00	12000		*Salix alba, Populus ciliata, P alba*

Contd...

31	Kungi	6.00	16000		*Salix alba, Populus ciliata, P alba*
32	Lalung	3.00	10000		*Salix alba, Populus ciliata, P alba*
33	Sagling-I	18.00	52500	50	*Salix alba, Populus ciliata*
34	Sagling-II	5.00	12500	55	*Salix alba, Populus ciliata*
35	Humbuk	5.00	12500	60	*Salix alba, Populus ciliata*
36	Hathoo	5.00	10000	50	*Salix alba, Populus ciliata*
37	Numbrang	5.00	10000	2	*Salix alba, Populus ciliata*
38	Sabling	2.00	5000	25	*Salix alba, Populus ciliata*
39	Kaza	10.00	18233	40	*Salix alba, Populus ciliata*
40	Kaza-new				*Salix alba, Populus ciliata*
41	Talk Langpa	10.00	10000	40	*Salix alba, Populus ciliata*
42	Kakti	4.00	6000	40	*Salix alba, Populus ciliata*
43	Rongtong	20.00	48000	41	*Salix alba, Populus ciliata*
44	Rangrik-I	12.00	16000	68	*Salix alba, Populus ciliata*
45	Rangrik Gompa	3.50	2200	40	*Salix alba, Populus ciliata*
46	Rongtong res	5.00	11500	30	*Salix alba, Populus ciliata*
47	Sumling-I	3.50	8500	58	*Salix alba, Populus ciliata*
48	Sumling-II	35.50	100000	40	*Salix alba, Populus ciliata*
49	Kee Gompa	12.00	12000	66	*Salix alba, Populus ciliata*
50	Kee-III	14.00	30500	85	*Salix alba, Populus ciliata*
51	Kee-I	8.25	4700	64	*Salix alba, S fragilis*
52	Kibber	1.00			*Melilotus officinalis*
53	Gate	15.00	8000	59	*Salix fragilis*
54	Kee-II	15.00	23463	40	*Salix fragilis, S alba*
55	Guinde	27.25	30000	45	*Salix fragilis*
56	Pangmo-I	5.00	14500	45	*Salix fragilis*
57	Chckzar	5.0	12500	50	*Salix fragilis*
58	Pangmo-II	6.00	11400	50	*Salix fragilis*
59	Pangmo-III	7.00	11500	50	*Salix fragilis*
60	Pangmo-IV	2.00	4600	70	*Salix fragilis*
61	Kaito-II	5.00	12500	40	*Salix fragilis*
62	Kaito-I	7.00	16000	70	*Salix fragilis*
63	Kaito-III	13.00	30000	65	*Salix fragilis*
64	Kaito-IV	4.00	9200	70	*Salix fragilis*

Contd...

65	Hansa-II	27.00	43000	50	*Salix fragilis*
66	Hansa-III	6.00	8000	5	*Salix fragilis*
67	Khulaska	15.00	40000	50	*Salix fragilis*
68	Kaimo-I	8.00	8000	50	*Salix fragilis*
69	Kaimo-II	20.00	16000	70	*Salix fragilis*
70	Kaimo-III	20.00	18000	50	*Salix fragilis*
71	Hansa-I	24.00	30000	60	*Salix fragilis*
72	Kaimo-IV	5.00	12000	50	*Salix fragilis*
73	Kaimo-V	5.00	12500	60	*Salix fragilis*
74	Kaimo-VI	4.00	10000	60	*Salix fragilis*
75	Chochong-II	18.00	36000	60	*Salix fragilis*
76	Losar-II	3.50	6500	60	*Salix fragilis*
77	Lossar-I	3.50	7000	60	*Salix fragilis*
78	Lossar-III	5.00	5600	50	*Salix fragilis*
79	Pindomax	30.00	65000	50	*Salix fragilis, S alba ,P ciliata*
80	Chhidang	13.00	5000	70	*Salix alba, Pinus wallichiana*
81	Gulling	10.00	42500	58	*Salix alba*
82	Tangti Gompa	12.00	21000		*Salix alba*
83	Kungri	5.00	10000	30	*Salix alba*
84	Teling	3.30	4200	71	*Salix fragilis*
85	Sagnam	15.00	17000	64	*Salix fragilis*
86	Rajgaon	5.50	12000	70	*Salix fragilis*
87	Mud	5.00	10000	100	*Salix fragilis*
88	Minsar	1.00	2200	22	*Salix fragilis*
89	Tabo-I	7.00	15500	35	*Salix alba, P ciliata, P armeniaca*
90	Tabo-II	7.00	15000	20	*Salix alba, P ciliata, P armeniaca*
91	Pharka	18.50	30000	0	*Salix alba, P ciliata*
92	Lari	3.00	5000	60	*Salix alba, P ciliata*
93	Tipta	7.00	15500	67	*P ciliata, , Salix alba, Robinia pseudoacacia*
94	Kharma	10.00	17550	82	*P ciliata, , Salix alba, Robinia pseudoacacia*
95	Lapcha	0.7	1800	66	*P ciliata , Salix alba, Robinia pseudoacacia*
96	Gue-II	7.00	22500	77	*P ciliata, , Salix alba, S fragilis*
97	Gue-I	5.00			*P ciliata, , Salix alba, S fragilis*
98	Kaurik	10.00	12500	80	*P ciliata, , Salix alba, S fragilis*
99	Hichech		17000	58	*Salix alba*

Source: Negi (1996)

Table 2: List of unsuccessful plantations carried out under DDP.

S No	Plantation Area	Area Planted (ha)	Species Planted
1	Gate	15.00	Willow
2	Kibber	1.00	Melilotus
3	Parka	18.50	Poplar
4	Khulaska	15.00	Poplar
5	Schling	2.50	Poplar
6	Chhidang	1.00	Kail
7	Lalung	3.00	Poplar
8	Poh-III	10.00	Juniper
9	Tang Gompa	12.00	Poplar, Willow
10	Hurling	3.00	Juniper
11	Poh ground	10.00	Juniper
12	Lara Ledang	20.00	Melilotus
13	Hull pasture	15.00	Melilotus
14	Takecha	20.00	Melilotus

Source: Negi (1996)

Table 3: Photosynthetic efficiency of traditional and cash crops.

S No	Crop/ Tree	Parameters					
		A	B	C	D	E	F
1	Wheat	1176.0	165.86	12.79	3.74	202.71	28.90
2	Barley	848.63	171.57	12.51	3.59	225.93	33.09
3	Potato	1041.99	164.09	14.96	3.91	276.76	34.42
4	Bean	997.89	176.20	8.44	4.15	256.57	33.94
5	Peas	896.64	145.01	9.96	5.54	200.99	33.40

Where

A = Photosynthetically active radiation (m mol/m^2/sec),

B = Co_2 intake (mol/m^2/sec)

C = Photosynthetic rate (m mol/m^2/sec)

D = Transpiration (m mol/m^2/sec)

E = Stomatal conductance (m mol/m^2/sec)

F = Relative humidity (%)

Table 4: Water use efficiency of traditional and cash crops.

S No	Crop	Photosynthesis	Transpiration	Water Use Efficiency
1	Wheat	12.79	3.74	0.0034
2	Barley	12.51	3.59	0.0035
3	Potato	14.96	3.91	0.0038
4	Peas	9.96	3.54	0.0028
5	Bean	8.44	4.15	0.0020

Table 5: Photosynthetic efficiency of various naturally growing plant species (grasses, herbs, shrubs and trees)

S No	Species / Weather Conditions	Parameters					
		A	B	C	D	E	F
1	Grasses	1683.50 (+25.22)	533.27 (+69.87)	25.35 (+12.38)	2.68 (+2.18)	101.33 (+102.9)	33.83 (+2.46)
2	*Chenopodium foliolosum*	1659.03 (+26.98)	117.20 (+33.86)	18.42 (+3.30)	4.79 (+0.87)	263.50 (+77.86)	37.00 (+0.82)
3	*Trigonella dioica*	1634.36 (+79.55)	182.34 (+41.88)	7.04 (+4.21)	6.71 (+3.01)	380.78 (+160.1)	36.74 (+3.51)
4	*Heracleum candicans*	947.90 (+713.98)	187.16 (+23.77)	13.41 (+6.58)	7.65 (+1.02)	355.54 (+42.80)	33.18 (+0.69)
5	*Clematis tibetana*	1717.93 (+9.82)	156.70 (+15.64)	16.34 (+3.96)	5.11 (+1.55)	233.70 (+77.08)	31.70 (+1.71)
6	*Mentha piperita*	1563.87 (+93.55)	202.45 (+31.21)	8.45 (+7.83)	3.92 (+0.54)	126.43 (+12.26)	28.78 (+1.00)
7	*Caragana jubata*	1606.63 (+132.73)	174.80 (+24.46)	12.79 (+3.51)	7.98 (+0.03)	297.40 (+21.22)	29.37 (+1.41)
8	*Thalictrum foliolosum*	1669.80 (+63.83)	179.30 (+21.55)	12.70 (+6.03)	8.15 (+0.58)	300.63 (+48.42)	30.40 (+1.27)
9	*Taraxacum officinale*	1444.60 (+104.19)	208.40 (+27.49)	9.35 (+8.27)	7.79 (+2.04)	333.63 (+61.91)	31.27 (+2.24)
10	*Dactylorhiza hatagirea*	1742.43 (+27.35)	212.30 (+10.05)	5.52 (+2.18)	5.82 (+3.16)	265.57 (+164.4)	29.43 (+1.23)
11	*Capparis spinosa*	1606.25 (+71.48)	217.80 (+73.19)	17.92 (+6.83)	9.09 (+1.03)	384.88 (+62.34)	26.90 (+0.95)
12	*Robinia pseudoacacia*	1710.80 (+49.38)	29.73 (+0.52)	8.40 (+1.69)	1.69 (+0.38)	32.40 (+13.72)	6.50 (+0.54)
13	*Alaianthus excelsa*	1727.34 (+67.85)	40.60 (+8.20)	7.33 (+11.55)	2.46 (+0.92)	58.96 (+26.54)	7.66 (+0.82)
14	*Populus euphretica*	1499.58 (+126.27)	30.44 (+3.15)	9.01 (+5.55)	1.11 (+0.26)	27.28 (+9.29)	5.72 (+0.53)

Contd...

Contd...

15	*Populus alba*	1057.05 (±777.84)	19.20 (±38.40)	15.13 (±3.25)	2.39 (±0.86)	63.30 (±25.49)	6.40 (±3.38)
16	*Hippophae rhamnoides*	1737.73 (±12.49)	28.77 (±40.68)	16.67 (±4.31)	2.73 (±0.81)	50.80 (±30.61)	7.97 (±1.79)
17	*Populus ciliata*	1578.82 (±229.64)	82.02 (±62.78)	9.21 (±2.89)	2.63 (±0.44)	80.82 (±18.49)	9.15 (±0.91)
18	*Salix viminalis*	1712.47 (±54.53)	47.67 (±67.41)	12.75 (±6.66)	0.87 (±0.64)	23.90 (±17.83)	5.77 (±0.45)
19	*Rumex nepalensis*	1619.37 (±21.37)	181.27 (±21.46)	17.27 (±2.31)	7.02 (±0.68)	417.53 (±61.59)	30.20 (±0.29)
20	*Cotoneaster bacillaris*	1522.87 (±44.61)	162.90 (±17.57)	7.50 (±1.71)	3.13 (±0.79)	139.53 (±52.25)	21.03 (±2.38)
21	*Tagetes minuta*	1637.47 (±34.52)	170.40 (±39.85)	15.05 (±5.06)	6.78 (±1.24)	321.13 (±47.05)	30.97 (±0.69)
22	*Arnebia guttata*	1498.37 (±177.61)	137.07 (±51.22)	10.41 (±5.99)	3.80 (±0.77)	129.60 (±36.19)	24.50 (±3.43)
23	*Myricaria legena*	1155.43 (±466.09)	110.53 (±62.13)	12.20 (±5.53)	3.35 (±1.03)	137.37 (±58.19)	22.30 (±2.05)
24	*Lilium polyphyllum*	1649.90 (±71.39)	43.23 (±61.14)	29.38 (±20.67)	3.47 (±1.43)	110.13 (±73.58)	21.10 (±0.85)
25	*Mentha longifolia*	1775.13 (±26.53)	58.47 (±82.68)	6.55 (±3.14)	1.39 (±0.46)	40.30 (±11.23)	19.43 (±0.80)
26	*Chenopodium album*	1462.20 (±379.60)	313.00 (±315.68)	18.04 (±10.04)	4.91 (±3.73)	103.67 (±112.9)	10.80 (±5.10)
27	*Arabidopsis himalaica*	1119.63 (±339.61)	225.60 (±16.17)	4.23 (±2.50)	4.96 (±1.40)	212.03 (±24.19)	23.43 (±3.22)
28	*Microula sikkimensis*	1779.97 (±8.97)	184.90 (±15.96)	9.05 (±2.48)	5.16 (±0.19)	137.63 (±87.11)	23.73 (±0.42)
29	*Ribes orientale*	1483.37 (±150.81)	178.23 (±34.04)	8.72 (±3.08)	4.00 (±0.29)	198.90 (±26.28)	24.77 (±0.79)

Where

A = Photosynthetically active radiation (m mol/m^2/sec),

B = Co_2 intake (mol/m^2/sec)

C = Photosynthetic rate (m mol/m^2/sec)

D = Transpiration (m mol/m^2/sec)

E = Stomatal conductance (m mol/m^2/sec)

F = Relative humidity (%)

Table 6. Water use efficiency of various naturally growing plant species (grasses, herbs, shrubs and trees).

Species	Photosynthesis	Transpiration	Water Use Efficiency
Grasses	25.35	2.68	0.0095
Chenopodium foliolosum	18.42	4.79	0.0038
Trigonella dioica	7.04	6.71	0.0011
Heracleum candicans	13.41	7.65	0.0018
Clematis tibetana	16.34	5.11	0.0032
Mentha piperita	8.45	3.92	0.0022
Caragana jubata	12.79	7.98	0.0016
Thalictrum foliolosum	12.70	8.15	0.0016
Taraxacum officinale	9.35	7.79	0.0012
Dactylorhiza hatagirea	5.52	5.82	0.0010
Capparis spinosa	17.92	9.09	0.0020
Robinia pseudoacacia	8.40	1.69	0.0050
Alianthus altissima	7.33	2.46	0.0029
Populus euphretica	9.01	1.11	0.0081
Populus alba	15.13	2.39	0.0063
Hippophae rhamnoides	16.67	1.73	0.0059
Populus ciliata	9.21	2.63	0.0035
Salix viminalis	12.75	0.87	0.0147
Rumex nepalensis	17.27	7.02	0.0025
Cotoneaster bacillaris	7.50	3.13	0.0024
Tagetes minuta	15.05	6.78	0.0022
Arnebia guttata	10.41	3.80	0.0027
Myricaria legena	12.20	3.35	0.0036
Lilium polyphyllum	29.38	3.47	0.0085
Mentha longifolia	6.55	1.39	0.0047
Chenopodium album	18.04	4.91	0.0037
Arabidopsis himalaica	4.23	4.96	0.0009
Microula sikkimensis	9.05	5.16	0.0018
Ribes orientale	8.72	4.00	0.0022

□□□

19

Agriculture Extension Through Local Institution in High Altitude Region of Ladakh

DORJEY ANGCHOK, M. PAL MURUGAN, A. K. KATIYAR
ELI PALJOR, HEMRAJ AND SHASHI BALA SINGH
Defence Institute of High Altitude Research, DRDO, Leh-Ladakh, J&K, India

INTRODUCTION

In India, risk prone cold-arid high altitude area constitutes the easternmost Trans-Himalayan part of Jammu and Kashmir (J&K) which includes Ladakh and Zanskar ranges and Lahaul Spiti region of Himachal Pradesh (HP). Ladakh, truly described as a cold arid desert constitutes of Leh and Kargil district. Leh district constitute an geographical area of 45,100 sq.km with an estimated population of 1,17,232 and population density of 3 per sq.km. Intensive sunlight, high evaporation rate, dry humidity (less than 40 %) strong wind velocity) and fluctuating temperature (+40°C to -40°C), minimum rainfall (8mm to 300mm) land locked condition (during winter months) characterize the general climatic condition of Ladakh. It is generally said that "In Ladakh, a man sitting in the sun with his feet in the shade can have sunstroke and frostbite at the same time".

Most of the land in Ladakh is a mountainous desert of rocks, sand and dust. Only small part about 8750 ha of land is used for cultivation and animal husbandry. By means of a system of terraces and irrigation canals, patches of land are made arable wherever possible. Because of heavy snowfall and low temperature during winter, cultivation season is restricted to about six months only during the summer season (April - September) and only 450 ha of land is double cropped in a year.

The present name of the region, Ladakh, is derived from the classical Tibetan word *la-daks,* meaning many passes. *La* means pass and *daks* meaning many; hence the name, Ladakh, a region with many passes.

Adventurous spirit, strength and stamina are required to inhabit such a region lying at an elevation varying between 2900 m to 5900 m above msl the highest region ever inhabited by human beings.

In the social aspect, Ladakh is characterized by a unique socio-cultural environment accompanied by harsh and inhospitable climatic conditions. In such a condition it is not desirable to 'copy and paste' the various extension models (most of them developed for conditions prevailing in western developed countries) "*as it is*" in this region. Similarly, as we have adopted in other disciplines, here in extension also we need to evolve our own strategies and action plan, appropriately suited to the conditions prevailing here in Ladakh.

Retrospection of Extension Service

Whatever limited extension activities, been carried out in these region by various organization have been based on the classical diffusion model, where it is assumed that once a technology is diffused and adopted by a farmer (mostly progressive resource rich), the technology in a social system will diffuse of its own like a contagious disease. The major assumption is that in a matter of time that all farmers will adopt the technology. But in reality it is not the case and consequently we call those farmers (who have not adopted the technology) as laggards / traditional / non-progressive / non-innovative / disadvantaged, without assessing, whether the technology is compatible with their (farmers') needs, interest, culture, skill, and the resources available with them. In other words we find fault in the farmer but not in the technology.

Therefore, in contrast to the above mentioned assumption, we need to look into the nature of the technology i.e. the appropriateness of the technology, whether it is in line with the needs (both felt and unfelt) of the farmer (both small and big) and its suitability for this cold desert. In this connection following measures can be taken up to improve upon such a situation:

- *Targeting approach:* Similar to the strategy adopted by the product marketing ventures, we have to approach and reach out the selected segment of the society for particular technology.(e.g. poultry Rearing - Muslim community in Ladakh)
- Increase awareness through various media (individual/group/ mass).
- Increase the countervailing power of the farmers through capacity building.
- Involve local farmers and local institution at every stage of technology generation process.
- Due to the mountainous physical environment of Ladakh, access to villages is not easy. Therefore the importance of secondary data is immense to carry out an effective extension activities (Districts/ blocks/ villages level data on population, area, distance, socio-economic status, crops grown, and other specific features; households- acreage, family size, major source of income, social status etc). Most of these data can be accessed from various state departments and the Hill Development Council of Ladakh through formal request.

Local Institution[1]: Their Role in Extension

The success of technological change in high growth regions should not blind us to the richness of the ethnic basis of local knowledge systems. Studies have shown that more diverse the environment and lower the population density greater is the need for social networking. The knowledge networks provide a platform for farmers to satisfy their curiosity about different innovations being tried by different people and not always successful.

Local institutions (LI) play a major role in sustaining the life of people in such conditions by designing strategies to deal with the inhospitable and tough environmental conditions. Institutions for resource management are a part of evolutionary cultural, religious, and social experience of any community. Diversified uses of a natural resource are essential for household survival in high-risk areas. One cannot analyse resource management institutions without understanding the conceptualization of the nature and repertoire of responses that a community evolves to adjust with changes in the natural phenomenon. The portfolio of activities, which is so evolved in an eco-specific manner, includes a combination of apparently rational and not so rational strategies of livelihood.

Keeping the above views in consideration an attempt has been made to generate a framework for extension activities through local institutions (Fig. 1) in context of risk prone high altitude areas. A number of studies have pointed out dangerous trends towards deforestation all over the world. The major cause of the current trend of natural resource degradation is the negligence and breakdown of local institutions, and top-down government policies. The current trends of deforestation, desertification and alarming high environmental damage can be halted and reversed by understanding the manner in which people utilize and manage the natural resources in rural areas through local institutions.

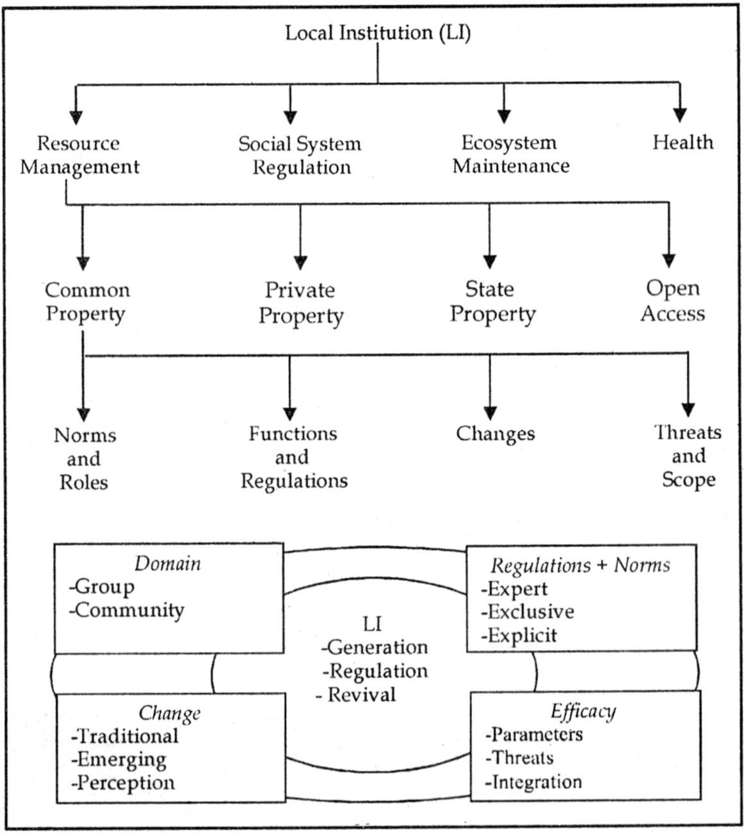

Fig. 1: Framework to Study Local Institution

LI play a major role in context of resource management by: (i) regulation of behaviour; (ii) maintain ecosystem productivity; (iii) maintain rate of use of resources ; (iv) influence allocation of resources; (v) ensures access to scarce resources; (vi) help in the reduction of uncertainty; (vii) sanctions (Coercion); (viii) scale allocation: through boundary, technology, time, by whom, by how, etc.

It is said that "Institutions are like grammar, and technology the word" i.e. the technology has to fit within the institutional grammar, otherwise it has no meaning, or produce distorted meaning. In other words institution to a great extent influences the nature of technology being generated and adopted.

Trench technology[2]: It is the most economical technology for vegetable cultivation during winter months in high altitude risk prone areas. The novel concept behind evolving this technology was to have a technology through which a very small and resource poor farmer can grow vegetables during winter. On the contrary, though being such a viable technology (as perceived by the technology generator) the rate of its adoption was quite less as compared to other technologies like **polyench**[3] (quite expensive compared to trench). Upon close interaction with local farmers it was found that this technology was not fitting into the local institution (the word was not fitting into the grammatical framework).

In Ladakh, in what property regime a resource fits depends on the time scale. A private resource (e.g. A field for wheat or barley cultivation) in summer becomes a common property resource **(CPR)**[4] in winter (open to all- animals of the village; free to graze any where), and fencing of these fields are avoided (through self discipline). Therefore, construction of a trench in such area for vegetable cultivation during winter is avoided because there is probability of animal intrusion causing damage to the crop and the structure. Adoption of this technology has been limited to, inside the compound of the household only.

In the past few decades it was a kind of fashion/craze among the agriculturist to study the Local Knowledge System **(LKS)**[5], to document and validate them for possible blending with modern scientific knowledge system. Till date whatever limited initiatives has been taken up in this potential area has been limited to the tangible/ material aspects of LKS (e.g. use of medicinal plants). In the process we ignored the intangible/non-material part of LKS, which we argue, was a mistake, similar to laying one line of a railway track. For smooth running, both the line should be run in parallel. The intangible part allows for the practice of those tangible parts to fit into the system and help plays a major role in sustaining the life of local people in Ladakh (e.g. a local healer in Ladakh goes to different terrains to collect materials for herbal medicine formulation and charges a very nominal fees (sometimes no fee is charged) upon medication). Unless the healer is purely philanthropic it would be difficult for him to sustain his own life, for which there are various institutions in a village where he is paid back (both tangible and intangible) by the villagers. To illustrate, he is one of the most prestigious person of the village (places high status during social gatherings and ceremonies) and during cultivation season the whole village pool labour and help him cultivate his field (the arrangement may vary - a group of household on rotation basis contributing in the healer's various farming activities). In other words, the strong social-networking system and the presence of local institutions help the healer in prediction of people's behaviour and vice versa so that both are mutually benefited.

Frame Work to Study Local Institutions

Altitude variation is the major factor deciding the type of agro-climatic region in high altitude areas like Ladakh. Such variations are found even within a single village. Since a single altitudinal level and eco-niche does not have the potential to provide the subsistence need of the villagers. All farmers require access to different eco-niches. To solve this problem many of the resources, which are necessary for survival, are held in common, where the rule of subtractability and excludability is followed. The availability of common property resource (CPR) helps people living in high-risk environment, to fulfill this need. The access to different microenvironment reduces the risk of bad years and effectively schedule labour, particularly for such activities as haymaking, at different altitudinal level in successive stages. Although, small agricultural plots can be effectively exploited, when owned individually, higher pasture can be more effectively utilised when owned in common. In the same way a single altitudinal level is not capable of providing subsistence needs, a single eco-niche activity of crop or animal rearing alone cannot provide such needs either.

In the context of CPR in Ladakh, we need to analyse institutions relatively more carefully and comprehensively than has been done by most theorist. We need to understand what institutions were/are there in the local community (literatures on CPR has given less importance to episodic institutions as against concurrent or durable institutions), and how specific institutional rules are generated and effect individual behaviour by allocating benefits from resources and structuring incentives.

There is immense need to identify the institutional diversity and infrastructures already existing/were exist in Ladakh, and see how they can be effectively used for better extension activities. Following are the details of the framework (Fig.01) to study LI especially in context of common pool resource management in Ladakh and other similar high altitude areas:

- *Management:* Among many functions of a local institution, management of resources that are held in common is one of them.
- *Norms and rules:* Institutions basically contain rules for effective management of the resource. These norms may be formulated by a group or a community and sometimes may be restricted to some class as expert, exclusive or explicit person/group.
- *Function and regulation:* How the institution work or function depends on the way the norms are laid down (the policy making process), which may be democratic or authoritative. Emphasis should also be given to understand the participation of different groups like representation of women and elderly people etc in the decision making process. It may be possible that in some issues there may be some exclusive rights enjoyed upon by some individual/group in the decision making process.
- *Changes* : Every institution changes with time, structurally or functionally. It is necessary to understand the factors behind those changes, which may be local, cosmopolitan and contextual and how people perceive those changes (desirable or undesirable).
- *Threats:* With changing time and space the institution come across various threats that puts in danger the effective functioning of the institutions. (e.g. In Ladakh, compared to few decades back, presently the status of *goba* (traditional village head) has degraded). People does not heed to his decisions because the individual has become more self sufficient (materially) and the norm-defaulters does not feel ashamed of paying penalties. Earlier the informal norms were very strong and heeded by the people, moreover social network was strong and efficacy of non-material sanctions was quite high.

- *Scope* : The scope for revival of an institution is always there. It may be in the form of learning from other institutions (working effectively) or add/delete some structural/ functional aspect of the institutions so that it is compatible with changing time and place. It is generally perceived that instead of evolving a new institution it is always desirable to revive an already existing old institution to lower "transaction costs", providing the decision-making structures, enforcement powers, experiences, and cultural expectations that might otherwise have to be created anew and at great economic and political expense. Moreover it maintains the credibility of the institution, which is major factor for effective functioning of an institution.

Identify Local Innovators

Innovations by the peasants whether in the field of technology or institutions require production of knowledge. But the reproduction of knowledge took place through the institutions, which has to be renewed over a period of time. Lots of traditional ecological knowledge has been retained through codes legitimized by religion or cultural institutions. It is being increasingly realized that the role of local institutions in generalizing respect for boundaries of common properties deserves greater attention.

Local farmers are very ingenious and innovative in designing new technologies or practices to cope up with the environment. Though their innovations may be at initial (rudimentary) stage, but with the integration of proper modern scientific inputs into these innovations, the technology so developed can be made effective and viable (e.g. A farmer from Gujarat got patent for his designing of a cotton stripper machine, with the help modern scientific inputs from Council of Scientific and Industrial Research (CSIR) & National Innovation Foundation (NIF). Similar programme can also be initiated in this region, through collaboration with various laboratories. It will not only help organizations but also the local farmers.

Emphasis should be on educating the farmers rather than just performing extension service. This will help farmers to decide of their own, what to adopt and what to discard out of their behaviour. These way organisations will not be an input supplier but a supplier of new ideas and innovations.

Incorporation of LI in Research and Extension Network

There is the need for developing a framework for incorporating LI into the Research and Extension System because:

(i) The transfer of technology (TOT) model fits badly with the needs and priorities of resource-poor farmers.

(ii) Agricultural extension programmes are still biased towards techniques and strategies, which are capital-intensive.

(iii) Resource-poor farmers (RPF) are scattered and are not able to make their needs and priorities readily known and felt.

(iv) The TOT model cannot easily handle the complex interactions of RPF farming; links between crops, especially with intercropping and multiple tiers; agro-forestry and livestock-crop-tree complementaries and the progressive adjustments required in the field in the face of seasonal and inter-annual fluctuations.

A framework for incorporating LI system into agricultural research and extension should have following features:

(i) Strengthening the capacities of regional research and extension organizations.

(ii) Identifying the need for extension scientists/social scientists in an interdisciplinary regional research team.

(iii) Formation of a sustainable technology development consortium to bring farmers, researchers, NGOs and extension workers together well ahead of the process of technology development

(iv) Understanding that it is impractical to depend entirely on research stations for innovations, considering the inadequacy of human resource capacity of the regional research system.

A comprehensive model for incorporating local knowledge in Research and Extension Network (REN) was developed and it fits well for the condition of Ladakh. Major features of the suggested model are:

(i) To set up a Center for Local Institution and Knowledge (CLIK), assigning broadly at least three-fold work, viz., (a) collection, (b) documentation and (c) preservation of LI and LK in kind of Knowledge Bank.

(ii) Other activities of CLIK should include coordination and monitoring, management of research outputs and related publications for developing literature on LI and LK, and dissemination of center's activities.

(iii) Curricular sanction for local knowledge in the school and undergraduate courses and integration in the pre-service training courses of village level agricultural extension workers to remove attitudinal barrier against LKS.

(iv) Support of general academia in addition to agricultural disciplines.

(v) Involvement of NGOs in the process.

(vi) In order to popularise LI and LK among the scientists, extension functionaries as well as the farming communities, adequate scope for publicity through electronic and print media should be provided.

(vii) The collected local agricultural knowledge should be subjected to preliminary screening and ready assessment in different professional forums before final recommendation.

Ladakh has a rich network of LI and LK, and therefore effort should be made to conserve it. In this context it is necessary in today's global scenario that we develop a viable strategy to protect ourselves from the exploitative global policies, and harness future opportunities. Moreover such a strategy is inevitable for the sustenance of this knowledge system in Ladakh which are on the verge of decadence.

It has been recognized all over the world especially western countries the so called modern scientific community that there lies tremendous scope through the integration of LK and LI into the mainstream modern knowledge system, to evolve eco-friendly and sustainable appropriate technologies and practices suited to a particular niche or an ecosystem.

In this regard, research organisations can initiate a major programme to document, validate, and integrate this local system, so that an cost effective appropriate technologies can be designed for Ladakh and similar other high altitude areas.

References

Angchok, D. 2000. Traditional Method of Weather Forecasting in Tibetan Astrology and its Relevance in Today's Agriculture. Unpublished M.Sc. Thesis, Department of Extension, Education, Banaras Hindu University, Varanasi, India.

Angchok, D. 2003. A Framework to Study Local Institutions: Their Role in CPR Management in Risk Prone High Altitude Areas. Paper Presented In: National Seminar on *Extension Strategy for Promoting Development Initiatives Among Farming Communities*, ISEE, Division of Extension Education, IARI, New-Delhi, 18-20 June 2003

Angchok, D. 2004. Appeasing Spirits, Protecting Nature. *Honey Bee*, IIM, Ahmedabad, 15 (4): 2-3

Atteh, David, O. 1989. Indigenous Local Knowledge as a Key to Local Level Development: Possibilities, Constraints and Planning. Paper Presented In: The Seminar on '*Reviving Local Self Reliance: Challenges for Rural/ Regional Development in Eastern and Southern Africa*', Arush, Tanzania, 21-24 Feb. 1989.

Box, L 1988. Networks and Nitpricks, Socio-Ecological Perspective on Agricultural Knowledge, Agri. University, Wageningen.

Buzdar, Nek 1988. Property Rights, Social Organization, and Resource Management in Northern Pakistan. *Working paper No.5*. Environment and Policy Institute, Honolulu, Hawai.

Edwards, Victoria M. and Steins, N. A. 1999. A Framework for Analysing Contextual Factors in Common Pool Resource Research. Journal of Environmental Policy & Planning. 1: 205-221.

Govt. of J&K .2005-06. District Leh at a Glance, LAHDC Leh, Govt. of J&K, District Statistical and Evaluation Agency Leh, Directorate of Economics and Statistics Planning and Development department.

Gupta, A. K 1991. Building Upon Peoples' Ecological Knowledge: Framework for Studying Culturally Embedded CPR Institutions. Paper Presented at the *Second Annual Conference of the International Association for the Study of Common Property*, Winnipeg, Manitoba, 26-29 September 1991.

Gupta, A. K. 1997a. Consumers can conserve: will you? Honey Bee, IIM, Ahmedabad, 8 (4): 1.

Gupta, A. K. 1997b. From Grassroots to Global: Linking Innovations, Institutions, Investors and Entrepreneurs. Honey Bee, IIM, Ahmedabad, 8 (2): 1-2.

Gupta, A. K. 1997c. Neem, Turmeric, Pongal and Patents: Traditional Knowledge and Contemporary Institutions. Honey Bee, IIM, Ahmedabad, 8 (3): 1-2, 14.

McCay, Bonnie J. 2000. Emergence of Self-Organized Cooperation. For Panel at *"Constituting the Commons," Conference of the International Association for the Study of Common Property*, Bloomington, Indiana, May 31-June 4, 2000.

Nandy Ashish. 1979. *Alternative Science*: Allied Publishers Private Limited, New Delhi, 1979, 115.

Ostrom, Elinor (1990). Governing the Commons: The Evolution of Institutions for Collective Action. New York: Cambridge University Press.

Ostrom, E; Gardner R and Walker J. 1994 . Rules, Games, and Common-Pool Resources. Ann Arbor: University of Michigan Press.

Parasar, B. 1994. Indigenous Agricultural Knowledge of Tribals in Eastern Ghat High Land Zone of Orissa. Unpublished Ph.D. Thesis, Department of Extension, Education, Banaras Hindu University, Varanasi, India.

Richards Paul 1989. Agriculture as Performance: Farmers first, London: Intermediate Technology Publication, 39-43.

Warren, D Michael. 1993. Indigenous Knowledge, Biodiversity Conservation and Development. Paper Presented in: International Conference on *Conservation of biodiversity in Africa: Local Initiatives and Institutional Roles*, Nairobi, Kenya, 30Aug-3Sept, 1993.

contd...

Endnotes

1 Institution: the organised system of practices and social roles developed about a value or series of values, and the machinery evolved to regulate the practices and administer the rules.

2 It is a very low cost greenhouse termed as an underground greenhouse preferably 30'x10'x3' of size. Trench is covered with transparent UV stabilized 200 micron thick polythene during day time to harvest maximum solar energy and extra cover of black polythene at night time to check the heat loss. With this technology even a small and marginal farmers can produce leafy vegetables during extreme winter months. For details contact: Director, Defence Institute of High Altitude Research, Leh.

3 It is semi underground double walled greenhouse working on the principle of zero energy chamber made from locally available unbaked bricks and wooden ballies. The air gap between two walls provides better insulation to check the heat loss. During extreme winters months it provides favorable environment for vegetable cultivation specially leafy vegetables and root crops of European origin.

4 One of the confusion of the tragedy of the common debate resulted from the use of terminology (Edwards & Steins, 1999). In particular, the crucial difference between a common pool resource and a common property resource is often ignored. A wide range of resource might be used in common, thus comprising a common pool from which individuals might draw 'resource units' (Ostrom, 1990). However common pool resource is a generic term adopted for resource used in common, which may or may not have formal and informal rights attached to them controlling their use. By terming a resource a common property, we are identifying that specific rights and duties re attached to it. Thus, common property resource forms a subset of common pool resources, being resources to which specific property rights for common usage are attached. Property rights are social institutions that have evolved as a means of establishing claims to a stream of benefits that might be derived from a resource system. A resource system with property rights attached to indicates the intension for some party to ensure that potential users observe any predetermined restrictions or prohibition concerning access to and use of the resource. The very existence of property rights can prevent the 'free for all' sometimes associated with resources used in common (Edwards & Steins, 1999).

According to the definition of Ostrom, Gardner, and Walker (1994, p. 6), a Common Pool Resource is characterized by two attributes: "the difficulty of excluding individuals from benefiting from a good and the subtractability of the benefits consumed by one individual from those available to others".

Typology of Goods

Appropriators			
Resource			
		Excludable	Non-excludable
	Subtractable	**PRIVATE**	**COMMOM POOL**
	Non-subtractable	**CLUB**	**PUBLIC**

Communal resources are resources exploited by many different users at the same or different times. When the rights to use resources are controlled by an identifiable group, and there exist rules defining who may and may not use the resource, and how, this is said to be a common property management system. Such a system depends on members of the group agreeing to limit their individual claims on a resource in the expectation that the other members of the group will do the same, and on the establishment of mechanisms to stop people cheating the system (Ostrom, 1990).

She further argues that disruption of common property regimes has often led to the creation of open access systems, where access to resources is entirely uncontrolled. This has resulted in a perception that common property regimes are equivalent to open access regimes. There is a misconception that only private property regimes can bring about responsible management and returns on investment.

5 "Indigenous" means native born, originating or born naturally in a country, and opposite to exotic. It also means innate and inherent. More specifically LKS refers to knowledge that is unique to a given culture and society. This knowledge is the information base of that society, codified in the language of the society and it facilitates communication and decision-making. While defining LKS, Atteh (1989) was of says that it is the untapped and unwritten body of language, which held in different brains, languages and skills in as many groups, languages, cultures and environments.

The term "Local Knowledge " is used synonymously with "Traditional " and Indigenous Knowledge ", to differentiate the knowledge developed by a given community from the International Knowledge System, sometimes called as "Western System", generated through universities, government research centers and private industries. IK refers to the knowledge of indigenous people as well as any other defined community. Local or traditional knowledge are the products of life long experience of the farmers, transferred from one generation to another.

□□□

20

Income Generation Through Sustainable Resource Management of Seabuckthorn in Nepal

SUSANNE VON DER HEIDE
HimalAsia Foundation, Institute for Intercultural Communication and Documentation, Kathmandu, Nepal

INTRODUCTION

Seabuckthorn project in Mustang initiated by the HimalAsia Foundation and Institute in 2003 together with local community groups/committees. The project served to safeguard medicinal plants and biodiversity of Nepal and in particular provided a sustainable income source for local people, since it was directly addressing the poverty faced by local communities.

Mustang District in North-Central Nepal, hemmed in partly by a chain of mountains, like the giant Annapurna and Dhaulaghiri, is divided into southern and northern Mustang. The Northern parts where the Lobas are living, who are ethnically and culturally related to the Tibetan sphere of influence, is also known as the former forbidden kingdom of Mustang and was opened to foreigners in 1992. In the southern part of Mustang the majority of inhabitants are Thakali and Baragaonli. They all make their living predominantly from agriculture, cattle breeding and small scale trade. Besides diversity, due to severe climate changes, Mustang is suffering from land degradation and soil erosion in different parts of the country, and this project helps to control this degradation by planting seabuckthorn on slopes and the degraded lands.

Seabuckthorn is a highly nutritious, multi-functional plant resource with ecological/ environmental, economic and social benefits. Out of the berries one can produce nutritious juice. This project has already started and juice is produce and sold by poor local inhabitants to local lodge owners that are selling it to trekkers and have also started to consume themselves particularly from the wild seabuckthorn forests in Mustang. The tea and jam is produced out of the berries which are sold to trekkers and shops in Kathmandu. Moreover, it is planned to process oil out of seabuckthorn for its high value for medicinal and cosmetic purposes. Already the selling of tea to the international market has started and seems to be very promising.

To meet the demands for land from which seabuckthorn can be harvested in the future (since the wild growing seabuckthorn forests in Mustang are far from each other and quite scattered and old making it difficult to harvest. HimalAsia has started to establish several seabuckthorn-nurseries in Mustang since 2003. This programme was initiated with the cooperation of village communities/Ama Samha groups, two monastic school committees (the Lekshey Choeling Chunmay Lobta Nunnery School and the Ku Tsab TerNga Monastery in Mustang) and with one Amchi family (doctors of Himalayan medicine), who also also formed a local committee.

It was planned to harvest the wild growing seabuckthorn forests that are found in certain areas of Mustang in a sustainable way, in order to bridge the gap of getting seabuckthorn-berries from the nursery. The collection from wild forests of seabuckthorn *tibetana* and *salicifolia* in Mustang has started in summer 2004 by forming women biodiversity forest user groups in collaboration with the local administrative bodies of the government. The women are mainly from low income groups and have been trained by HimalAsia to carry out this work without over harvesting the wild forests.

In the mean time a certain amount of seabuckthorn plants have been raised in the nurseries to cultivate them on community land, in order to harvest enough seabuckthorn berries for the juice, tea and oil production for the national and international market.

In regard of that there is infrastructural need to put up gabions as a measure for protecting from a meandering river, that area for about 25 hectares of community land for the cultivation of seabuckthorn plants.

It was also necessary to provide training to the Amar Samha women groups of the concerned villages for 1) proper planting process, 2) proper harvesting procedures 3) processing knowledge and 4) and knowledge of upgrading the existing value chain towards a more sustainable export delivery system. Also it is necessary to procure oil filter machine for the community in order to enhance processing of seabuckthorn berries, which will be empowered the women group with a proper system to provide a sustainable income source for the community in the future. As a project activity there was a need to promote and educate people about the value of seabuckthorn and its products to improve possibilities of trade within and outside of the country. Moreover a workshop with stakeholders, to determine the challenges in market promotion will be organised with a view to concentrate on the marketing of seabuckthorn products.

Increasing local awareness via trainings and workshops will enhance the seabuckthorn value chain and support towards possibility of increasing the number of seabuckthorn trees (which will avoid mono-cultures as the roots of seabuckthorn provides enough nitrogen to supports the growth of other vegetables and medicinal herbs) in different communities of Upper and Southern Mustang in Nepal to use it as an income source for the locals. This approach could serve as a model that can be replicated in other areas of Nepal in which seabuckthorn plants are growing.

Objectives

Since many of the local population in Mustang have migrated from relatively arid area, which on the other hand has a highly developed traditional culture, dating hundreds of years, an initiative to improve the livelihood of such a distinct and deprived society will help safeguard this unique culture with a possibility of increasing income for the people of Mustang.

HimalAsia is hoping that with this project a market in Nepal for seabuckthorn products will be established where the local cooperatives can directly benefit and small and medium sized enterprises are established to further add value to the process of seabuckthorn production. In this regard it is important to ensure local market development for seabuckthorn juice. The oil of seabuckthorn can also be used to treat patients with burns and scars in local hospitals more efficiently and less costly as with usual medicine/treatment (A project in this regards has been initiated recently already from HimalAsia in cooperation with the Sushma Koirala Hospital). The envisioned project will also try to see if such marketing measures can be used to further strengthen market possibilities for promoting seabuckthorn at the international level. On the other hand it is also envisaged that international companies will buy the products from the local cooperatives.

The berries of seabuckthorn can be harvested for processing after five years of plantation. If proper planting material is used the period can be reduced to 3 to 4 years and this will be a reliable cash crop for the local community for decades. Currently, HimalAsia is training locals to take care of the seeds and plants. Besides, this with the support of this program, they will get further training in the coming years and learn how to prepare juice out of the berries, and markets facilitating locals to be able to sell this juice to foreign trekkers, and to the local population by appropriately highlighting the nutritional value of the juice. The program will support awareness campaigns based on the experience of HimalAsia in the region. The programs to promote Seabuckthorn tea will be explored (having already sold the first samples to Trekkers). Additionally, the activities will be diversified by aptly mixing the berries and seabuckthorn leaves for special Amchi remedies and with other medicinal herbs will be developed.

HimalAsia has been able to produce the first batches of extracted seabuckthorn oil together with the Research Centre for Applied Sciences and Technology (RECAST), Tribhuvan University which is used by Sushma Koirala Hospital to treat burns and scars of the patients successfully.

The two species of seabuckthorn existing in Mustang (*Hippophae tibetana* and *H. salicifolia*) are growing in the Himalayan area at an altitude between 2500 and 4000 meters. The quality/properties of the seabuckthorn juice and of the oil are one of the best in the world (very high amount of Beta Carotene, Vitamin C and E, etc.) HimalAsia has comprehensively tested all relevant properties in German Laboratories. This gives us the opportunity to introduce a very unique product with tremendous properties to cure multiple complications related especially to health. Since seabuckthorn is unlike any other medicinal herbs of the Himalaya and it is not threatened by extinction, so it can serve as a particularly suitable cash crop plant.

The Mustang region and other similar areas in the Himalayas suffer from a high level of land degradation, desertification and soil erosion.

As seabuckthorn plants are extremely hardy and exhibit vigorous root growth, they could play a significant role in ecological rehabilitation, and it is hoped that this project will help the people of the region to promote seabuckthorn as a high value product for the European market, giving them an opportunity to strengthen their livelihood, in addition to mitigate many environmental concerns as listed above and especially the outburst of glacier-lakes is expected and also the further melting of glaciers in e.g. the Dhaulagiri and Neelgiri/Annapurna mountain ranges.

The project will directly address the poverty faced by the local communities - especially the local women forest user group members belonging usually to very low caste groups by simultaneously supporting the establishment of seabuckthorn plantations as well as the collection of seabuckthorn in the wild forests in a sustainable way and developing a market for the plant products for locals and as high value commodity internationally.

The project social objectives are:

- Income from the seabuckthorn nurseries / the cultivation of the seabuckthorn plants on plantations are used directly to provide a sustainable source of income for the Lekshey Choeling Chunmay Lobta Nunnery School and the Ku Tsab TerNga Monastery, providing education to young children.
- The project alleviates the problem of malnutrition through the development of seabuckthorn products rich in vitamins A and C, E and omega-3 fatty acids, minerals and other active substances.
- The project also addresses the question of gender equality, given the involvement of several women's co-operatives in the collection, cultivation and sale of seabuckthorn products. The project also provides revenues for the education of young women at the Lekshey Choeling Chunmay Lobta Nunnery School.
- Furthermore, the project could play an important role for the local hospitals that are having patients with burns, the seabuckthorn oil can be used to treat patients more efficiently and less costly. The Sushma Koirala/Interplast Hospital in the city of Kathmandu is having project on seabuckthorn to carry out more scientific research in this regard.

Strategy

HimalAsia Foundation had already started to measure whether it is beneficial for the women groups to collect seabuckthorn and produce different products out of that. However, there is a need to strengthen the quality standards in terms of processing quality. Quality concerns will be addressed by implementing the above mentioned training and capacity building exercises.

The community groups are very happy that HimalAsia is buying the berry pulp for making tea and Amchi medicine, because this is an unexpected additional income which helps them to raise their living standard significantly. Also HimalAsia has started the basic certification process (seabuckthorn is weighed and it is marked from which area it has been collected, also involved women are giving their names and signatures on papers which are kept by the chairperson of the women collecting groups). With such a system in place everybody gets a transparent view from where the pulp comes, which batch number it is, etc. Such certification process helps towards the traceability of the product. The activities envisioned through this program will further strengthen such activities so that a proper certification system can be built upon for the future.

In the long run beside the seven Village Development Committees (VDCs) with around 871 households (approximately 4890 inhabitants) who are already involved in the Seabuckthorn Programme in Mustang, and have partly provided the Community Land for further cultivation of seabuckthorn plants from nurseries, more VDC's in Mustang are willing to be included into the Seabuckthorn Programme of HimalAsia, which will provide them a good income source in future.

Beneficiaries from the Project - Expected Qualitative Outputs

As mentioned above already low income groups will benefit from the project. They are the ones who are already selling the seabuckthorn juice to the local lodge owners and are earning good money as well by selling pulp to HimalAsia. Besides this, they will earn more in future when they have the opportunity to sell more quality products made from Seabuckthorn to a much organized market within and outside Nepal.

Since most of the locals involved are women biodiversity groups, these groups of women will gain from the project. Besides, the Nunnery School in Tsarang, which has 28 little nuns will profit from the project since it will provide an opportunity to earn direct income measures and become self sustainable in the future. For that reason a seabuckthorn nursery has been established at their nunnery. HimalAsia is currently supporting the school - but it is intended that the school will be self sustainable after implementing proper harvesting techniques of the berries in the nursery.

The same is relevant for the young monks (age between 6 -16 years) school in Thini, that is also supported by HimalAsia. Moreover, this school was established after the Foreign Office of Germany financially supported the restoration together with HimalAsia of the Monastery where the monk's school is now located. In front of this Monastery the seabuckthorn nursery has been established to raise income in future for the monks and make them self-sustainable.

Impact of the Project on Disadvantaged Groups, Women and Children

As mentioned above low income groups specially from the so called occupational castes of Mustang (Sarki, Dami, Kami) from Lower Mustang and the Gurung low cast groups from Upper Mustang will benefit from the project. Besides, the Nunnery school project from Tsarang with its 28 nuns and the young monk's school project from Thini will benefit. And also the Amchi family with the gardeners, which also belong to the lower income groups will benefit. Besides, the women biodiversity forest user groups from the different villages in southern and upper Mustang will benefit.

All these groups/people will be empowered to increase their incomes and to have a better live standard in future. They are/will be also in future trained (HimalAsia has already started with that programme) how to conduct the different measures as mentioned above.

The project deems it important, that women are consulted and involved in the decision-making process and has a fair share of resources and benefits from the project. Besides, the current certification processes which is mostly handled by women, in addition to the preparation of juice, tea and Amchi medicine.

All these measures related to quality control, harvesting, processing, and marketing will further strengthen the possibility of producing quality products, diversifying the export of products made from Seabuckthorn, encouraging more people to be engaged in seabuckthorn production process (increasing employment possibilities), and increasing the income level of the deprived group of Mustang.

Reference

Seabuckthorn programme in Mustang, Nepal - Conservation of Bioderversity and Poverty Eradication, In: V.Singh (Ed.): Seabuckthorn a multipurpose wonder plant; Vol III - Advances in Research and Technology; Proceedings of the International Conference on Seabuckthorn in Beijing 2005. Delhi 2007, 519-523.

□□□

21

Storage Studies of Biodiesel and its Physico-chemical Properties Under High Altitude Cold Desert Conditions of Ladakh

SYED BASHARAT MUSTAFA AND SHASHI BALA SINGH
Defence Institute of High Altitude Research, DIHAR, DRDO, Leh-Ladakh, J&K, India

INTRODUCTION

The world is presently confronted with the twin crises of fossil fuel depletion and environmental degradation. Indiscriminate extraction and lavish consumption of fossil fuels have led to reduction in underground-based carbon resources. The search for alternative fuels, which promise a harmonious correlation with sustainable development, energy conservation, efficiency and environmental preservation, has become highly pronounced in the present context.

The fuels of bio-origin can provide a feasible solution to this worldwide petroleum crisis. Gasoline and diesel-driven automobiles are the major sources of various undesirable emissions. Scientists around the world have explored several alternative energy resources, which have the potential to quench the ever-increasing energy thirst of today's population. Various biofuel energy resources explored include biomass, biogas, primary alcohols, vegetable oils, biodiesel, etc. These alternative energy resources are largely environment friendly.

Biodiesel is a fuel composed of mono-alkyl esters of fatty acids derived from vegetable oils or animal fats. Biodiesel is the product obtained when a vegetable oil or animal fat is chemically reacted with an alcohol to produce fatty acid alkyl esters. A catalyst such as sodium or potassium hydroxide is required. Glycerol is produced as a by-product. Biodiesel is designated B_{100}, and meets the requirement of ASTM (American Society for Testing & Materials) D 6751, for use in compression ignition (diesel) engines.

It can be used either pure or in blends with diesel fuel in unmodified diesel engines as it easily mixes with normal diesel. Biodiesel is the pure, or 100 percent, biodiesel fuel and is referred as B_{100} or "neat" biodiesel. Biodiesel blends are referred to as Bxx where xx indicates the amount of biodiesel in the blend (i.e., a B_{20} blend is 20 percent by volume biodiesel and 80 percent by volume petrodiesel).

Thus, biodiesel is an alternative fuel for diesel engines that is receiving great attention world wide. Although it attracts the most attention because it is renewable, and also reduces some exhaust pollutants. It is also attractive because it can be produced easily from common feed stocks.

Production of Biodiesel

Biodiesel fuel can be produced from oilseed plants such as rape seeds, sunflower, canola and *Jatropha curcas*

The Biodiesel Reaction

Oil or Fat		Glycerin (By Product)		Bio diesel
CH_2OCOR'''		CH_2OH		$R'''COOR$
|	Catalyst	|		|
CH_2OCOR'' + 3 ROH	----------->	CH_2OH	+	$R''COOR$
|		|		|
CH_2OCOR'		CH_2OH		$R'COOR$

Environmental Implications of Biodiesel

Biodiesel is environment friendly. The overall reduction in carbon dioxide cycle is about 78%, similarly emission of other harmful gases like hydro carbons, aromatics, sulphur dioxide etc. are far less compared to petrodiesel.

Average B_{100} and B_{20} Emissions (In Percentage) Compared to Normal Diesel

Emission	B_{100}	B_{20}
Carbon monoxide	-48	-12
Total unburned hydrocarbons	-67	-20
Particulate matter	-47	-12
Sulphates	-100	-20
Air toxics	-60 to -90	-12 to -20

Biodiesel as Engine Fuel

- Wear of vital moving parts of 20% biodiesel -operated engines [B_{20}] is about 30% lower compared to the neat diesel-operated engines
- Carbon deposits on cylinder, a piston top & injector tip of biodiesel operated engines is far less than diesel fuelled engines.
- Ash content in the lubricating oil for biodiesel operated engines is less than diesel fuelled engines.
- Biodiesel fuelled engines produce lower amount of wear debris.
- 20% biodiesel B_{20} improves the peak thermal efficiency of the engine by 2.5% and reduces brake specific energy consumption substantially.

For use of biodiesel, no changes are needed in the existing engines or vehicles and hence avoid huge investment in developing new type of engines. Thus a diesel engine can perform satisfactorily on biodiesel blends without any engine hardware modifications. Long term endurance test using biodiesel has proved that biodiesel can be used for substituting mineral diesel in long run.

Storage of Biodiesel

Long term storage of bio-diesel leads to increase in viscosity due to auto-oxidation and polymerization. Adequate data on long-term storage of biodiesel and blends are not available. Based on the experience of producers, biodiesel can be stored up to a maximum of 6 months period. As a mild solvent, biodiesel tends to dissolve sediments normally encountered in old diesel storage tanks. Brass, teflon, lead, tin, copper, zinc etc oxidize biodiesel and create sediments. The existing storage facilities and infrastructure for petrol and diesel can be used for the biodiesel with minor alteration.

Since biodiesel is made using biological based products, the temperature at which it is stored is more critical than in the case of petroleum diesel. The moulds can grow in biodiesel if it is stored at high temperature. On the contrary it will thicken and could become difficult to dispense if stored at very low temperature.

Underground storage is preferred in cold climates but it can be stored in open with proper insulation, heating and other equipments should be installed. Low temperature can cause biodiesel to gel. Additives can be used for low temperature storage and pumping. The biodiesel and its blends should be stored at temperature at least 15^{0}C higher than the pour point of the fuel. While splash blending the saturated compounds can crystallize and separate out to cause plugging of fuel lines and filters. Condensation of water in the tank should be avoided as hydro carbon degrading bacteria and mould can grow and these use biodiesel as food.

Usage of Diesel in High Altitudes of Ladakh

Ladakh is the northern most part of India stretching over 360 Km from North West to east and widest distance 335 Km from Karakoram pass to Rohtang pass is characterized by prevailing adverse eco environment. Having an area of 92,000 sq. Km, it lays approximately between 33°and 35° North latitudes and 76° to 79° East longitude at an altitude of 8,500 to 15,000 ft. This sparsely populated trans-Himalayan land of about 2.35 Lakh people is an icy barren landscape and the world's highest inhabitant plateau. Glistering white snow capped peaks and glaciers dominate in this most inhospitable terrain of its kind in the world.

Ladakh remains cut off from rest of the world during winter months for about six months and hence diesel fuels have to be stored for a long period of time to avoid shortage of fuel during winter. In this region mercury drops to 25 to 30°C below the freezing point during winter. Due to very low temperature, the usage of petroleum diesel in the vehicles leads to many practical problems due to its high compression ratio and also results in numerous eco-environmental hazards (air pollution etc.)

Different types of diesel are used in different defined temperature zones in these altitudes

Diesel Used in High Altitudes

S. No.	Temperature range	Nomenclature (type of fuel)
1	Above O°C	100% DHPP (Diesel High Pour Point)
2	O°C to -18°C	80% ATF (Aircraft Turbine Fuel), non aviation, High Grade Kerosene + 20% DHPP (Diesel High Pour Point)
3	-18°C to -60°C	98% ATF (Aircraft Turbine Fuel) + 2% OM (oil mineral)

Bio-diesel in High Altitudes of Ladakh

Storage Studies

Storage studies on Biodiesel and its blends with winter diesel have been carried out in open environmental conditions in Ladakh in winter months (under subzero temperature) and also under controlled conditions. It is observed that B_{100} stored under 0°C gets frozen, while blends of biodiesel i.e. B_1 to B_{20} (1-20%) doesn't freeze up to -17°C with winter diesel and under controlled conditions, biodiesel blends with winter grade petro-diesel (B_{15} to B_{100})gets frozen when stored at -25°C and below

-17°C

B5 B20 B100 DHPP'W'

B5 B20 B100 DHPP'W'

Open Environmental Conditions

Physicochemical Analysis

Physicochemical analysis of the biodiesel blends with petro diesel (winter diesel) was carried out for cloud point, pour point and kinematic viscosity at -5°C.

Cloud Point

A fuel property that is particularly important for the low temperature operability of diesel fuel is the cloud point. The cloud point is the temperature at which a cloud of wax crystals first appears in a liquid upon cooling. Therefore, it is an index of the lowest temperature of the fuel's utility under certain applications. Operating at temperatures below the cloud point for a diesel fuel can result in fuel filter clogging due to the wax crystals. The cloud point is determined by visually inspecting for a haze in the normally clear fuel, while the fuel is cooled under carefully controlled conditions. The cloud point is an important property for biodiesel since biodiesel fuels typically have higher cloud points, i.e. crystals begin to form at higher temperature, than standard diesel fuel. This feature has implications on the use of biodiesel in cold weather applications.

Cloud Point Apparatus

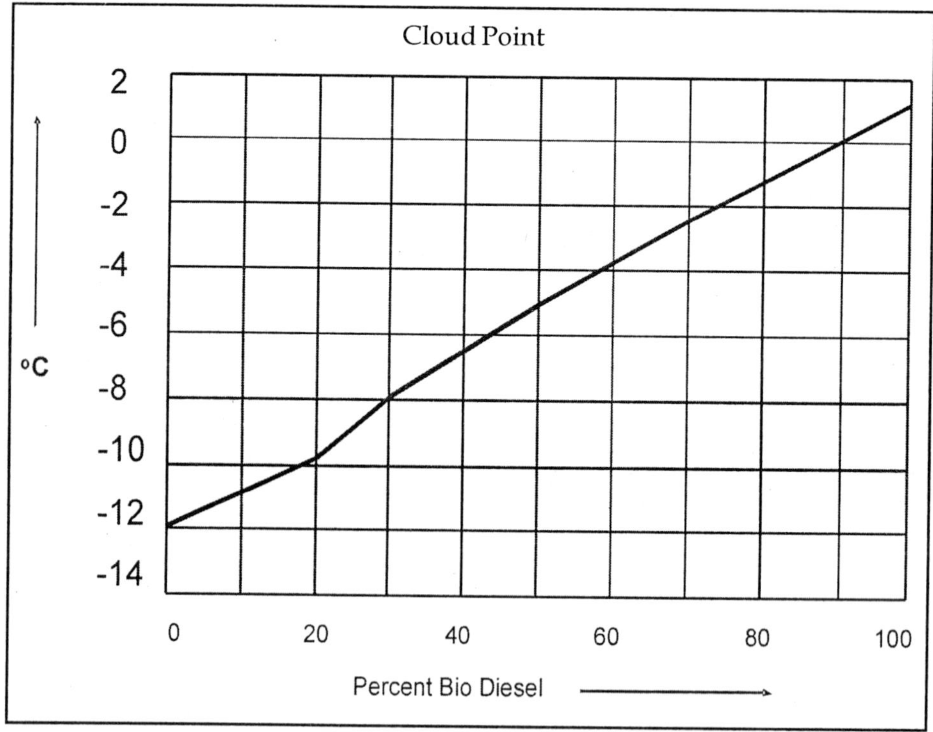

Pour Point

It is a measure of the low temperature performance of diesel/ biodiesel fuels. The pour point is the lowest temperature at which a fuel sample will flow. Therefore, the pour point provides an index of the lowest temperature of the fuel's utility for certain applications.

The pour point also has implications for the handling of fuels during cold temperatures. A summary of the procedure steps is 1) the sample is cooled at a specified rate, 2) the sample is examined at 3°C intervals for flow, and 3) the lowest temperature at which sample movement is observed is noted. The repeatability of the pour point test is <3°C and the reproducibility is <6°C.

POUR POINT APPARATUS

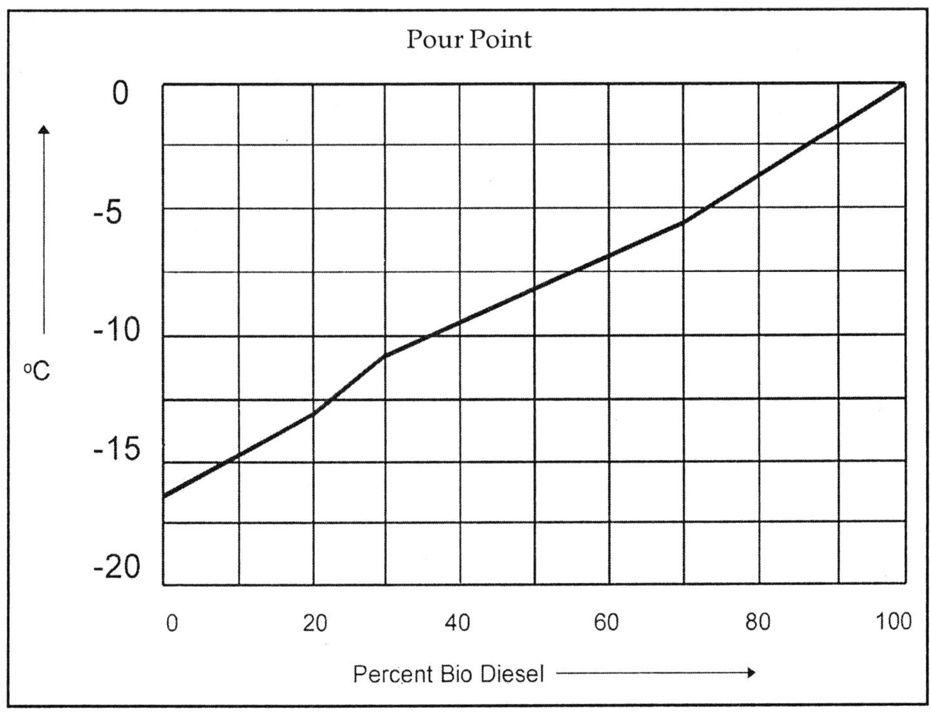

Kinematic Viscosity

Viscosity is the ease with which a fluid will flow. Technically it is the ratio of the shear stress to the shear rate for a fluid. The kinematic viscosity (í) is the resistance to flow of a fluid under gravity.

Therefore, the kinematic viscosity of a fluid is related to the dynamic viscosity through the density (ñ), i.e., í = ç/ñ. The procedure for measuring kinematic viscosity in diesel or biodiesel fuels is "Measured for a fixed volume of liquid to flow under gravity through the capillary of a calibrated viscometer under a reproducible driving head and at a closely controlled and known temperature." The repeatability of the test is within 0.5- 1.0%, while the reproducibility is 1-2%. The test is usually conducted by measuring the time required for a volume of liquid to flow under gravity.

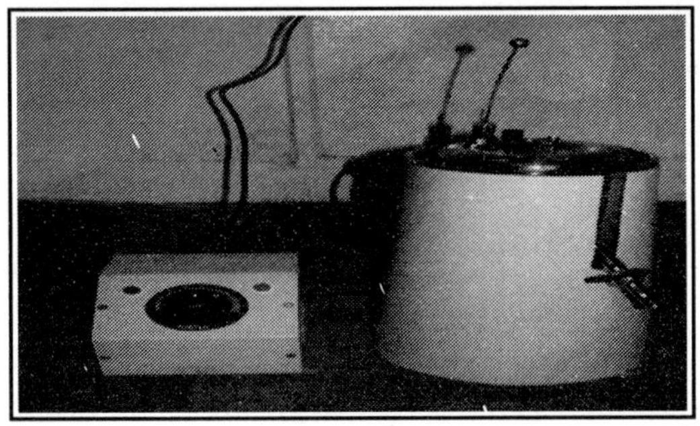

VISCOMETER

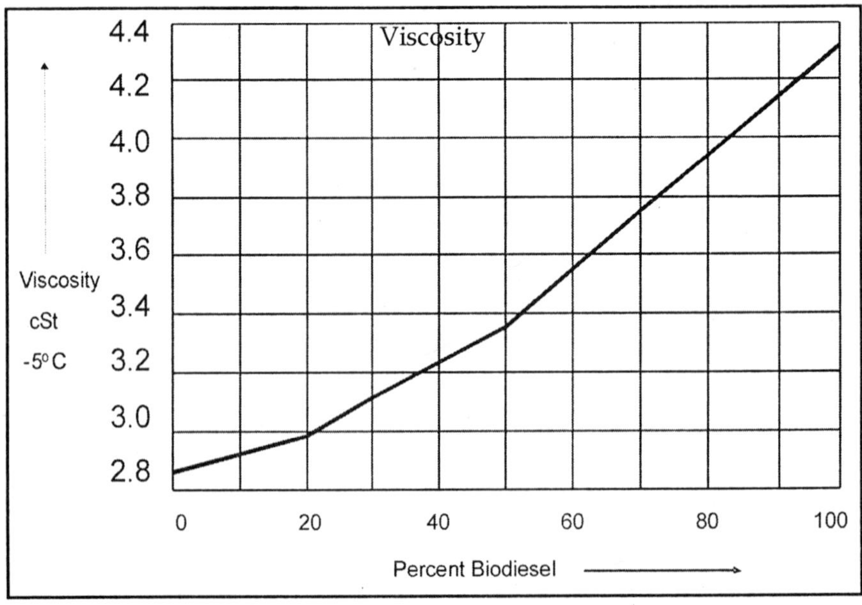

Performance Evaluation

Four vehicles were run with biodiesel blends (with winter diesel) starting from 1% up to 17.5% biodiesel i.e. B_1 to $B_{17.5}$ under subzero temperature in high altitudes of Ladakh. Following observations were recorded:

(a) Knocking and chocking not observed during performance test of biodiesel blends in MT vehicles in Leh in the temperature range of -17°C to 0°C.

(b) The feed pump, atomizer & spray nozzles are functioning smoothly and no obstruction or blockade have been observed in the fuel flow.

INFERENCES/CONCLUSIONS

- Viscosity increases (almost varies linearly) as the concentration of biodiesel increases.
- Cloud & pour point vary mostly in a linear fashion with the amount of biodiesel blended with petroleum diesel fuel (winter diesel).
- Proved the feasibility of biodiesel utilization in vehicles during winter up to -17° C.

❑❑❑

22

Biotechnological Approaches for Enhancing Crop Productivity and Farmers' Income

K.R. KOUNDAL
Indian Agricultural Research Institute, New Delhi-110012, India.

Classical Plant Breeding vs. Plant Biotechnology

Agricultural production in India got a major boost with introduction of dwarf varieties of wheat and rice in the mid-sixties. The introduction of these varieties led to a dramatic increase in the yields of the two crops. However, now the yields of rice and wheat have been stagnating. The productive agricultural areas in the North, due to continuous rice-wheat cultivation are encountering serious problems of water depletion, deficiency of micronutrients in the soil and increase in the use of pesticides, fungicides and herbicides to control pests, pathogens and weeds. In India, we are faced with twin problem of increasing population and decreasing available land for agriculture. India's long term economic prospectus depends heavily on the agricultural sector which contributes a quarter of the GDP and provide livelihood to two-thirds of the population. To achieve the desired growth rate it may not be wise to depend entirely on conventional methods of developing and transferring the technologies.

Agricultural production is becoming more and more dependent on agrochemicals, thereby increasing input costs and causing significant damage to the environment and human health. Both farmers and consumers are at the receiving end, farmers by exposure to agrochemicals and consumers due to residues of agrochemicals in the consumed food. While there is self-sufficiency in cereal grains at present, the yields and productivity of dry land crops, mostly grain legumes and oil seeds remain low and no major breakthroughs in productivity enhancement and yield stabilization have been achieved. India's population is expected to reach approximately 1.5 billion by 2050. It is estimated that around 300 million (roughly 30 percent) of India's population suffers from malnutrition. Due to socio-economic reasons, women and children are more at the receiving end in terms of micronutrient deficiency and low calorie intake. Nutritional security for everyone would require more extensive availability of grain legumes, edible oils, fruits and vegetables, milk and poultry products. The challenges of malnutrition, enhanced productivity and crop diversification can be met by better resource management and by breeding more productive, more nutritious and at the same time less resource input demanding crops.

Conventional technologies of agriculture are inadequate to meet the formidable challenges listed above. Some of the limitations of conventional breeding are: i) lack of germplasm resources for some of the major pests and pathogens of crops, ii) the new plant types evolved for higher productivity are more vulnerable to pests and diseases, iii) problems in sourcing genes from wild relatives, iv) lack of nutritional qualities in major cereals crops, v) the methodology of plant breeding is based on phenotypic selection and vi) plant-environment interactions affect the selection process. Advances in modern biology, especially biotechnology, offer many advantages over traditional techniques of plant breeding. Biotechnology offers possible solutions to previously intractable problems and difficult targets such as drought tolerance. The ability of scientists to tailor make genes have provided unlimited opportunities specially for crop and livestock improvement and have given hope for meeting the challenge of combating protein and energy malnutrition. The vast opportunities thrown open as a result of new technology are immense.

Application of Biotechnology for Crop Improvement

The genetic improvement of crop plants using classical plant breeding involves the introduction of required characteristics into plant by genetic crossing during sexual reproduction. However, only two sexually compatible plant species can be crossed. Recent advances in recombinant DNA methodology have given birth to a new technology called genetic engineering, which is allowing plant breeders to produce plants with new gene combinations. This has at least two major advantages (i) there is no crossability barrier between donor and the recipient and (ii) there is no co-transfer of unwanted characters thereby reducing the time required to develop new improved varieties. GM crops are genetically modified by manipulating the plant's own genes and by introducing genes from unrelated organisms for developing resistance to biotic and abiotic stresses.

Genetically modified (GM) crops could produce higher yields in countries where soil conditions are poor and food is in shortage. Crop yields are reduced to considerably low levels because of biotic stresses (e.g. pathogens and insect-pests attack) and post harvest losses. Similarly drought, salinity and temperature are major abiotic stresses affecting crop productivity. There is a great potential for transgenic technology applications to counter biotic and abiotic stresses and to improve the nutritional quality of foods. Concerted efforts must be undertaken to isolate genes and promoters from indigenous crop plants. It is important to isolate the genes and promoters which can be expressed in plants for strengthening the development of transgenic crops resistant to biotic and abiotic stresses and post harvest losses. It is important to remember that biotechnology tools complement and extend the traditional methods to enhance agricultural productivity.

Several genetically modified crops have been released for commercial cultivation in many developed countries. Some of the examples include: improved resistance to herbicide, improved resistance to insect-pests, improved post-harvest characteristics, improved oil quality, manipulation of seed storage proteins, improved iron and vitamin A content in rice. The major genetically modified crops which are commercially grown include insect resistant cotton, potato and corn, herbicide resistant soybean, canola and cotton. The multiple benefits of transgenic crops include flexibility in terms of crop management, decreased dependency on insecticides and herbicides, cleaner and high quality end products. Genetic engineering technology combined with tissue culture approaches should be used to increase the production of crops and provide access to food for all.

In recent years, development of transgenic crops expressing a variety of novel traits such as insect resistance, disease resistance, herbicide tolerance, hybrid production, improved oil quality etc have led to large scale cultivation of GM crops which currently occupy 114.3 million hectares (282.4 million acres) on a global scale. It is noteworthy that in 2007 India, registered the highest proportion of increase in Bt cotton area to 6.2 million hectares. Recently, GM cotton has been released for commercial cultivation which is resistant to cotton bollworm by the public sector. In 2007, the US, followed by Argentina, Brazil, Canada, India, and China continued to be the principal adopters of biotech crops globally, with 57.7 million hectares planted in the US (50 percent of global biotech area) of which approximately 37 percent are stacked with two or three traits. The stacked GM crops, currently deployed in the US, Canada, Australia, Mexico, South Africa and the Philippines are important and growing future trend, which meets the multiple constraints of farmers. GM soybean continued to be the principal biotech crop in 2007, occupying 58.6 million hectares (51 percent of global biotech area), followed by maize (35.2 million hectares at 31 percent), cotton (15 million hectares at 13 percent) and canola (5.5 million hectares at 5 percent of global biotech crop area). The increasing collective impact of the five principal developing countries is an important continuing trend with implication for the future adoption and acceptance of biotech crops worldwide. Each of the five countries have benefited in a different way from biotech crops.

Substantial benefits have been reaped by the farmers worldwide by cultivating GM crops. In the near future we may expect many more GM crops, which have been modified for better availability of vitamins, iron, micronutrients, quality proteins and oils. These developments will be supported by programs aimed at unraveling the genome sequences of important crops such as maize, cotton and tomato. Great opportunities thus exist for Indian scientists to do prospecting for novel genes in a variety of biological systems and establish intellectual property rights on their discoveries. Several scientific groups have utilized the transgenic technology to answer basic questions in plant molecular biology and to develop transgenic crops, which now are in the advanced stages.

In the global arena, the genomics of rice has already provided a wealth of information. Functional genomics seeks to analyze the cellular and organismal functions of the genes and the manner in which gene products interact to control those functions. There could be myriad positive implications of genomics with respect to food, nutrition and environmental security of the nation. Therefore, the five major goals given below deserve consideration that there is high probability that crop biotechnology can deliver these promises in the near future.

(i) Increasing global crop productivity to improve food, feed and fiber security in sustainable crop production systems that also conserve biodiversity.

(ii) Contributing to the alleviation of poverty and hunger.

(iii) Reducing the environmental footprint of agriculture.

(iv) Mitigating climate change and reducing greenhouse gases (GHG).

(v) Contributing to the cost-effective production of biofuels.

However, when introducing a new gene into the food chain, there is a need to adopt appropriate safeguards to protect human and animal health. There are several areas of concern regarding GM crops. These include: (i) toxicity (ii) allegenicity (iii) carcinogenicity (iv) food intolerances (v) use of antibiotic markers and (vi) nutritional value. Acceptability or non-acceptability is established by determining whether a novel food is substantially equivalent to an endogenous conventional food in terms of nutritional properties, toxin and allergen content, the type of processing (industrial or domestic) that the food might undergo, and consumption by human beings.

In the U.S. regulatory system involves the public in the decision making process. Under this regulatory framework, the U.S. Department of Agriculture's (USDA), Animal and Plant Health Inspection Service (APHIS) ensures that new biotech or transgenic plant varieties are safe to use in agriculture as conventional varieties. The U.S. Food and Drug Administration (FDA) consults with the developers of transgenic plants to ensure that new crops and foods produced from these are safe to consume as conventional foods. The Environmental Protection Agency (EPA) ensures the safe use in the environment of new plant pesticide substances (such as Bt gene in corn or cotton) and new user of herbicides in conjunction with transgenic plants. Each agency involved respect and value the role and independence of others.

Similarly, the Indian regulatory agencies have established appropriate policies, laws regulations and enforcement mechanisms for genetically modified crops/foods whether for testing, export and import or release on the commercial scale. Laws, regulations and guidelines on research, release containment and monitoring of transgenic plants, and food safety assessment are developed by the Department of Biotechnology, Government of India, in consultation with other concerned organizations. Besides these, we need to build scientific expertise, infrastructure and resources for adequate assessment of the possible risks during commercialization of genetically modified crops. We must also discuss appropriate policies for proper labeling of food products. This essentially provides consumers with a choice either to either use or not use GM foods. The EU and Japan are in favor of including GMO related information on products labels on a mandatory basis. The EU has introduced the labeling rule in 1997 and also has aggressively disputed the consensus in the international scientific community that the new technology is no more than an extension or refinement of earlier genetic techniques and the associated risks are basically the same as for other products.

The challenge for the international community and the leading biotech crop developing countries of India, China, Argentina, Brazil and South Africa, which have already benefited from biotech crops, is to openly share their experience and knowledge with the legion of developing countries that have yet to have first-hand experience with biotech crops. There is no substitute for sharing the collective experience of a "national team of practitioners" who have been involved in a successful national crop biotech program such as Bt cotton in India or China, or biotech maize in South Africa or the Philippines. The national team of practitioners includes agronomists, biotechnologists, economists and farmers who have been directly engaged with all aspects of biotech crops. Both the pros and the cons must be frankly shared so that there is no need for newcomers to the technology to reinvent the wheel. One key question that must be asked to the team sharing the experience is what have been the lessons and learning of the first generation of biotech crop adopters that can be shared with second generation adopters so that the latter can gain from the experience.

The most important constraint to the development of GM crops in most developing countries that deserves highlighting is the lack of appropriate cost-effective and responsible regulatory systems that incorporate all the lessons of a dozen years of regulation. Current regulatory systems in most developing countries are usually unnecessarily cumbersome, and in many cases it is difficult to implement the system to approve products. The current regulatory systems were designed more than ten years ago to meet the initial needs of industrial countries dealing with a new technology and with access to significant resources for regulation. With the accumulated knowledge of the last few years, it is now possible to design appropriate regulatory systems that are responsible, rigorous and yet not onerous, requiring only modest resources that are within the means of most developing countries - this should be assigned top priority. Today, unnecessary and unjustified stringent standards designed to meet the needs of resource-rich industrial countries are denying the developing countries timely access to products such as golden rice.

Clear and understandable consumer information is a very important part of the public acceptance process. Besides media, research organizations and scientific institutions concerned with crop improvement must also take up the responsibility for promoting awareness among the public about the applications of genetic engineering in agriculture and their potential benefits as well as risks and constraints. Media serve as a bridge between science and society and should be enabled to play an important role in fostering the technological upgrading of Indian agriculture, thereby enhancing our agricultural competitiveness.

An effective communication strategy must be developed and a cohesive mechanism established to ensure that messages are consistent with national policies on agricultural biotechnology and also that all target groups are reached. Education and development communications must receive high priority. Field research may be required to ensure that the concerns of various groups of population are understood and addressed to see that the messages are evidence based, simple and effective. Transparency in communication should be the bottom line. The issues in regard to release of GM crops are not understood correctly owing to the lack of information on this subject even amongst the otherwise well-informed members of the public. The golden triangle of partnership between various Research Institutes and Universities; ICAR and the SAUs; and the private sector can help in bringing the transgenic crops from the laboratory to the field at a much faster pace. Steps will have to be taken so that there is a swing from 'mistrust' to 'trust' between the public and private sectors, especially in terms of IPR (spell out) related issues and benefit sharing. There are a number of intellectual property rights/patent issues related to plant regeneration, plant transformation, gene sequences, promoters, vectors and screening techniques which need to be kept in mind while commercializing transgenic crops. There is a need to develop appropriate mechanisms and approaches to overcome the intellectual property barriers for the successful commercialization of transgenic crops developed so that farmers are able to reap the benefits of biotechnology. An information campaign needs to be conducted to generate public awareness on the benefits and risks associated with biotechnology and the social, ethical, economic, scientific, environmental and health issues which are addressed by regulatory bodies.

CONCLUSIONS

It is essential that we improve food production and distribution. GM crops could produce higher yields in countries where soil conditions are poor and the population is desperate for food. Countries like India with 70 percent rural base cannot ignore agricultural biotechnology. It is important that we enhance farm productivity per unit land, water and capital without harming the ecosystem. Crop yields are reduced to considerably low levels because of biotic (e.g. pathogens and insect pests attack) and abiotic (drought, salinity, temperature) stresses affecting crop productivity. Major network projects should be initiated to usher in transgenic technology to combat biotic and abiotic stresses and to improve the nutritive quality of foods.

Furthermore, it is vital that the public sector and the private sector forge partnerships that will allow equal advantages to both parties in achieving the objectives of global food security. The government must ensure continued safe and effective testing and introduction of transgenic crops and implement regulatory programs that inspire public confidence. The public should be well informed and engaged in a dialog about the impact of the technology on environment, food safety, sustainability and food security. Concerted efforts must be undertaken to investigate human health and environmental effects, both positive and negative, of GM technologies. It is now possible to design appropriate regulatory systems that are responsible and requiring only modest resources that are within the means of most developing countries - this should be assigned top priority. A rational regime for biotechnology is imperative; otherwise, society might either face uncontrolled risks or loose out on major benefits which may improve and enrich peoples' lives. With biotechnology, particularly GM crops, the potential to achieve food security and alleviate hunger in the third world is huge. We must ensure that society will continue to benefit from the vital contribution that plant breeding offers, using both conventional and biotechnology tools. This approach, I believe will be the most cost effective, environmentally safe and sustainable way to ensure global food security in the future.

References

Brar, D.S., Ohtani, T. and Uchimiya, H. 1995. Biotechnology and genetic engineering reviews. 13, 168–179.

Cook, R.J. 2000. Agricultural Biotechnology and the Poor. *In* Proceedings of an international conference, Washington DC, pp 123–130. GI Persely & M.M. Lantia (Eds).

Corner, T. 1993. Food safety issues relating to genetic engineering of crop plants. Agriculture Science. 6, 36–41.

Gelvin, S.B. 1998. The introduction and expression of transgenes in plants. Current Opinion in Biotechnology. 9, 227–231.

Grover, Anil. and Pental, Deepak. 2003. Breeding objectives and requirements for producing transgenics for major field crops of India. Current Science. 84, 310-320.

James, C. 2007. Global review of commercialized transgenic crops: 2007. ISAAA Briefs. No. 37.

Kahl. G. and Writer, P. 1995. Plant genetic engineering for crop improvement. World Journal of Microbiology & Biotechnology. 11, 449–460.

Koundal, K..R. and Rajendran, P. 2003. Plant insecticidal proteins and their potential for development transgenics resistant to insect pest. Indian Journal of Biotechnology. 2, 110-120.

Koundal, K.R. 2001. Transgenic for the betterment of mankind. The Botanica. 51, 46–51.

Koundal, K.R., Dash, P.K. and Kansal, Rekha. 2003. Plant defense proteins: Mechanism and potential for pest control through genetic manipulation. J. Plant Biol. 30(2), 2111-2227.

Koundal, K.R. and Rajendran, P. 2003. Plant insecticidal proteins and their potential for developing transgenics resistant to insect-pests. Indian Journal of Biotechnology. 2, 110–120.

Koundal, K.R., Santha, I.M., Mehta, S.L. and Malik, V.S. 1989. Plant genetic engineering prospective of improving crop productivity. *In* Recent Advances in Plant Biochemistry. ICAR Publication, New Delhi.

Lacy, P. 2003. Developing the full Arsenal: Fighting hunger with biotechnology. AIS. Review. 23, No. 1 Winter Spring.

Qaim, M. and Zilberman, D. 2003. Yield effects of genetically modified crops in developing countries. Science. 299, 900–902.

Revised guidelines for research in transgenic plants and guidelines for toxicity and altergenicity evaluation of transgenic seeds plants and plant parts. DBT, DST, Govt. of India, August 1998.

Vain, P. 2007. Thirty years of plant transformation technology developments. Plant Biotechnol. J. 5(2), 221.

□□□

23

Molecular / Biotechnological Approaches for Sustainable Agriculture

PARAMJIT KHURANA
Department of Plant Molecular Biology, University of Delhi
South Campus, New Delhi, India

INTRODUCTION

Forty percent of the world's land used for agriculture is seriously degraded and to meet the nutritional needs of the growing population, cereal production in general and wheat production in particular will need to be increased by 40 percent in the next 20 years. Wheat is one of the world's most important food grain which provides more than 20 percent of the calories and protein in human diet, and the staple food for over 40 countries, accounting for 35 percent of the world's population. Increase in wheat productivity has been limited by using conventional breeding methods. Biotechnological tools for sustainable food supply are thus needed to meet these requirements in the near future.

Wheat has a huge diversity of dough quality and other visco-elastic characteristics, which are essential for its use in a range of food products. One of the major limitations in conventional breeding involves barriers for gene transfer due to incompatibility and species differences. Progress in biotechnology has resulted methods to insert desired foreign gene(s) to overcome problems of sexual incompatibility and species barriers between organisms. This technology helps breeders and molecular biologists to introduce only the gene of interest with more selective modification and represents a significant advance. In addition transgenic plants have many other advantages, *viz.*, the potential to resist biotic and abiotic stresses, added nutritional quality, more herbicide and pesticide resistance. Recent genetic engineering technology allows not only transfer of a single gene, but a couple of genes in a much more precise, controllable and predictable way than is achievable with conventional breeding. Recently, genomic efforts are supplanting efforts at improving the major crop plants by integrating genetic engineering and molecular breeding efforts. With this perspective, this article summarizes our efforts of the last decade at improving wheat for higher yield and a better value-added crop of global importance.

Wheat, a major food crop has remained a challenge for biotechnologists due to being recalcitrant to *in vitro* propagation. The key factor in the successful generation of transgenic plants is the optimization of conditions for regeneration of plantlets after several rounds of selection. Recently, Chauhan *et al, (2007)* developed an efficient genotype independent, in vitro regeneration system for nine popular Indian wheat cultivars of *Triticum aestivum, Triticum durum* and *Triticum dicoccum.* This protocol can be used for raising transgenic plants for functional genomics and analysis of agronomically important traits in the three species of wheat.

Monocotyledonous plants, including important cereals were earlier thought to be recalcitrant to *Agrobacterium*-mediated gene transfer but this has changed in the last few years. Tissue cultures of wheat have generally been raised from embryogenic explants such as immature embryos, scutella, their calli, and are regularly employed as the target tissues for various transformation strategies due to their high regeneration efficiency. However, tissues of non-embryogenic origin such as leaf or basal segments have been rarely employed for transformation studies owing to their low regeneration response. Our work reveals the unusually high morphogenic potential of the basal leaf segment of various wheat cultivars. Of particular interest is its extreme sensitivity to short-term treatments of 2, 4-D for rapid induction of somatic embryogenesis. Wheat leaf base segments treated with 2, 4-D for 24 h developed somatic embryos without an intervening callus when cultured on basal medium for less than 2 weeks. This response is auxin specific and genotype independent. The rapid induction of somatic embryogenesis in wheat leaf bases under easily manipulable culture conditions provides a unique opportunity to monitor the molecular events associated with auxin-mediated induction of somatic embryogenesis and also to study the associated developmental events.

Traditionally, immature embryo derived-calli have been found efficient for *in vitro* regeneration in wheat, but more recently, mature embryo derived tissues such as callus and leaf bases have also shown promise for *in vitro* propagation and transformation studies. Regeneration from mature and immature embryos of bread and emmer wheat was standardized and transgenics from emmer wheat were successfully raised for the first time through *Agrobacterium*-mediated transformation. An efficient and reproducible method of *Agrobacterium*-mediated transformation of mature embryos of hexaploid bread wheat (*Triticum aestivum*) and tetraploid pasta wheat (*Triticum durum*) was reported subsequently. Optimization of this method facilitated the introduction of *bar* gene as a selectable marker conferring herbicide resistance as well as potato proteinase inhibitor gene (*pin*2) for insect resistance in wheat.

Somatic embryogenesis forms the basis of cellular totipotency in higher plants wherein the developmental restructuring of somatic cells towards the embryonic pathway occurs. It involves a broad repertoire of gene expression, regulated by various factors which are either independent or interdependent. Somatic embryogenesis is the chief mode of differentiation in cereal tissue cultures and wheat leaf base treated with 2, 4-D (10 μM) for one day and subsequently transferred to basal medium for the initiation of somatic embryogenesis and plant regeneration offers a unique system for understanding this phenomenon. The induction of somatic embryogenesis is a multifactorial event, but induction by plant hormone auxin is generally considered to be the most important factor. The cellular responses to auxin involve changes in gene regulation and stimulation of the transcription of numerous genes. The *Aux/IAA* genes are best characterized among early auxin-responsive genes. By screening the available databases, at least 15 ESTs have been identified from wheat which exhibit high sequence identity with *Aux/IAA* homologues in other species. One of these *Aux/IAA* genes, *TaIAA1* is light sensitive, tissue specific, and is induced within 15-30 min of exogenous auxin application. In addition to auxin-induced gene, the *TaIAA1* gene is induced by brassinosteriod and also it's transcript levels increase in the presence of divalent cation Ca^{2+}.

Three *Somatic Embryogenesis Receptor Kinase (TaSERK)* genes belonging to a small gene family of receptor-like kinase and involved in somatic embryogenesis were isolated and characterized from wheat. It was shown that somatic embryogenesis is associated with high *SERK* expression via a calcium-mediated pathway. *SERK* expression could be enhanced with application of auxin, epibrassinolide and abscissic acid suggesting a crosstalk between auxin and BR signaling. Identification and comprehensive analysis of *SERK*-like (*SERL*) gene family from rice *(Oryza sativa)* as well as the significance of SPP motif and the adjoining region in *SERK* and *SERL* proteins was performed. This work involved the identification of *SERL* gene family from rice, analysis of their structure, duplication, chromosomal distribution and phylogenetic relationship. The real-time polymerase chain reaction (RT-PCR) analysis demonstrated that *SERK* and *SERL* genes are expressed differentially in developing seed, various organs/tissues, under light or dark, and are regulated by auxin/brassinosteroid treatment in rice, indicating that they may perform specific as well as redundant functions in plant development.

In order to unravel the molecular basis of somatic embryogenesis, cDNA library was made from the regeneration proficient wheat leaf base segments treated with auxin. In total, 1440 clones were sequenced. By reverse northern analysis, a total of 57 clones were found to be upregulated during somatic embryogenesis. A substantial number of genes involved in hormone response, signal transduction cascades, defense, anti-oxidation, programmed cell death/senescence and cell division were identified and characterized partially. Analysis of data of select genes suggests that the induction phase of somatic embryogenesis is accompanied by the expression of genes that may also be involved in zygotic embryogenesis. The developmental reprogramming process may in fact involve multiple cellular pathways and unfolding of as yet unknown molecular events. Thus, an interaction network draft using bioinformatics and systems biology strategy was constructed. This work is the outcome of a systematic and comprehensive analysis of somatic embryogenesis associated interactome in a monocot leaf base system.

Transgenics for Stress Tolerance

Nematodes are the principal animal parasites of plants causing global losses to agriculture estimated at $100 billion each year. The major plant parasitic nematode infesting Indian wheat is the cereal cyst nematode (CCN) *Heterodera avenae*. In India, *Heterodera avenae* causes severe damage to cereal crops such as barley and wheat in Rajasthan resulting in heavy economic losses. CCN infestation in wheat root results in decline of root and shoot growth and also a concomitant reduction in transpiration. To some extent, crop rotation and nematicides can reduce the soil-borne nematode population. However, both for economic as well as for environmental reasons, the best way to avoid crop damage is to use cultivars resistant to the cereal cyst nematode. It is thus imperative to impart resistance to a broad range of plant parasitic nematodes. Methodologies aimed towards the development of transgenic plants with improved nematode resistance encompass strategies such as anti-invasion and migration, feeding-cell attenuation and anti-nematode feeding.

The plant proteinase inhibitors (PI's) bind with proteinases present in the nematode gut and thereby hinderthe nematode's proteolytic activity and impede the development of feeding pests. Plant PI's are generally categorized into four types based on the active amino acid in the active center, i.e., cysteine, serine, metallo- and aspartyl proteases. A potato proteinase inhibitor, PIN2 (a serine protease), was introduced into wheat, *T. durum* PDW215, *via Agrobacterium*-mediated mature embryo co-cultivation approach. The expression of the *PIN2* was regulated by its own promoter and had the *Act1* gene intron present within for enhanced expression levels of the promoter, together with the 3′-terminator sequence. Greenhouse trials of selective T_1 transformants have been carried out by inoculating the progeny with the nematode cysts at germination stage. The transgenic lines showed enhanced resistance, vigor, yield and proteinase inhibitor activity as compared to the control. Additionally, nematode fecundity, i.e. the number of eggs and larvae produced, in the transformants was lower as compared to a resistant *T. aestivum* var. Raj MR. Current efforts are focused on understanding the molecular interactions between plants and the nematodes which would further open new avenues for environment friendly nematode control.

Drought is another major constraint reducing crop yields worldwide. To obtain tolerance against abiotic stress, introduction of the ABA-responsive barley gene, HVA1, a member of group 3 late embryogenesis abundant (LEA) protein gene into *Triticum aestivum* has been initiated to overcome drought stress in commercial cultivars of Indian wheat through transgenic technology. This is being combined with double haploid production techniques for overexpression of the *HVA1* gene. Preliminary results indicate that gene insertion is stable and no gene silencing has occurred in any of the double haploid lines.

Wheat Genomics

Grass genomes are labile, rapidly evolving entities and their structural and functional relationships are complex. Rice sequence based maps are used for locating genes of interest in wheat; however it requires extensive cross-species mapping, sequencing, and analysis at the BAC level. Gene-by-gene comparison of the sequence homology of 2,932 predicted rice genes from the long arm of chromosome 11 was done with wheat ESTs and their location was specified on wheat chromosome bins. This comparative analysis showed significant conservation of individual gene sequences between rice and wheat. Furthermore, it was shown that rice chromosome 11 and wheat chromosomes of homoeologous group 4 share a common origin. While there exists a definite conservation of gene sequences, there is no obvious conservation of the gene order. Lack of extensive colinearity between rice and wheat genomes suggests that there have been many insertions, deletions, duplications and translocations that make synteny comparisons much more complicated than earlier thought.

Future Endeavors

In the last decade preventive medicine has undergone a great advance. Research has demonstrated that nutrition plays a crucial role in the prevention of chronic diseases. In recent years cereals and its ingredients have been considered as functional foods and nutraceuticals because of providing dietary fibre, proteins, energy, minerals, vitamins and antioxidants required for human health. Wheat is one of the most important cereal crops in the world and with the advent of transgene technology and practical amenability of cereals it should be possible to introduce advantageous traits in different wheats to augment their nutritive value and consumer acceptance.

Carotenoids (vitamin A) and tocochromanols (vitamin E) are the two major groups of lipid-soluble antioxidants in photosynthetic tissues. In our lab, we are trying to develop genetically enhanced wheat varieties having high carotenoid and tocochromanol content. Different genes of carotenoid and tocochromanol biosynthetic pathway are isolated and over-expressed under the control of constitutive and tissue specific promoter, in hexaploid and tetraploid wheat varieties.

In India, wheat is grown under sub-tropical environment during mild winters, which warms up towards grain filling stages of the crop. Of the 25 million hectare under wheat in India, heat stress at grain filling stage is of particular relevance to the 4.5 m ha late sown wheat in North West Plain Zone. Heat stress induces decrease of the duration of developmental phases leading to fewer and smaller organs, reduced light processes related to carbon assimilation. Tolerance to heat stress is a complex phenomenon and no single trait fully explains why some wheat varieties are able to yield better even under heat stress environments. To increase our understanding of the physiological and molecular basis of plant stress tolerance we have started unraveling the transcriptome of high temperature stress in wheat at different developmental stages such as seedling, unfertilized flower, and developing grain. Identification of genes and alleles, understanding their expression and regulation under heat stress would aid in molecular breeding programs to produce superior wheat cultivars.

References

Chauhan, H., Desai, S.A. and Khurana, P. 2007. Comparative analysis of the differential regeneration response of various genotypes of *Triticum aestivum*, *Triticum durum* and *Triticum dicoccum*. Plant Cell Tiss. Org. Cult. *91*, 191-199.

Chugh, A. and Khurana, P. 2002. Gene expression during somatic embryogenesis-recent advances. Curr. Sci. 83, 715-730.

Chugh, A. and Khurana, P. 2003a. Regeneration via somatic embryogenesis from leaf basal segments and genetic transformation of bread and emmer wheat by particle bombardment. Plant Cell Tiss. Org. Cult. 74, 151-161.

Chugh, A. and Khurana, P. 2003b. Herbicide-resistant transgenics of bread wheat (*T. aestivum*) and emmer wheat (*T. dicoccum*) by particle bombardment and *Agrobacterium*-mediated approaches. Curr. Sci. 84, 78-83.

Khurana, J.P., Chugh, A. and Khurana, P. 2002. Regeneration from mature and immature embryos and transient gene expression *via Agrobacterium*-mediated transformation in emmer wheat (*Triticum dicoccum* Schuble). Indian J. Exp. Biol. 40, 1295-1303.

Khurana, P., Chauhan, H. and Desai, S.A. 2007. Wheat, In : Kole C and Hall TC (eds) A compendium of transgenic crops. Willey Blackwell (In Press).

Mahalakshmi, A. and Khurana, P. 1995. *Agrobacterium*-mediated gene delivery in various tissues and genotypes of wheat (*Triticum aestivum* L.). J. Plant Biochem. Biotech. 4, 55-59.

Mahalakshmi, A. and Khurana, P. 1997. *Agrobacterium*-mediated cereal transformation: A critical appraisal. Ind. J. Exp. Biol. 35, 416-426.

Mahalakshmi, A., Khurana, J.P., Khurana, P. 2003. Rapid induction of somatic embryogenesis by 2,4-D in leaf base cultures of wheat (*Triticum aestivum* L.). *Plant Biotech.* 20, 267-273.

Patnaik, D. and Khurana, P. 2001. Wheat Biotechnology: a minireview. Elect. J. Biotechnol. 4, 1-29.

Patnaik, D. and Khurana, P. 2003. Genetic transformation of Indian bread (*T. aestivum*) and Pasta (*T. durum*) wheat by particle bombardment of mature embryo-derived calli. BMC Plant Biology. 3, 5.

Patnaik, D., Vishnudasan, D. and Khurana, P. 2006). *Agrobacterium*–mediated transformation of mature embryos of *Triticum aestivum* and *Triticum durum.* Curr. Sci. 91, 307-317.

Singh, N., Raghuvanshi, S., Srivastava, S.K., Gaur, A., Pal, A.K., Dalal, V., Singh, A., Ghazi, I.A., Bhargav, A., Yadav, M., Dixit, A., Batra, K., Gaikwad, K., Sharma, T.R., Mohanty, A., Bharti, A.K., Kapur, A., Gupta, V., Kumar, D., Vij, S., Vydianathan, R., Khurana, P., Sharma, S., McCombie, W.R., Messing, J., Wing, R., SasakI, T., Khurana, P., Mohapatra, T., Khurana, J.P. and Tyagi, A.K. 2004. Sequence analysis of the long arm of rice chromosome 11 for rice-wheat synteny. Funct. Integr. Genomics. 4, 102-117.

Singla, B., Chugh, A., Khurana, J. and Khurana, P. 2006. An early auxin-responsive *Aux/IAA* gene from wheat (*Triticum aestivum*) is induced by epibrassinolide and differentially regulated by light and calcium. J. Expt. Bot. 57, 4059-4070.

Singla, B., Khurana, J.P. and Khurana, P. 2007b. Characterization of the SERK genes from wheat (*Triticum aestivum)* and their possible involvement in somatic embryogenesis. *PCR* (In Press).

Singla, B., Tyagi, A.K., Khurana, J. P. and Khurana, P. 2007a. Analysis of expression profile of selected genes expressed during auxin-induced somatic embryogenesis in leaf base system of wheat (*Triticum aestivum)* and their possible interactions. Plant Mol. Biol. 65, 677-692.

Vishnudasan, D. and Khurana, P. 2005. New paradigms towards appraising parasitic nematode infestation with special reference on cereal cyst nematode (*H. avenae*). Physiol. Mol. Biol. Plants. 11, 1-18.

Vishnudasan, D., Tripathy, M.N., Rao, U. and Khurana, P. 2005. Assessment of nematode resistanc in wheat transgenics expressing potato proteinase inhibitor (*PIN2*) gene. Trans. Res. 14, 665-675.

❑❑❑

24

Unraveling of Rice-Wheat Synteny Using Single Copy Genes

N. K. SINGH
National Research Centre on Plant Biotechnology, Indian Agricultural Research Institute, New Delhi, India

INTRODUCTION

High throughput decoding of genomes has added a new dimension to our understanding of evolution in plant species. But so far only a limited number of plant genomes, e.g. *Arabidopsis* and rice have been sequenced to completion. Comparisons with other plant species are based largely on the available genetic and EST information. Such comparative studies are important for the practical utilization of the knowledge of genomics. Large genomes appear to have evolved through extensive duplications and divergence of genes and non-coding repetitive sequences. Duplications are present in all genomes as tandem arrays of individual genes and chromosome segmental duplications resulting from aneuploidy and polyploidy. Transposition of DNA segments to multiple locations in the genome is also a common mode of duplication. Duplications and rearrangements pose a real challenge in the analysis of gene synteny and colinearity, which is required for the transfer of genome information from model species to other economically important crop species. Except for a well-established colinearity among the closely related members of tribe Triticeae (wheat, barley and rye), previous studies have shown definite but limited conservation of gene order among rice, maize and wheat. Results of our study is reviewed on filtering out the noise due to genome wide duplications on rice-wheat synteny and colinearity.

Gene Prediction and Annotation of the Rice Genome

The sequence of 3343 BAC/PAC clones in the minimum tiling path of 12 rice chromosomes were downloaded using the "Batch Entrez" tool of NCBI (www.ncbi.nlm.nih.gov) in August 2003. Each clone was checked for its orientation with respect to its neighbours and incorrect orientations were corrected by reverse complementation to bring all the BAC/PAC sequences in a uniform North-South orientation. Gene prediction was performed on individual BAC/PAC clones using FGENESH (www.softberry.com) trained for monocots and gene information including cDNA sequence (CDS) and protein sequence were extracted in separate set of files using an in-house Perl script "fgenex". The predicted CDS were used for mega BLASTX search with nr-protein database to check for similarity with known proteins (www.genome.ad.jp). The mega BLASTX output was saved in text format to serve as an input for another Perl script "annotate", to automatically classify these genes into three functional categories viz. exact/putative, unknown and hypothetical, as per the International Rice Genome Sequencing Project (IRGSP) guidelines (www.rgp.dna.affrc.go.jp). The overlap between neighboring BAC/PAC clones were removed manually. Each gene from the 12 rice chromosomes was assigned a unique number to indicate its chromosomal location and relative position, e.g. gene numbers 01-0001 to 01-6737 were assigned to the 6,737 genes predicted for the rice chromosome 1 according to their physical order in the chromosome beginning from the short arm end. This database is available at website www.nrcpb.org/rgp/VanshanuDhan.html.

Sequence Homology Between Rice and Wheat Genes

Total 56,298 genes were predicted and CDS of these genes were compared against 39,813 wheat EST contigs developed by the International Wheat SNP Mapping Consortium (build 3, Dec 02, http://wheat.pw.usda.gov/ITMI/2002/ Wheat SNP), using an optimized set of search parameters. All the rice CDS showing matches with the wheat EST contigs at a cutoff bit score of 100 were considered significant, and the output was tabulated in Excel sheets using an in-house Perl script "mblastex". In addition, the 56,298 rice CDS were also compared with a database of 3,792 wheat EST contigs (ver. Aug. 03) containing bin-mapped ESTs (http://wheat.pw.usda.gov). An AQL query, "aql_retrieve", was used to extract the bin information and other details for these contigs directly from the Grain Genes database of bin-mapped wheat ESTs. This information was used as a base to calculate different statistics regarding the conservation of gene synteny between rice and wheat at different cutoff bit scores and to find out the rice gene copy numbers by sorting and filtering tools in the Microsoft Excel. The location of each rice gene homolog was plotted on to the 21 wheat chromosomes and their bins using a combination of automated and manual steps, and cross-checked at least three times by different individuals.

An optimized BLAST search of the CDS of 56,298 rice genes with the consensus sequences of 39,813 wheat EST contigs showed 24,209 matches at a cutoff bit score of 100. A further search against 3,792 bin-mapped wheat EST contigs revealed that 8,834 rice genes matched with these bin-mapped wheat EST contigs. Genes from each of the 12 rice chromosomes mapped predominantly to a syntenic wheat group. The proportion of rice genes mapping to their syntenic wheat groups increased from 28.4 per cent at 100 cut off bit score, to 33.9 per cent and 38.2 per cent at cut off bit scores of 200 and 300, respectively, but the total number of matches dropped concomitantly from 8,834 to 5,840 and 4,238, respectively. Therefore, an intermediate cutoff bit score of 200 was used as a compromise between the need for higher stringency and allowing for retention of distant matches between rice and wheat genes for further analysis. A graphical depiction of the location of each of the 5,840 rice gene homologs on to 21 wheat chromosomes revealed many interesting patterns of rice-wheat synteny (Fig. 1). We compared individual rice genes against a set of presumably non-redundant wheat EST contigs, hence did not expect more than one match for a single rice gene. However, more than one rice gene may hit the same wheat contig due to presence of gene duplications in the rice genome. Thus 5840 rice genes matched with only 1896 wheat contigs, as quite often the same wheat EST contig did match with two or more rice genes. Since these matches were at a very high stringency (bit scores 200 or above and E values 10^{-40} or less), it reflected the existence of extensive gene duplications in the rice genome. The multiple rice genes matching with same wheat EST contig were quite often located consecutively in a rice chromosome, indicating predominance of tandem gene duplications. The homology plot showed 79.7% of the 5840 rice genes mapping to unique wheat groups, but only 33.9% of the genes mapped to the respective syntenic wheat groups. In all, four distinct patterns of synteny were observed. The first pattern showed single largest proportion of the rice genes (~25%) mapping to a syntenic wheat group while the remaining 75% genes distributed over all the six non-syntenic wheat groups. The second pattern was typical of the regions of rice chromosomes involved in an ancient polyploidy event where two-thirds of the rice genes were located on just two wheat groups (dual synteny) and one-third were distributed on the remaining five wheat groups. The third pattern clearly identified chromosomal translocations within the specific wheat genomes and therefore they must have originated after the divergence of wheat from rice. The fourth pattern represented regions of rice genome that show no specific synteny with wheat.

Evidence for an Ancient Polyploidy in the Rice and Wheat Genomes

After completion of the genome sequence of *japonica* rice variety 'Nipponbare', nine large segmental duplications were identified covering sixty two per cent of the predicted rice transcriptome. These duplications are also present in the genome of *indica* rice line '93-11' and their origin has been dated to a time about 70 mya, well before the divergence of rice and wheat, through an ancient polyploidy event in the ancestral grass genome. Much earlier than these genomic studies, Nandi (1936) based on his detailed cytological investigations concluded that *Oryza sativa* is a balanced allopolyploid originating from an ancestral grass species with basic chromosome number five. Since rice and wheat diverged from a common ancestor long after this duplication event, the signs of this ancient polyploidy must be present in the wheat genome also. Segmental duplications corresponding to the ancient polyploidy were clearly visible in our rice-wheat homology plots showing location of 5840 rice gene homologs in wheat. The rice chromosome segments corresponding to the ancient duplication showed high number of gene matches with two different wheat groups (dual synteny), which was at the same time consistent with the pattern of rice-wheat synteny for the 12 rice chromosomes (Fig. 1 B). A pattern of ancient duplication and post-duplication conservation of lineages between rice and wheat were present for all the large segmental duplications in the rice genome. In the ancient duplicated segments, 64 per cent of all rice gene homologs mapped to the two syntenic wheat chromosome groups, while the remaining 36% genes were distributed over 5 non-syntenic wheat groups. This provided independent evidence, in addition to the high Ks values of the duplicated gene pairs in rice, that the origin of these segmental duplications must predate the divergence of rice and wheat. Subsequent diploidization of these duplications in wheat was evident from the fact that ninety six percent of the mapped rice gene homologs in these regions were retained in only one of the two syntenic wheat groups and were deleted or diverged beyond recognition from the second wheat group. Existence of ancient polyploidy in rice genome has been a subject of debate but it is consistent with our observations showing corresponding duplications in the wheat genome also in the form of dual synteny patterns.

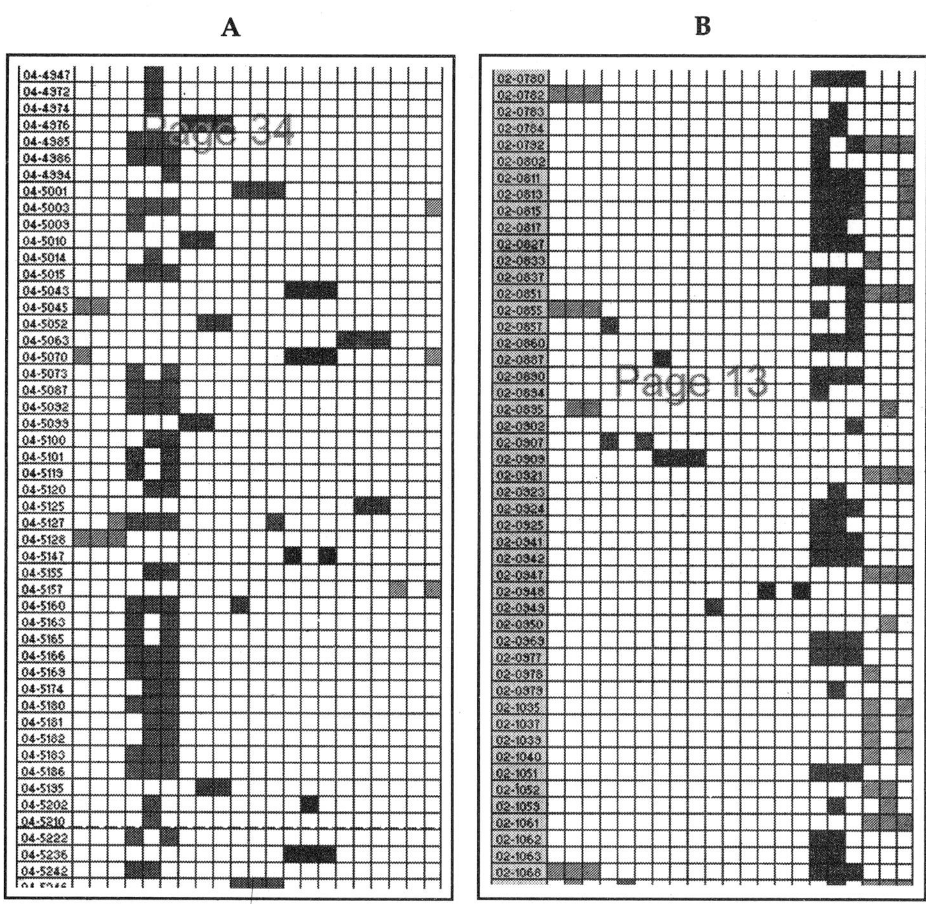

Fig. 1: Different patterns of conserved synteny between rice and wheat. **A.** Synteny of rice chromosome 4 with a single wheat chromosome group (red color); **B.** Dual synteny of rice chromosome 2 with wheat groups 6 (brown color) and 7 (orange color)

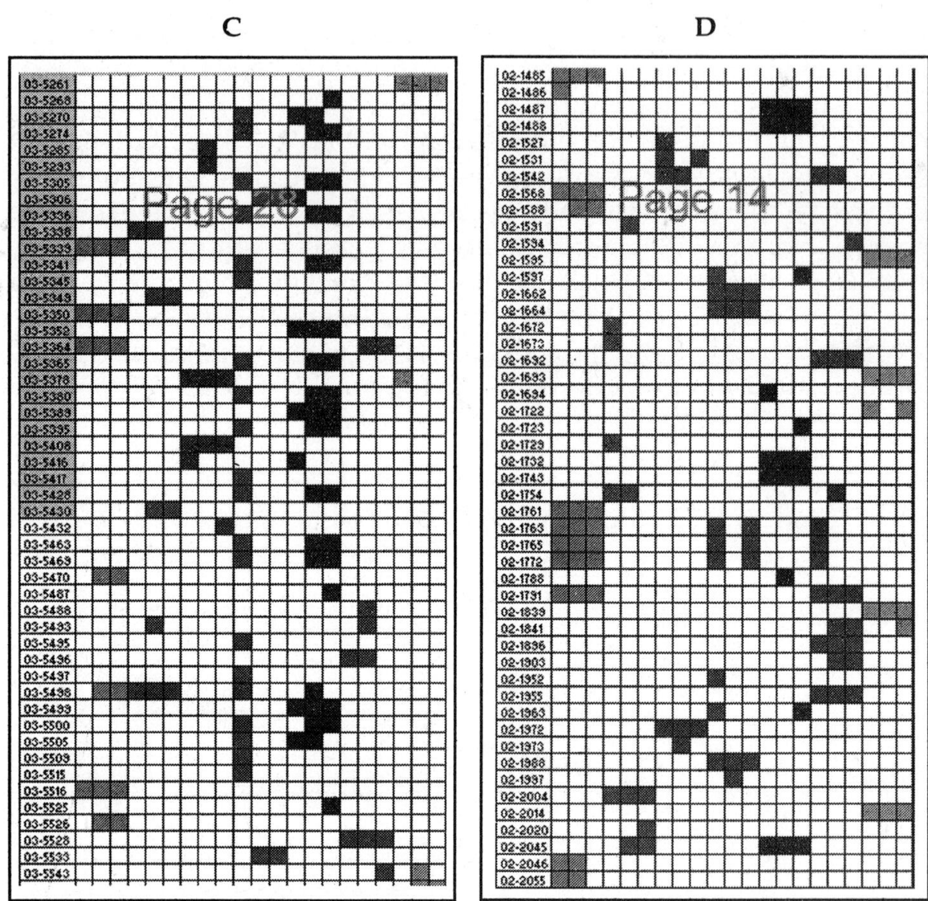

Fig. 1: Different patterns of conserved synteny between rice and wheat. **C.** Translocation of synteny between wheat chromosome 5A (first blue column) and 4A (first green column); **D.** Lack of synteny with any particular wheat chromosome.

Conserved Synteny and Colinearity of Single Copy Rice Genes with Wheat

The rice genes showing unique matches with a wheat EST contig were identified as single- copy (SC) genes. Two or more rice genes matching with a single wheat contigs represented a case of duplicated or multi-copy (MC) rice genes. Earlier studies used single-bin wheat genes for the analysis of wheat-rice synteny but these were not necessarily single copy wheat genes due to common occurrence of tandem duplication of genes. The identification of true SC genes is possible for rice only due to the availability of complete genome sequence information. Frequency distribution of copy number of genes in a genome follows a power law equation due to an inherent control on the amplification of gene copies. In a power law distribution as the number of genes in a gene family increases the frequency of such gene families in the genome will decrease. The frequency distribution of genes with different copy numbers among the 4659 rice genes mapping to unique wheat groups was consistent with this pattern. The highest frequency of 1063 was for the SC genes, and then there was stepwise reduction in the frequency of genes with higher copy numbers. While only 27.2 per cent of the MC genes mapped to a syntenic wheat group, the proportion was more than doubled to 58.7 per cent for the SC rice genes, showing striking conservation of synteny of the SC genes.

With the SC rice genes, each rice chromosomes showed synteny with only one wheat group, except for rice chromosome 11 which showed significantly high number of matches with wheat groups 4 and 5. This appeared to be due to an exchange of distal chromosomal segments between wheat groups 4 and 5 in the region corresponding to the long arm of rice chromosome 3. This exchange was common to all the three wheat genomes and therefore must have occurred quite early in the progenitor of the A, B and D genomes of wheat but after the divergence of rice from wheat. There was random distribution of a large number of gene homologs from each rice chromosome to all the non-syntenic wheat groups, thus highlighting the fluid nature of the plant genomes. A plausible explanation for this is background movement of genes in the genome through infrequent transposition. Role of transposition in the breakdown of synteny was further indicated by disproportionately high movement of genes near the wheat centromeres from a syntenic wheat group to nearly all the non-syntenic wheat groups. The breaks in synteny due to transposition were associated with the wheat centromeres rather than rice centromeres (Fig. 2). Wheat chromosomes have large amount of repetitive sequences near their centromere and a detailed characterization of these may elucidate the exact mechanism of this transposition.

Based on the conserved order of SC rice genes across the wheat chromosome bins, we reconstructed how the karyotypes of rice and wheat relate to a common ancestral grass genome karyotype (Fig. 3). Nandi (1936) through his extensive cytological study on several hundred Prophase I nuclei concluded that both aneuploidy (two separate duplications of individual rice chromosomes) and polyploidy (created by hybridization between two related species) have

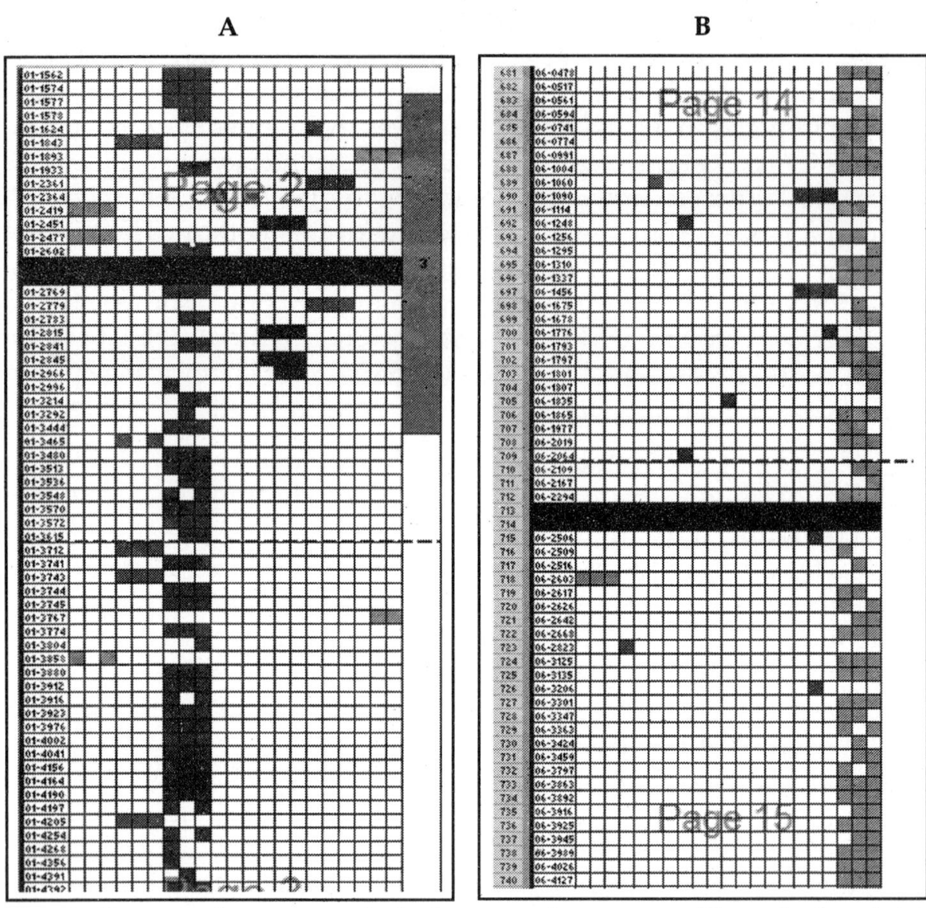

Fig. 2: Conserved synteny of single copy rice genes with a single wheat chromosome group. **A.** rice chromosome 1 with wheat group 3 (magenta color); **B.** rice chromosome 6 with wheat group 7 (orange color). Note heavy transposition of genes located near wheat centromere (orange vertical bar) to non-syntenic wheat groups as compared to no loss of synteny between rice chromosome 6 and wheat group 7 near region corresponding to only rice centromere (black horizontal bar)

contributed to the origin of *Oryza sativa,* and he suggested a haploid chromosome formula for rice as AA_1A_1, BB_1B_1, CC_1, DD_1, EE_1. The simplest explanation as to how multiple ancestral chromosomes of rice are arranged in single wheat chromosome group is by fusion of the ancestral chromosomes leading to the evolution of larger size wheat chromosomes. Presence of seven basic chromosomes in wheat in contrast to twelve chromosomes in rice could be explained by two centromeric fusions in wheat chromosomes of homoeologous group 1 and 4, and seven translocations, one each in wheat homoeologous group 1, 2 and 4, and two each in group 5 and 7 (Fig. 3). It is well known that the failure of meiotic chromosome pairing in the interspecific hybrids (polyhaploids) leads to chromosome breakage and fusion. Cereals with divergent basic chromosome numbers e.g. 7 in wheat, barley and rye, 10 in sorghum, maize and sugarcane and 12 in rice could have easily originated in such polyhaploid(s). Such chromosomal rearrangements have been observed experimentally in synthetic polyploids of mustard and wheat. Although the outcome of synteny studies are prone to be influenced by the choice of genes, our conclusions are based on a comprehensive analysis of all the available genes rather than any selected set of genes, and therefore these conclusions are more consistent and justified.

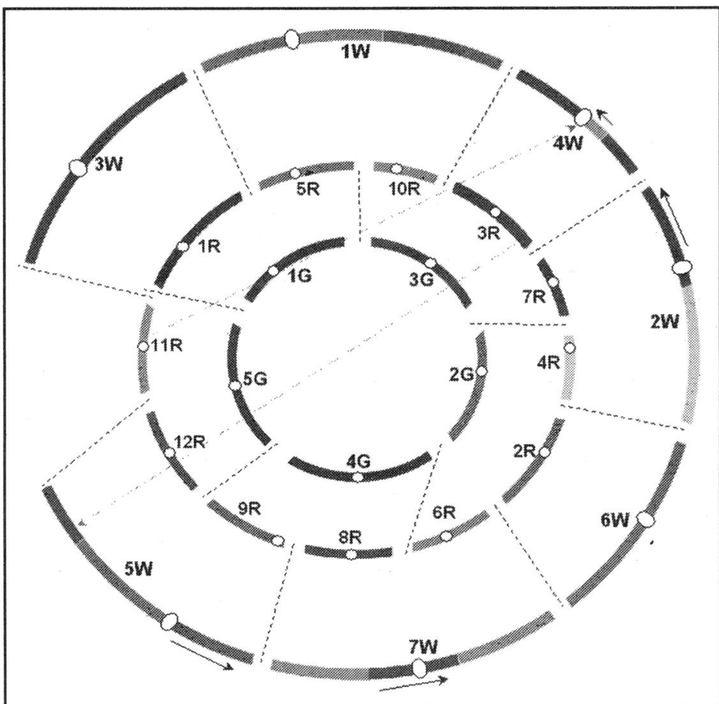

Fig. 3: Origin of 12 rice chromosomes (1R-12R) and 7 wheat chromosome groups (1W -7W) from an ancestral grass genome with 5 chromosomes (1G-5G) by whole genome duplication, aneuploidy (for 2G and 3G), chromosome fusions and translocations. The orientation of all chromosomes is clockwise with respect to the rice chromosome pseudomolecules, except in parts of wheat chromosomes 2W, 5W and 7W as indicated by arrows in anti clock direction.

Validation of the Synteny of Single Copy Rice Genes in Wheat

High resolution molecular maps are an essential resource for the structural and functional genomics of crop plants. Assignment of large number of ESTs to specific wheat chromosome bins will reduce the time and cost of developing high resolution maps in wheat. Based on the conserved colinearity of 1063 mapped SC rice genes in wheat, we predicted the chromosome bin location of 6178 unmapped SC rice gene homologs. To validate these predictions, PCR primers were designed using the Primer 3 software (http://frodo.wi.mit.edu) for a set of 213 wheat EST contigs predicted to map in the telomeric bins of 21 wheat chromosomes, and used genomic DNA from Chinese Spring wheat and its telomeric deletion lines for validation of these perditions. One hundred and sixty four of these primer pairs amplified DNA fragments in the expected size range. Sixty nine of these amplified a single DNA band, and in forty nine instances the single amplified band could not be assigned to the predicted wheat bin, probably due to lack of polymorphism or in some cases due to their location on non-syntenic wheat chromosomes. Nonetheless, 58 genes (35.4% of the total amplified genes) were located in the predicted telomeric bins of the seven wheat chromosome groups. This is a conservative estimate, as more primers are likely to be validated to their predicted locations using CAPS (cleaved amplified polymorphic segments) and SNP (single nucleotide polymorphism) genotyping of the single monomorphic PCR products. Only 1244 of the 7,540 SC rice gene homologs are presently bin-mapped in wheat. Mapping of remaining 6178 rice gene homologs in wheat will be greatly expedited based on their conserved synteny and colinearity as described here.

Rice and Wheat Hared a Long Common Lineage after Ancient Polyploidy

The coding sequence of a random set of orthologous pairs of rice and wheat genes, showing homology at cutoff bit scores of 200-400, 401-600, 601-800, 801-1000 and more than 1000, were extracted. A total of 116 such representative gene pairs, including 22-24 genes for each of the five bit score intervals and 8-10 genes from each of the twelve rice chromosomes, were aligned using ClustalW in MEGA3 software. The meg files were then analyzed using DnaSP 4.0 software (www.ub.es/dnasp/) for synonymous (Ks) and non-synonymous (Ka) substitution rates. We plotted the frequency distribution of Ks values for the 116 gene pairs to see if the bit score cutoff has any influence on the Ks values. The mean Ks values were indeed lower for the highly conserved gene pairs with higher bit scores but the modal Ks values were not much affected by the cutoff bit scores (Fig. 4), and hence provide a better criterion for the phylogenetic dating of duplication events. The modal Ks values for the 116 syntenic SC rice genes and their wheat orthologs were in the range of 0.2 to 0.3, giving an estimated time of divergence for rice and wheat between 15 to 23 mya using the molecular clock of Muse (2000). It is consistent with the idea that the common ancestor of present day rice and wheat must have followed a common lineage for a long time, somewhere in the range of 20 to 50 million years before their divergence.

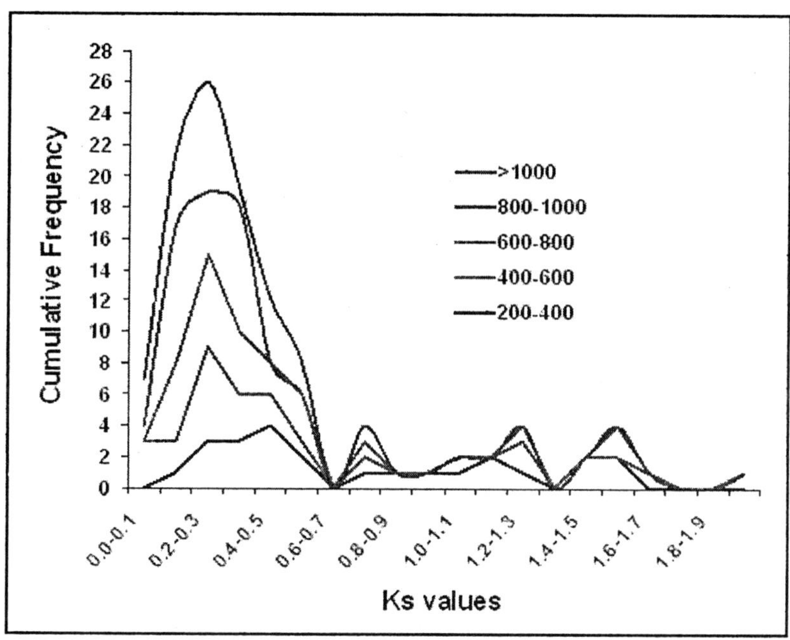

Fig. 4: Dating of the diversification of rice and wheat from a common ancestral grass genome based on synonymous substitution (Ks) statistics between 116 conserved rice-wheat gene pairs showing different levels of homology (bit score values 200 to >1000). Please note the conserved modal Ks value in the range of 0.2-0.3, corresponding to a divergence time of 15-23 mya using evolutional clock of Muse (2000)

Molecular Function of the Single Copy Genes

Based on the presence of protein functional domains identified by BLASTX search we classified the SC rice gene homologs of wheat into two categories: (i) with exact/putative function (ii) with unknown functions; none of the genes belonged to hypothetical categories as each gene showed homology with expressed wheat genes. Significantly, 50.2 per cent of the bin-mapped SC genes have unknown function as compared to only 29.4 per cent of the bin-mapped MC genes showing unknown function. The proportion of genes with unknown function was still higher (54.0 per cent) in the unmapped SC genes. This shows that the SC genes have lagged behind in both their mapping and functional characterization, probably due to their poor representation in the mRNA populations. The conserved sequence and single copy constitution of these genes point towards their vital role in the plant, which is also supported by the annotated function of some of these genes. The SC rice genes have maintained their single copy status despite the prevailing forces of gene duplication, including tandem duplication, background duplication and polyploidy. More than 85 percent of the SC rice genes mapped to a unique wheat group, indicating that duplicate copies resulting from the ancient polyploidy have largely been deleted or diverged beyond recognition in the wheat genomes. More than one-third of the expected triplicate sets of the 1063 SC rice gene homologs (3189 genes for the three genomes in bread wheat) have already been eliminated or diverged beyond recognition. The bread wheat has only 2149 homologs of the SC rice genes including duplicates and triplicates. A rapid mechanism for the elimination of duplicated loci has been described in synthetic wheat polyploids, but other duplicates may take longer for elimination, or may even be tolerated and evolve new functions. Single copy genes are present in almost all the sequenced genomes despite increasing evidence that complex species have evolved by duplication and divergence of genes and genomes. The inferred deletions of SC rice gene homologs in the wheat need not have occurred after the formation of polyploid wheat and many of these could have occurred during independent evolution of the diploid progenitor genomes of bread wheat over a period of 2.5 to 7 mya. Tandem duplication of genes (a kind of paralogs) is much more frequent than the segmental duplication. For example tandem duplications have played a key role in the rapid divergence of disease resistance genes in rice. The tandem copies of genes are likely to evolve the fastest, as this represents a more fundamental feature of all the sequenced genomes and is also present within the large duplicated segments of the chromosomes. The orthology must have its origin in the duplication of chromosome segments by translocation, aneuploidy or polyploidy, whereas transpositions are responsible for moving the individual genes from their orthologous positions (ancestral gene context) to new paralogous positions. After their divergence from a common ancestor the daughter species are likely to retain the orthologous set of genes in the same order (colinearity), depending on the duration of their separation or adaptation to specific ecological niche which may lead to the loss of colinearity.

ACKNOWLEDGEMENTS

Work was supported by the Indian Council of Agricultural Research, and Department of Biotechnology, Government of India. We are thankful to NCBI, *Softberry.com*, USDA and RGP for making their databases and genome analysis tools publicly available. We are grateful to Dr. Dave Matthews for his help with wheat EST data sets and Drs. Bikram Gill and Li Qi for kindly providing the DNA samples of wheat deletion lines.

References

Devos, K.M., Beales, J., Ogihara, Y., Doust, A.N. 2005. Comparative sequence analysis of the phytochrome C gene and its upstream region in allopolyploid wheat reveals new data on the evolution of its 3 constituent genomes. Plant Mol. Biol. 58, 625–641.

Huynen, M.A., van Nimwegen, E. 1998. The frequency distribution of gene family sizes in complete genomes. Mol. Biol. Evol. 15, 583-589.

IRGSP. 2005. The map-based sequence of the rice genome. Nature. 436, 793-800.

Lynch, M., Conery, J.S. 2000. The evolutionary fate and consequences of duplicated genes. Science. 290, 1151-1155.

Muse, S.V. 2000. Examining rates and patterns of nucleotide substitution in plants. Plant Mol. Biol. 42, 25-43.

Nandi, H.K. 1936. The chromosome morphology, secondary association, and origin of cultivated rice. J. Genet. 33, 315-336.

Paterson, A.H., Bowers, J.E., Chapman, B.A. 2004. Ancient polyploidization predating divergence of cereals, and its consequences for comparative genomics. Proc. Natl. Acad. Sci. USA. 101(26), 9903–9908.

Singh, N.K., Dalal, V., Batra, K., Singh, B.K., Chitra, G., Singh, A., Ghazi, I.A., Yadav, M., Pandit, A., Dixit, R., Singh, P.K., Singh, H., Koundal, K.R., Gaikwad, K., Mohapatra, T., Sharma, T.R. 2007. Single-copy genes define a conserved order between rice and wheat for understanding differences caused by duplication, deletion, and transposition of genes. Funct. Integr. Genomics. 7, 17–35.

The Rice Chromosomes 11 and 12 Sequencing Consortia. 2005. The sequence of rice chromosomes 11 and 12, rich in disease resistance genes and recent gene duplications. BMC Biol. 3, 20.

Yu, J., Wang, J., Lin, W., Li, S., Li, H., et al. 2005. Genomes of *Oryza sativa*: a history of duplications. PloS. Biol. 3(2), e38.

❑❑❑

25

Development of Genome Resources and Understanding Molecular Biology of Rutin Biosynthesis in a Nutraceutical Food Crop (Buckwheat) of High Altitudes

R.S. CHAUHAN*[1], N. GUPTA[1], S. SHARMA[1] AND J. C. RANA[2]
[1]Department of Biotechnology & Bioinformatics, Jaypee University of Information Technology, Waknaghat, Solan-173215, H.P. India, [2]National Bureau of Plant Genetic Resources, Regional Station, Phagli, Shimla, H.P. India

NTRODUCTION

Buckwheat is a dicot plant belongs to the family Polygonaceae.)ut of 14 species that are known to exist, two viz. *Fagopyrum tataricum* tartary buckwheat) and *F. esculentum* (common buckwheat) are ultivated types and all others occur as wild or escape in the cultivated ields. In India it is grown in North-Western and North-Eastern Iimalayan regions above 1800m elevation. The two species which re of highest economic importance are *F. esculentum;* generally called ommon buckwheat and *F. tataricum* or tartary buckwheat, the latter eing more important in high altitudes. The cultivation of common uckwheat is limited to approximately 500-2500m in altitude, whereas ırtary buckwheat can be grown above 2,500m in Himalayas. uckwheat contains high rutin (a flavonoid) content and due to high nutritional value of the crop it is considered as an important sourc for functional food. Besides antioxidant properties, rutin strengthen capillaries and, therefore, helps people suffering from arterioscleros (the hardening of the arteries) or high blood pressure. The tartar buckwheat contains 10x higher rutin content compared to commo buckwheat, thus making it an ideal food crop with industrial potentia Buckwheat has a small genome size (~450Mbp), diploid, self pollinate and short life cycle (~60 days) thus making it an ideal model pla species for understanding various aspects of biosynthesis an accumulation of flavonoids. Buckwheat also possesses othe pharmacological properties such as it ameliorates spatial memory contains anti-microbial peptides, possesses therapeutic role in huma leukemia, suppresses gall stone formation, treats peridontitis and gu bleeding, retards memory carcinogenesis, and more importantl flavonoids in buckwheat provide protection from UV-B radiations a high altitudes. In spite of Himalayas being the richest source o buckwheat genetic diversity no efforts have been made till date t identify genotypes rich in medicinal and nutritional components. Th genome resources are very limited in buckwheat, thereby, limiting th utilization of crop for industrial applications as well as its geneti improvement.

It is the only field crop containing highest amount of flavonoid especially rutin and quercitin. Rutin possesses well establishe medicinal values including pharmacological, vasconstrictiv spasmolitic and positive inotropic effects in addition to its ability t reduce hemorrhaging in people with high blood pressure. Tartar buckwheat seeds contain more rutin (about 0.8-1.7% dw) than commo buckwheat seeds (0.01% dw). Rutin and other flavonoids derived fro buckwheat are in increasing demand in the food, pharmaceutical an cosmetic industries due to their biological and physiological activiti including antioxidant, anti-inflammation and anti-hypertension.

Current Status of Genome Resources in *Fagopyrum* Species

A significant amount of research has been carried out on in th functionalities and properties of buckwheat proteins, flavonoid flavones, phytosteroles, thiamin-binding proteins, and other ra compounds in buckwheat. However, the availability of genom resources such as a good linkage map, different classes of molecul markers, EST libraries, large-insert DNA library, etc. is limited. The are fragmentary reports on understanding the species relationship b using molecular markers such as PCR-based DNA fingerprinting in *Fagopyrum*. Preliminary efforts have been made to develop a genetic map for *F. esculentum* and *F. homotropicum* on the basis of 223 and 211 AFLP markers, respectively. The map of *F. homotropicum* has eight linkage groups with 211 AFLP markers covering 548.9 cM. Microsatellite markers have been developed in common buckwheat by sequencing 2785 clones from the libraries and it was shown that 1483 clones contained microsatellites which were enriched for $(CT)_n$ and $(GT)_n$ repeats. Fifty-four primer pairs were evaluated for their ability to detect variations in common buckwheat populations and utilized in seven related *Fagopyrum* species including *F. tataricum*. SSR markers have been developed in tartary buckwheat. One bacterial artificial chromosomal (BAC) library have been constructed from a wild buckwheat species, *F. homotropicum*.

Although tartary buckwheat has several desirable properties compared to common buckwheat but the presence of adhering seed coat renders it underutilized commercially because of its bitter taste. The genetics of easy dehulling characteristic of *F. tataricum* has shown that the trait is controlled by a single recessive gene. We are attempting to genetically map the loose-hull trait for its eventual incorporation into genetically superior genotypes through molecular breeding.

Development of Molecular Markers in *F. tataricum*

Due to the lack of a well developed linkage map of buckwheat and seeing the availability of a limited number of molecular markers, we looked for *in silico* alternatives for rapid identification of additional molecular markers by utilizing ESTs available in other plant species belonging to taxonomic order, Caryophyllales to which *Fagopyrum* species belong (Table 1). All ESTs available in a particular plant species were downloaded from the database of TIGR (www.tigr.org) and simple sequence repeats (SSRs) were identified by using PGG Bioinformatics tool. Primer pairs were designed for their amplification by using Primer 3. Only 13 SSRs were successfully amplified on *F. tataricum* thereby indicating a poor transferability of SSRs (Table 2). Fifty four SSRs reported in *F. esculentum* were also used in *F. tataricum* to see the extent of transferability.

Comparative Genomics of Genes Involved in Rutin Biosynthesis

Rutin biosynthetic pathway has been elucidated in various plant species. The pathway has 9 genes which are phenylalanine ammonia lyase (PAL), cinnamate-4-hydroxylase (C4H), 4-coumaryl CoA ligase, Chalcone synthase (CHS), Chalcone isomerase (CHI), Flavonol synthase (FLS), Flavanone-3-hydroxylase (F3H), Flavanone-3'-hydroxylase (F3'H) and glucosyl/rhamnosyl transferase (Figure 1).

Table 1: Plant Species Belonging to Caryophyllales from which EST Sequences were Retrieved

Family	Plant Species	Ests
Aizoaceae (Ice plant family)	*Mesembryanthemum crystallinum*	27,191
Amaranthaceae (Goosefoot family)	*Beta vulgaris* *Suaeda salsa*	25,834 973
Plumbaginaceae (Leadwort family)	*Limonium bicolor* *Plumbago zeylanica*	4,686 1,701
Tamaricaceae (Tamarix family)	*Tamarix androssowii* *Tamarix hispida*	4,627 17,082

Table 2: SSRs Identified in Different Plant Families Amplified on *Fagopyrum Tataricum*

Primer Sets	Ssr Motif	Product Size	Plant
>TA8860_161934 >EG551447 >EG550581 >EG550404 >EG549488	(cag)17 (ga)28 (ca)4(ac)3a(ac)17 (tc)20 (ct)17	300-400 bp 300-400 bp 100-200bp 200 bp 500 bp	***B. vulgaris*** (Goosefoot family)
>TA3587_3544 >BE034161 >BE034182 >BE034334 >BE035077	(tac)14 (ga)55 (cca)15 (tc)21 (ag)38	400 bp 200bp 200bp 300bp 200 bp	***M. crystallinum*** (Ice plant family)
>TA209_76149 >CB818246 >TA304_126914	(gaa)7 (gaaag)7 (caa)5	500 bp 400 bp 200bp	***P. zeylanica*** (Leadwort family)

Phenylalanine ammonia-lyase (PAL, EC 4.3.1.5) catalyzes the first step in the phenylpropanoid pathway and is considered an important regulation point between primary and secondary metabolism.

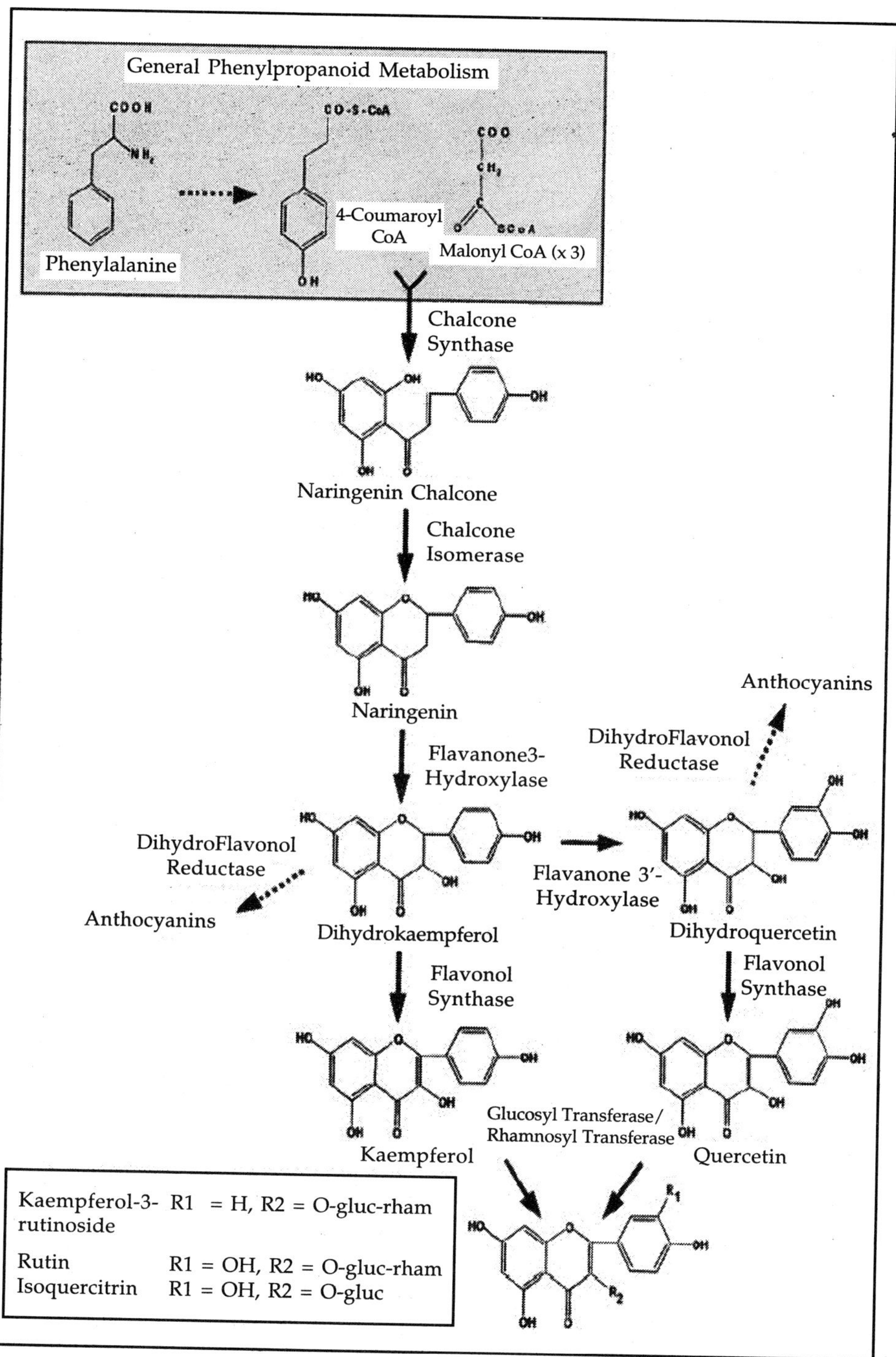

Fig. 1: Rutin Biosynthetic Pathway in Plants

Chalcone synthase (CHS), a key enzyme in flavonoid biosynthesis, catalyses sequential decarboxylative condensations of *p*-coumaroyl-CoA with three malonyl-CoA molecules. Chalcone synthase biosynthesis is specific to the formation of flavonoids pigments and isoflavonoids. Chalcone synthase gene has been found to be critical in rutin biosynthesis because down regulation of CHS in tomato fruit by RNA interference (RNAi) resulted in suppression of CHS and subsequently reduced accumulation of Flavonoids.

Chalcone isomerase, an enzyme in the isoflavonoid pathway in plants, catalyzes the cyclization of chalcone into (2*S*)-naringenin. Chalcone isomerase gene activity appears to be critical as in tomato peel tissue, the expression of a sequence encoding the *P. hybrida* chalcone isomerase leads to a large increase in the level of quercetin-glycoside accumulation. CHI-suppression by RNA interference (RNAi) showed reduced pigmentation and change of flavonoid components in flower petals of transgenic *Nicotiana tabacum.*

Flavanone 3-hydroxylase (F3H) catalyses the hydroxylation of flavanones at 3 position of C ring leads to the formation of dihydroflavanols. The full-length cDNA(1071 bp) and genomic DNA sequences of F3H gene were isolated from *Ginkgo biloba* for the first time. In addition F3H gene expression appears to be vital in the regulation of the flavonoid pathway as it is coordinately regulated with CHS and CHI in petunia. The strong correlation between the concentration of catechins and *Camellia sinensis* F3H expression indicates its critical role in catechin biosynthesis.

Flavonol synthase (*FLS*) catalyse the formation of kempferol and quercitin from dihydrokempferol and dihydroquercitin respectively. Knock-out alleles of genes involved in flavonoid biosynthesis were generated in *Arabidopsis*. One mutant line containing an *En-1*(maize transposable element) insertion in the flavonol synthase gene (*FLS*) showed drastic reduced levels of kaempferol. Allelism tests with other lines containing *En-1* insertions in the flavanone 3-hydroxylase gene (*F3H*) demonstrated that *transparent testa 6* (*tt6*) encodes flavanone 3-hydroxylase.

Glucosyl/rhamnosyl transferases catalyse the glycosylation of quercetin following the rhamnosylation of isoquercitrin. Among these enzymes, uridine-diphosphoglucose (UDPGlc), flavonoid 3-O-glucosyltransferases (3GT) have been studied in different plants. The 3GT has been purified and characterized from buckwheat cotyledons.

It has been studied that the expression pattern of a UGT gene, *UGT89C1*, was found to be highly interrelated with known flavonoid biosynthetic genes which is identified as flavonol 7-*O*-rhamnosyltransferase. This was done by using transcriptome co-expression analysis accessible on the ATTED-II public database. Search for rutin biosynthesis genes in the GenBank revealed that there are variable copy number for these genes in the genomes of different plant species (Table 3). The nucleotide and protein sequences of genes involved in rutin biosynthesis was retrieved from different plant species and the multiple sequence alignments (MSA) were done in order to find out extent of sequence similarity. Multiple sequence alignment for protein and DNA sequences for each gene from different plant species revealed conservation of gene structure and sequence in identity (Table 4). Out of 9 genes involved in rutin biosynthesis, two genes CHS and glucosyl transferase have been identified in *F. esculentum* and *F. tataricum and F.esculentum*, respectively.

Table 3: Rutin Biosynthesis Gene Copy Number in Different Plant Species

Gene	Plant Species/ Gene Copy Number
Chalcone synthase	*Citrus*/ 2 (CHS 2 in embryogenesis) *Poplar* (6) *Arabidopsis* (1) *Soyabean* (7) *Walnut*/2 (98% identity) *C. sinensis*/ 2 (CHS 2 plays key role) *V.vinifera* (4) *Petunia*/ 8 complete and 4 partial (Koes *et al.*,1987;1989)
PAL (Phenylalanine ammonia-lyase)	*V.vinifera*/ 13 (1, 2, 3 are major ones) *Poplar* (5) *Arabidopsis*/ 4 (1 and 2 are major) (Olsen *et al.*, 2007)
Cinnamate 4 hydroxylase (C4H)	*C. sinensis*/ 2(C4H 2 plays major role) *Arabidopsis* (1) *V.vinifera* (1) *Poplar* (2) *Ipomea purpurea*/ 6 (Durbin et al.,1995;1999; Fukada-Tanakaet al, 1997)
4Coumarate CoA ligase	*Arabidopsis*/ 14 genes, 4 are major (*At4cl1, At4cl2, At4cl3,* and *At4cl5*).

Contd...

	At4CL3 plays role *Raspberry* (4) *V.vinifera* (1) *Poplar* (5)
CHI (Chalcone isomerase)	*C.sinensis* (1) *Petunia hybrida, Phaseolus vulgaris and Glycine max* (4) *Arabidopsis* (1) *Poplar* (1) *V.vinifera* (1)
F3H (Flavanone 3-hydroxylase)	*Poplar* (3) *Arabidopsis* (1) *V.vinifera* (1)
FLS (Flavonol synthase) 2-oxoglutarate-Dependent dioxygenase family.	*V.vinifera* (4) *Poplar* (4) *Arabidopsis*/6 (AtFLS1 is major) Daniel K. Owens *et al*,. 2008
Glucosyl transferase	*V.vinifera* (1) *Poplar* (1) *Arabidopsis* (1)

Table 4: Sequence identity in rutin biosynthesis genes from different plant species

Gene	**Plant Species**				
Chalcone Isomerase					
	Lotus japonicus	*A. thaliana*	*Citrus sinensis*		
Glycine max	85%	68.90%	82.20%		
Flavanol-3-Hydroxylase					
	Brassica juncea	*Raphnus sativus*	*A. thaliana*	*Vitis vinifera*	*Dacus carota*
Brassica rapa	97.10%	97.90%	93%	83.50%	81.70%
Flavanol Synthase					
	Nicotiana tabacum	*Vitis vinifera*			
Solanum tuberosum	86.40%				
Citrus		71.30%			
Phenylalanine Ammonia Lyase					
	with all plant species(mentioned in table 5)				
Glycine max	~85.1%				
Cinnamate-4-Hydroxylase					
	A. thaliana	*B. oleracea*	*Brassica juncea*		
Brassica rapa	95.60%	85.90%	4.90%		
4-Coumarate CoA Ligase					
	B. oleracea				
Brassica rapa	53.80%				
~6% among different plant species					

Rutin Content Estimation in *F. tataricum* Genotypes

A collection of *F. tataricum* genotypes (408 no.) is available with the National Bureau of Plant Genetic Resources, Regional Research Station (Phagli, Shimla, Himanchal Pradesh, India). The genotypes have been collected from remote areas of Jammu & Kashmir, Himachal Pradesh, Uttaranchal, Sikkim and North-Eastern hills of India. Rutin content analysis in seeds on 200 *F. tataricum* accessions in our laboratory revealed significant variations ranging from 0.6 per cent to 2.0 per cent (dw) compared to 0.07 per cent (dw) in accessions of *F. esculentum* (Fig. 2). This observation is analogous to previous studies wherein rutin content variation has been reported, though on limited number of genotypes. The *Fagopyrum tataricum* genotypes showing contrasting variation for rutin content are being used in the identification of DNA sequence variations in genes involved in the rutin biosynthesis. The identification of SNPs in high vs. low rutin content genotypes will be of great importance in molecular breeding of buckwheat for high rutin content. The high rutin content genotypes are also being used in identifying regulatory genes controlling rutin biosynthesis through differential display analysis. We are also

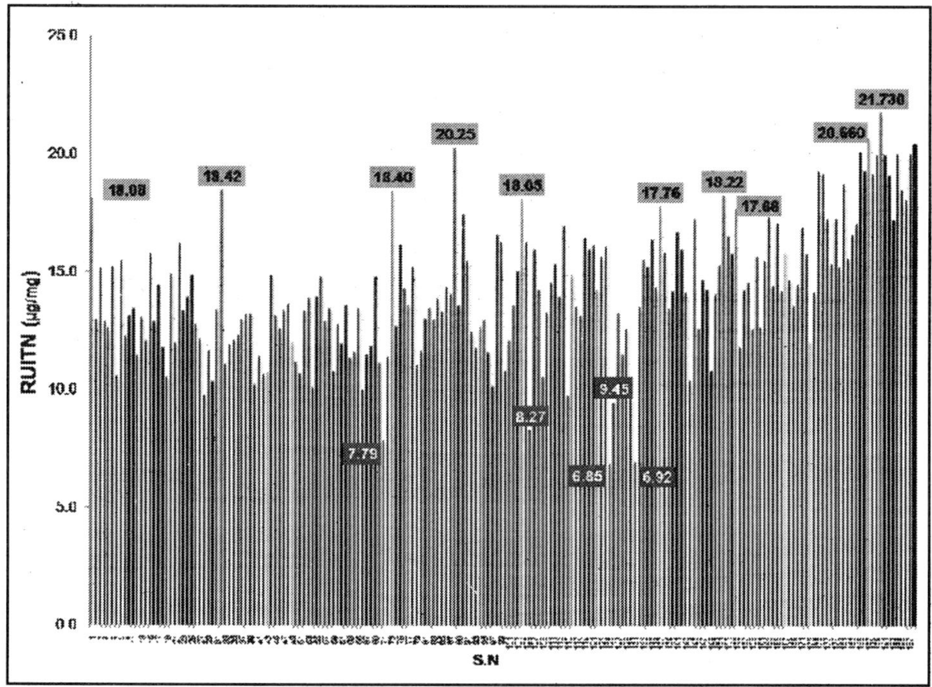

Fig. 2: Variation in Rutin Content Among Different Accessions of *F. Tataricum*

constructing a BAC library from high rutin content, easy-hull genotype with the ultimate goal of cloning useful genes. The BAC library would be very useful in pursuing genomics of *F. tataricum*

ACKNOWLEDGEMENTS

The authors are thankful to the Department of Biotechnology, Ministry of Science & Technology, Govt. of India for providing financial assistance to pursue the above mentioned studies. We are also thankful the Management of the Jaypee University of Information Technology for providing infrastructural support in the Department of Biotechnology & Bioinformatics.

References

Bonafaccia, G. and Fabjan, N. 2003. Nutritional comparison of tartary buckwheat with common buckwheat and minor cereals. Zb. Bioteh. Fak. Univ. Ljublj. Kmet. 81, 349-355.

Colliver, S., Bovy, A., Collins, G., Muir, S., Robinson, S., de Vos, C.H.R. and Verhoeyen, M.E. 2002. Improving the nutritional content of tomatoes through reprogramming their flavonoid biosynthetic pathway. Phytochem Rev. 1, 113–123.

Eggum, B. O., Ivan, K., and Branka, J. 1981.Chemical composition and protein quality of buckwheat *(Fagopyrum esculentum Moench).* Plant Foods Hum Nutr. 30, 175-179.

Fabjan, N., Rode, J., Kosir, I. J., Wang, Z., Zhang, Z., and Kreft, I. 2003. Tartary buckwheat (*Fagopyrum tataricum* Gaertn.) as a source of dietary rutin and quercetin. J. Agric. Food Chem. 51, 6452 -6455.

Hrazdina, G., Lifson, E., and Weeden, N.F. 1986. Isolation and characterization of buckwheat (*Fagopyrum esculentum* M.) chalcone synthase and its polyclonal antibodies. Arch Biochem Biophys. 247(2), 414-419.

Jiang, P., Burczynski, F., Campbell, C., Pierce, G., Austria, J.A. and Briggs, C.J. 2006. Rutin and flavonoid contents in three buckwheat species *Fagopyrum esculentum, F. tataricum,* and *F. homotropicum* and their protective effects against lipid peroxidation. Food Res. Intern. 40(3), 356-364.

Joshi, B.D. and Paroda, R.S. 1991. Buckwheat in India. National Bureau of Plant Genetic Resources, Regional Station, Phagli, Shimla, India.

Konishi, T., Iwata, H., Yashiro, K., Tsumura, Y., Ohsawa, R., Yasui, Y. and Ohnishi, O. 2006. Development and Characterization of Microsatellite Markers for Common Buckwheat. Breed. Sci. 56, 277-285.

Li, Y., Shi, T., and Zhang, Z. 2007. Development of microsatellite markers from tartary buckwheat.Biotechnol Lett. 29, 823–827.

Nagano, M., Jotaro, A., Campbell, C., Kawaskl, S. and Adachi, T. 2001. Construction a BAC Library of the Genus *Fagopyrum*. Proceeding of the 8th ISB.

Nishihara, M., Nakatsuka, T. and Yamamura, S. 2005. Flavonoid components and flower color change in transgenic tobacco plants by suppression of chalcone isomerase gene. FEBS J., Elsevier B.V.

Pelletier, M. K. and Shirley, B. W. 1996. Analysis of Flavanone 3-Hydroxylase in Arabidopsis Seedlings (Coordinate Regulation with Chalcone Synthase and Chalcone Isomerase). Plant Physiol. 111, 1339-1345.

Rana, J.C. 2004. Buckwheat Genetic Resources Management in India. Proceedings of the 9th ISB.

Schijlen, E. G.W.M., Vos Ric de, C.H., Stefan, M., Jonker, H. H., Rosin, F.M., Molthoff. J. W., Tikunov, Y. M., Angenent, G. C., Tunen, A., van, J. and Bovy, A.G. 2007. RNA Interference Silencing of Chalcone Synthase, the First Step in the Flavonoid Biosynthesis Pathway, Leads to Parthenocarpic Tomato Fruits. Plant Physiology. 144, 1520-1530.

Sharma, T. R., and Jana, S. 2002. Species relationships in Fagopyrum revealed by PCR-based DNA fingerprinting. Theor. Appl. Genet. 105, 306–312.

Shen, G., Pang, Y., Wu, W., Deng, Z., Zhao, L., Cao, Y., Sun, X. and Tang, K. 2006. Cloning and Characterization of a Flavanone 3-Hydroxylase Gene from *Ginkgo biloba*. Biosci. Rep. 26, 19–29.

Singh, K., Rani, A., Kumar, S., Sood, P., Mahajan, M., Yadav, S.K., Singh, B. and Ahuja, P.S. 2008. An early gene of the flavonoid pathway, flavanone 3-hydroxylase, exhibits a positive relationship with the concentration of catechins in tea (*Camellia sinensis*). Tree Physiol. 28, 1349–1356.

Suzuki, T., Kim, S., Yamauchi, H., Takigawa, S., Honda, Y., and Mukasa, Y. 2005. Characterization of a flavonoid 3-O-glucosyltransferase and its activity during cotyledon growth in buckwheat (*Fagopyrum esculentum*). Plant Sci. 169, 943–948.

Verhoeyen, M. E., Bovy, A., Collins, G., Muir, S., Robinson, S., de Vos, C. H. R. and Colliver, S. 2002. Increasing antioxidant levels in tomatoes through modification of the Favonoid biosynthetic pathway. J. Exp. Bot. 53, 2099-2106.

Wang, Y. and Campbell, C.G. 2007. Tartary buckwheat breeding (*Fagopyrum tataricum* L. Gaertn.) through hybridization with its Rice-Tartary type. Euphytica. 156, 399-405.

Wisman, E., Hartmann, U., Sagasser, M., Baumann, E., Palme, K., Hahlbrock, K., Saedler, H., and Weisshaar, B. 1998. Knock-out mutants from an *En-1* mutagenized *Arabidopsis thaliana* population generate phenylpropanoid biosynthesis phenotypes. Proc. Natl Acad Sci U S A. 13. 95(21), 12432–12437.

Yasui, Y., Wang, Y., Ohnishi, O., and Campbell, C. G. 2004. Amplified fragment length polymorphism linkage analysis of common buckwheat (*Fagopyrum esculentum*) and its wild self-pollinated relative *Fagopyrum homotropicum* .NRC Research Press.

Yonekura-Sakakibara, K., Tohge, T., Niida, R., and Kazuki, S. 2007. Identification of a Flavonol 7-*O*-Rhamnosyltransferase Gene Determining Flavonoid Pattern in *Arabidopsis* by Transcriptome Coexpression Analysis and Reverse Genetics. J. Biol. Chem. 282, 14932-14941.

□□□

26

Development of Short Duration Quality Protein Maize Hybrids Suitable for Hills through 'Marker Assisted Selection'

H.S. GUPTA, P.K. AGRAWAL, AND V. MAHAJAN
Vivekananda Parvatiya Krishi Anusandhan Sansthan (ICAR), Almora, India.

ABSTRACT

Quality protein maize (QPM) possesses roughly twice as much usable protein as normal maize grown in the tropics. The QPM protein in general contains 55 percent more tryptophan, 30 percent more lysine and 38 percent less leucine than that of normal maize. The biological value of normal maize protein is 40 percent, while that of QPM is 90 percent. With QPM, the niacin availability is higher due to higher tryptophan and lower leucine content and carotene utilization. The QPM maize can also be transformed into edible products without any deterioration of its quality and acceptability. The nutritional and biological superiority of QPM has been studied in animals (rats, pigs) and humans (infants, small children, adults). In Columbia, QPM was used to bring children suffering from kwashiorkor back to normalcy. It can also be used as ingredient in the preparation of composite flours to supplement wheat flour for bread and biscuit preparation. Composite flours have been mixed with 10 percent maize flour in many countries such as Brazil, Zambia, Zimbabwe and Ghana. QPM can play an increasing role in reducing protein supplementation of animal feeds. That is why it has been widely adopted for cultivation in developing countries to fight protein malnutrition. In India, QPM was released for commercial cultivation almost a decade ago by introducing QPM lines from CIMMYT. However, all these inbred lines have longer growth duration and thus, give rise to QPM hybrids of full season maturity. Those hybrids are not suitable for the hills and many rain-fed regions of peninsular India. Utilizing 'marker assisted selection' (MAS) methodology at VPKAS (spell out), the QPM trait was transferred to two early maturing Indian maize inbreds that were, in turn, crossed to give rise a short duration QPM hybrid, Vivek QPM 9, with 30 percent higher lysine and 40 percent more tryptophan while retaining the same level of productivity as Vivek Maize Hybrid 9, the elite non-QPM hybrid. It has also been found suitable for cultivation under organic farming. We advocate this hybrid for large scale adoption in the hills and mountains of India and value added products for the nutritional security.

Key words : Maize, QPM, MAS, SSR, Vivek QPM 9

INTRODUCTION

Maize is grown in over 7.0 million ha area in India and production is 15.1 m tons (Anon., 2007). It is used both as food and feed. However the protein of normal maize has a biological value of 40 percent that of milk and is poor as a food as well as feed. It therefore, needs to be supplemented with proteins from legumes and animal products to be valuable as food. In normal maize protein, the essential amino acids like lysine, tryptophan and threonine are in reduced quantities, lysine being the most limiting followed by tryptophan. In contrast, QPM protein has nearly twice the quantity of lysine and tryptophan making it 90 percent equivalent to milk protein. Therefore, it is imperative to incorporate QPM in our food and feed for imparting better nutrition and alleviation of protein-malnutrition (Bressani, 1991, Bressani, 1992, Graham *et al.,* 1990, Graham *et al.,* 1980, Mertz *et al.,* 1964, Mertz *et al.,* 1965).

The QPM protein in general contains 55 percent more tryptophan, 30 percent more lysine and 38 percent less leucine than that of normal maize. In QPM, the niacin availability and carotene utilization is higher due to higher tryptophan and lower leucine content. The QPM can also be transformed into edible products without any deterioration of its quality and acceptability. The nutritional and biological superiority of QPM has been studied in animals (rats, pigs) and humans (infants, small children, adults) (Bressani, 1991, Graham *et al.*, 1980, Krivanek *et al.,* 2007, Glover *et al.,* 1987, Paes *et al.,* 1994, Mertz *et al.,* 1965, Lopez-Pereira 1993, Graham *et al.*, 1990). In Columbia, QPM was used to children suffering from kwashiorkor back to normalcy. It can also be used as ingredient in the preparation of composite flours to supplement wheat flour for bread and biscuit preparation. Composite flours have been mixed with 10 percent maize flour in many countries such as Brazil, Zambia, Zimbabwe and Ghana. QPM can play an increasing role in reducing protein supplementation of animal feeds. That is why it has been widely adopted for cultivation in developing countries to fight protein malnutrition.

Soon after the discovery of the nutritional benefits of *opaque* 2 (*o*2) mutations that doubled the quantity of lysine and tryptophan, several breeding programs were initiated to incorporate this gene aiming at conversion of normal endosperm populations and inbreds to *o*2 versions through direct backcrossing. However, the soft endosperm of the *o*2 maize coupled with reduced grain yield and susceptibility to diseases and insect-pests did not allow its widespread adaptation. During this process of conversion, partially hard endosperms (vitreous) were noticed. Meanwhile, Paez *et al.*, in 1969 reported endosperm modification in *o*2 kernels with much less negative pleiotropic effects. This opened the door for introducing the selection of hard endosperm modifications in the *o*2 breeding scheme. Thus, the approach for selection of *o*2 plants coupled with hard endosperm at the phenotypic level became a viable approach. The resultant genotypes with elevated lysine and tryptophan as compared to the normal maize and without the negative effect of soft endosperm were termed as Quality Protein Maize (QPM) by International Wheat and Maize Centre (CIMMYT). QPM, thus looked and performed like normal maize, except that their nutritional value got elevated.

The most important aspect in developing acceptable QPM lines is to combine the nutritional advantages offered by the *o*2 mutation with the *o*2 modifiers that imparts the acceptable agronomic traits like grain yield, resistance against biotic stresses and grain vitreousness. It was noticed that the Caribbean and Cuban flints, with *opaque* 2 in general, tended to exhibit a higher frequency of modified kernels and those materials provided considerable hope for the development of QPM genotypes. For the recovery of the QPM lines, selection for kernel modification had to be practiced at all the stages, while maintaining the protein quality. During 1980's CIMMYT took initiatives to convert a number of non-QPM genotypes to QPM genotypes, representing tropical, subtropical and highland maize germplasm with different maturity durations. In a short span of 5 to 6 years, CIMMYT could convert many normal germplasm into QPM which were as good as their non-QPM counterparts for grain yield and other agronomic traits.

At this stage handling of QPM materials in the homozygous *o2* background was emphasized. Simultaneously 10 QPM advanced populations were identified and tested through the 'International Progeny Testing Trials' for their evaluation and dissemination. Thereafter, CIMMYT initiated development of QPM hybrid in 1985 since hybrid maize occupies a significant percentage of the total maize area and the grain yield of hybrids in maize is significantly higher. They identified a number of lines with better combining ability (Vasal *et al.*, 1993a, Vasal *et al.*, 1993b). In many cases the test-hybrids outperformed the local hybrids. India is one of the first few countries to focus on *o2* maize and released three *o2* composites namely Shakti, Rattan and Protina in 1970 and one modified superior *o2* composite 'Shakti 1' in 1997. Later India released six QPM hybrids, all of which were developed from the CIMMYT QPM inbreds as parental lines and are of full season maturity with maturity duration of 120 to 130 days (Anon., 2008; Table 1). When these hybrids are grown in hills, their duration increases further and does not fit to the cropping systems. Besides, short duration QPM hybrids are also better suited over a large rainfed area in the peninsular India. The present investigation was undertaken following marker assisted selection (MAS) (Babu *et al.*, 2005, Gibbon *et al.*, 2005, Krivanek *et al.*, 2007) to develop short duration QPM hybrid through conversion of normal inbred to QPM inbreds followed by hybridization.

Here, we recovered QPM hybrid with 4.25 percent lysine (percent of total protein). Another important aspect of this investigation was the development of QPM hybrids within 4 years (two generations per year) instead of 6 to 8 years following conventional approach.

MATERIALS AND METHODS

Vivek Maize Hybrid 9 is an extra early elite maize hybrids that matures in 85-90 days yielding 50-55 q/ha of maize grain. The hybrid was released for commercial cultivation in zone I (Himalayan states; Jammu & Kashmir, Himachal Pradesh, Uttarakhand) as well as Zone IV (Peninsular India; Andhra Pradesh, Tamil Nadu, Karnataka and Maharashtra) by the Central Seed Subcommittee on Crop Standard and Notification in the year 2000. The parents of this hybrid are CM 212 and CM 145.

DNA-marker technology and MAS were used to convert the two inbreds, CM 212 and CM 145 to QPM version. Two QPM lines *viz.*,

CML 180 and CML 170 obtained from CIMMYT, Mexico were used as donors. Three SSR markers, *viz.*, *phi* 057, *phi* 112 and *umc* 1066 located as internal repetitive elements within the *opaque*2 gene, were used for transferring *opaque*2 gene from CML 180 to CM 212 and CML 170 to CM 145. The background selection was done with approximately 100 polymorphic SSR markers distributed over the entire genome. All the selected plants were self-pollinated and upon harvest the kernels were subjected to selection for kernel modification under transmitted light using a light box. The kernels of each selected self-pollinated BC2F2 ear were grouped under three classes of modification *viz.*, less than 25 percent, 25–50 percent and more than 50 percent opaque. The biochemical analyses for total protein content in the endosperm and tryptophan concentration in endosperm protein in each class of kernel modification were carried out according to standard procedures. During the process, the grain samples were de-germed after removing the pericarp and finely ground. The resulting flour was defatted and total nitrogen content was determined by Microkjeldhal procedure and percentage of protein was calculated by multiplying the N content with a factor of 6.25. Quantitative analysis of tryptophan was done using colorimetric estimation. Since tryptophan and lysine are tightly linked, tryptophan quantity was routinely measured to estimate the quantity of lysine and tryptophan. Lysine was measured with HPLC at the National Bureau of Plant Genetic Resources, New Delhi. The resultant QPM inbreds, VQL 1 and VQL 2 (Fig. 1 and 2) were crossed to get the QPM hybrid, Vivek QPM 9.

RESULTS

The agronomic and biochemical attributes of the reconstituted QPM hybrid (Vivek QPM 9) and the non-QPM hybrid (Vivek Maize Hybrid 9) are presented in Table 2. The QPM hybrid, Vivek QPM 9 was found to contain 40 percent higher tryptophan and 30 percent more lysine than the non-QPM hybrid, Vivek Maize Hybrid 9. The appearance of the kernel is yellow, semi-flint with cap as in Vivek Maize Hybrid 9 with straight rows and low tip sterility (Fig. 3). Thereafter, the hybrid was tested for multilocation trial under AICRP on maize and yield trials carried out in the state of Uttarakhand under organic conditions.

Vivek QPM 9 was evaluated during *Kharif* 2005 and 2007, under the AICRP on Maize at seven locations in Zone I (comprising of Jammu and Kashmir, Himachal Pradesh, Uttarakhand and Meghalaya) and thirteen locations in Zone IV (comprising of Andhra Pradesh, Maharashtra, Karnataka and Tamil Nadu). The average productivity of the QPM hybrid (5,843 kg ha^{-1} in Zone I & 5,435 kg ha^{-1} in Zone IV) was at par with Vivek Maize Hybrid 9 (5,931 kg ha^{-1} & 5,404 kg ha^{-1} in zone IV) in both the zones over two years (Tables 3 and 4). In addition, Vivek QPM 9 showed equal level of resistance to Triticum blight, one of the most important diseases of maize crop (Table 2). Based on performance in Zone 1 and Zone IV, Vivek QPM 9 was selected for release for commercial cultivation in Zone I and Zone IV. Similarly Vivek QPM 9 was also evaluated in the state of Uttarakhand during *Kharif* 2006 and 2007. The mean yield of Vivek QPM 9 for two years was found to be 4,575 kg ha^{-1}, which is at par with that of Vivek Maize Hybrid 9 (4,422 kg ha^{-1}) (Table 4). The productivity of Vivek QPM 9 was numerically superior to Vivek Maize Hybrid 9 at all the six locations over two years. Considering the yield performance and better quality protein, it was released by the 'State Varietal Release Committee' (SVRC) of the Uttarakhand state for commercial cultivation under organic farming. The F_1 hybrid seeds of Vivek QPM 9 are being distributed in Uttarakhand and other parts of the country. Few villages in Uttarakhand have also been identified where QPM hybrids are introduced to convert them in to QPM maize villages.

DISCUSSIONS

The development of QPM involves three genetic systems *viz.*, i) recessive homozygous allele of the *o*2 gene, ii) modifiers for kernel hardness, and the iii) amino acid modifiers, which affect the relative level of lysine and tryptophan content in grain endosperm. The recessive *o*2 allele is the essential component of the QPM. This gene codes a transcription factor, which is a regulator for zein synthesis. Zeins, in particular alpha-zeins are the most abundant protein in maize endosperms which are known to be poor in amino acids tryptophan and lysine. The *o*2 allele when present in homozygous condition reduces production of zeins, particularly the á-zeins and triggers increase in the level of lysine and tryptophan. As a consequence, in the grain with *o*2 mutation a dramatic reduction in this enzyme leads to a corresponding increase in free lysine in the endosperm. The second component of QPM breeding is the hardness of the grains. Increased level of the gamma zein is likely to contribute to the recovery of hard endosperm. Many more genes are likely to be associated with the grain hardness and thus, they need to be understood. However this character can be visually selected under a light box where white light is projected through the grains and completely hard seeds are selected. The third component is the set of amino acid modifiers which affect the relative content of lysine and tryptophan. The lysine content of normal maize is around 2 percent whereas it is approximately 4 percent of the total QPM protein, with a range of 1.6 percent to 2.6 percent in normal maize and 2.7 percent to 4.5 percent in QPM (Krivanek *et al.*, 2007).

In the present study, we monitored the presence of *o*2 gene through the markers located within the gene. Kernel hardness was monitored through light box and level of tryptophan was measured through biochemical methods. Selection of desirable heterozygous and homozygous states of *o*2 gene was easily and quickly detected in F_1, BC_1, BC_2 and BC_2S_2. The selected plants with more than 90 percent genome of the recurrent parent were evaluated for tryptophan level in their seeds. This was done to select plants with hard endosperm and high tryptophan.

During the investigation, we noted similar effect and found that despite the presence of *opaque*2 in homozygous state (*o*2/*o*2) in VQL 1 and VQL 2, the level of enhancement in tryptophan was significantly different in both the QPM inbreds (0.42 in CM 212 and 0.52 in VQL 1; 0.55 in CM145 and 0.58 in VQL 2). This significant difference in the level of increase in tryptophan in the two recipient (of *opaque*2 gene) inbred lines (CM 212 and CM 145) is due to the fact that we did not continuously monitor tryptophan level during the selection. The results of the present investigation are, thus, in conformity with that of Krivanek *et al.*, 2007. Following the strategy developed in our laboratory, we recovered inbred lines homozygous for opaque 2 gene with more than 90-95 percent recurrent parent genome coupled with higher tryptophan within a short span of three years.

Mertz *et al.* 1964, first reported that the lysine content of *o*2 was 3.3 to 4.0 g per 100g of endosperm protein, which was more than twice that of normal maize endosperm (1.3g lysine/100g endosperm protein). The superior quality and digestibility of QPM protein has been demonstrated by many researches (Bressani, R. 1991, Bressani, R. 1992, Graham *et al.*, 1990, Graham *et al.*, 1980, Mertz *et al.*, 1964, Mertz *et al.*, 1965). The QPM protein in general contains 55 percent more tryptophan, 30 percent more lysine and 38 percent less leucine than that of normal maize. Furthermore, the biological value of normal maize protein is 40 percent, while that of *o*2 maize protein is 90 percent. Only 37 percent of the common maize protein intake is utilized compared to 74 percent of the same amount of *o*2 maize protein. The nitrogen balance index for skim milk and *o*2 maize protein is 0.80 and 0.72 respectively, which indicates that the protein quality of QPM is 90 percent the quality of milk protein. Besides, around 24g of normal maize per kg of body weight is required for nitrogen equilibrium, compared to only around 8g for QPM. In QPM protein, the niacin availability is higher due to higher tryptophan and lower leucine content and also the carotene utilization. The QPM maize can also be transformed into edible products without any deterioration of its quality and acceptability.

In the present investigation, we recovered QPM hybrid with 44 percent higher tryptophan, 30 percent more lysine and 12 percent less isoleucine. Another important outcome of the conversion process was the yield and pest resistance attributes were retained. During the present investigation, the development of QPM hybrids took only 4 years by taking two crops per year. On the contrary, conventional breeding will require 6 to 8 years out of which 12 generations will be needed just for conversion of normal inbreds to QPM inbreds because each backcross has to be followed by selfing for recovery of the *o*2*o*2. Availability of a combination of molecular probes that would allow selection of endosperm modifiers in *o*2 genotypes prior to selection of agronomic characters would facilitate rapid and efficient conversion of non-QPM inbred lines into QPM lines. The utility of three maize SSR markers present within the o2 gene in molecular-marker assisted selection for *o*2 has been successfully used for conversion of non-QPM to QPM lines at CIMMYT, Mexico, VPKAS, India and elsewhere in Asia and Africa.

For a country like India, with diverse climatic and soil situation, we need to develop a number of QPM composites and hybrids. There is a need to emphasize composites equally as hybrids, since the hybrid seed production and distribution are not feasible in many parts of the country. The limiting factor for QPM is the unavailability of suitable seeds to the farmers and the lack of location specific QPM cultivars. There is also a need to create awareness among consumers and industry on the use of QPM in food and feeds. There is also an urgent need to involve private players for the seed production and, link the industry with the farmers so that the farmers are assured of fair price for their produce. We at VPKAS have initiated a move to develop a linkage between seed producers, farmers and the industry.

References

Anon. 2008. Bulletin published by the AICRP on Maize, Directorate of Maize Research, New Delhi, pp. 21.

Anon. 2007. Fertilizer Statistics Eds Chanda, T.K., Kuldeep, S. and Robertson, The fertilizer association of India, New Delhi.

Babu, R., Nair, S.K., Kumar, A., Venkatesh, S., Sekhar, J.C., Singh, N.N., Srinivasan, G., Gupta, H.S. 2005. Two generation marker-aided backcrossing for rapid conversion of normal maize lines to quality protein maize (QPM). Theor. Appl. Genet. 111, 888-897.

Bressani, R. 1991. Protein quality of high-lysine maize for humans. Cereal Food World. 36, 806-811.

Bressani, R. 1992. Nutritional value of high-lysine maize in humans. In Quality Protein Maize (Ed Mertz, E. T.). Am. Assoc. Cereal Chemists. 205-225.

Gibbon, B.C., Larkins B.A. 2005. Molecular genetic approaches to developing quality protein maize. Trends Genet. 21, 227-233.

Glover, D.V. and Mertz, E.T. 1987. In Nutritional Quality of Cereal Grains: Genetic and Agronomic Improvement (Eds Olson, R. A. and Frey, O. J.). Am. Soc. Agron. Madison, WI, pp. 183-236.

Graham, G.G., Lembake, J. and Morales, E. 1990. Quality Protein Maize and as the sole source of dietary protein and fat for rapidly growing young children. Pediatriccs. 85, 85-91.

Graham, G.G., Placko, R.P. and Maclean, W.C. 1980. Nutritional value of normal, *opaque2* and, *sugary* 2 and *opaque* 2 maize hybrids for children and infants 2. Plasma free amino acids. J. Nutr. 110, 1070-1074.

Krivanek, A.F., De Grote, H., Gunaratna, N.S., Diallo, A.O., Friesen, D. 2007. Breeding and disseminating quality protein maize (QPM) for Africa. African Journal of Biotechnology. 6, 312-324.

Lopez-Pereira, M.A. 1993. Economics of Quality Protein Maize as a feed stuff. Agribusiness. 9, 557-568.

Mertz, E.T., Bates, L.S., Nelson O.E. 1964. Mutant gene that changes protein composition and increases lysine content of maize endosperm. Science. 145, 279-280.

Mertz, E.T., Vernon, O.A., Bates, S. and Nelson, O.E. 1965. Protein value of Colombian *opaque* -2 corn for young adult men. Science. 148, 1741-1744.

Paes, M.C.D. and Bicudo, M.H. 1964-1994.Nutritional Perspectives of quality protein maize. In Quality Protein Maize. Proceedings of the International Symposium on Quality Protein Maize (Eds Larkins, B.A. and Mertz, E.T.), Sete Lagaos, Brazil, pp. 65-78.

Paez, A.V., Helm, J.L. and Zuber, M.S. 1969. Lysine content of *opaque* -2 maize kernels having different phenotypes. Crop Sci. 9, 251-252.

Vasal, S.K., Srinivasan, G., Pandey, S., Gonzalez, C.F., Crossa, J. and Beck, D.L. 1993a. Heterosis and combining ability of CIMMYT's quality protein maize germplasm: I. Lowland tropical. Crop Science. 33, 46-51.

Vasal, S.K., Srinivasan, G., Gonzalez, C.F., Beck, D.L. and Crossa. 1993b. Heterosis and combining ability of CIMMYT's quality protein maize germplasm: II. Subtropical. J. Crop Sci. 33, 51-57.

Table 1: QPM Hybrids Released for Commercial Cultivation in India

Name	Pedigree	Year of Release	Centre's name	Maturity group
Shaktiman 1	(CML 142 x CML 150) x CML 186	2001	Dholi	Full season
Shaktiman 2	CML 176 x CML 186	2004	Dholi	Full season
HQPM 1	HKI 193-1 x HKI 163	2005	Uchani	Full season
Shaktiman 3	CML 161 x CML 163	2006	Dholi	Full season
Shaktiman 4	CML 161 x CML 169	2006	Dholi	Full season
HQPM 5	HKI 163 x HKI 161	2007	Uchani	Full season
Vivek QPM 9	VQL 1 x VQL 2	2007	Almora	Extra early

Table 2: Major Agronomic Traits of Vivek QPM 9

Traits	Vivek QPM 9
Plant Height (cm)	195
Reaction to turcicum blight (scale 1-5)	1.0
Days to silking	51.0
Protein Content (%)	8.46
Tryptophan (%)	0.83
Lysine	4.19
Lysine (accepted range in QPM)	2.7-4.5%
Iron (ppm)	37.0
Zinc (ppm)	29.0
Kernel hardness	Hard
Grain yield (t/ha)	6.2

Table 3: Grain Yield (kg/ha) of QPM hybrid (FQH 4567) in Zone I and Zone IV

Hybrid	Zone I			Zone IV		
	2005	**2007**	**Mean**	**2005**	**2007**	**Mean**
FQH 4567	6118	5567	5843	3531	7338	5435
Vivek 9	6360	5502	5931	3502	7306	5404
CD 5%	638	998		488	863	

Table 4: Grain Yield (Kg/ha) of Vivek QPM 9 in Uttarakhand under Organic Condition

Location	Vivek QPM 9	Vivek Maize Hyb. 9	C.D. (5%)	CV %
Kharif 2006				
Hawalbagh	4503	3965	1249	15.67
Majhera	4633	4800	1160	19.60
Thal	1944	1204	NS	27.80
Ranichauri	6020	5740	1073	12.15
Mean (4)	4275	3927	-	-
Kharif 2007				
Hawalbagh	4738	4493	731	9.74
Majhera	5611	6333	808	727
Mean (2)	5175	5413	-	-
Weighted Mean (2 years)	4575	4422	-	-
Rank	1	3		
Superiority of FQH 4567		3.46%		

CM 212

Normal

QPM version

Fig. 1. Normal and QPM Version (VQL 1) of CM 212

CM 145

Normal **QPM version**

Fig. 2: Normal and QPM Version (VQL 2) of CM 145

(a) **(b)**

Fig. 3: A Crop of Vivek QPM 9 at Maturity (a) and the Cobs (b)

❑❑❑

27

Investigating Two Medicinal *Scutellaria* Species of Himalayan Origin

NIRMAL JOSHEE[1], AGNES M. RIMANDO[2], PRAHLAD PARAJULI[3], GOPAL S. RAWAT[4], AND ANAND K. YADAV[1]

[1]Agricultural Research Station, Fort Valley State University, Fort Valley, GA, USA; [2]USDA-ARS, Oxford, MS, USA; [3]Karmanos Cancer Research Institute, Wayne State University, Detroit, MI, USA, [4]Wildlife Institute of India, Dehradun, Uttarakhand, India.

ABSTRACT

Our current research emphasis is on medicinal, nutraceutical, and biofuel plants. We have been working on the bioactive flavonoids from *Scutellaria* for their anti-inflammatory and anticancer properties. There is very little scientific evidence for the anticancer bioactivity of *Scutellaria* species, most of which has come from studies on the root extracts of a limited number of species, namely, *Scutellaria baicalensis* and *S. barbata*, among few. Leaf and stem extracts of *Scutellaria*, although known to be substantially rich in flavonoid content and other active ingredients have rarely been studied for their potential biomedical activities. Among the leaf, stem, and root extracts examined, the leaf extracts of two Himalayan species, *S. angulosa* and *S. scandens*, showed consistent, dose-dependent, antiproliferative and pro-apoptotic activities against various malignant cell lines. Here we report development of micropropagation protocols for rapid mass multiplication and conservation. This study addresses concerns related to the need to pay closer attention to issues of sustainable and ethical harvests, as the majority of plants entering the trade are still being sourced from the wild.

Key Words : Scutellaria, Micropropagation, Conservation, Flavonoids, Bioactive Compounds

INTRODUCTION

The green consumerism, as pointed out by Duke (1993), has not yet reached the pharmaceutical industry, where the "green" pharmaceutical might be defined as that product whose major ingredient compound is still commercially extracted from higher plant(s). Recent research trends into the African medicinal plants show that while ethnobotanical surveys continue, there is an appreciable increase in research activity in the area of bioactivity of natural products. For example, between 1981 and 2002, five percent of the 1,031 new chemical entities approved as drugs by the US Food and Drug Administration (FDA) were natural products, while another 23 percent were natural-product-derived molecules (Clardy and Walsh, 2004). Many species of *Scutellaria* have been extensively used for their medicinal properties in China, India, Korea, Japan, several European countries and North America. Thus far, more than 50 phenolic compounds and several essential oils like monoterpenoids, sesquiterpenoids, phenylpropane derivatives or iridoid glycosides, have been reportedly isolated from many *Scutellaria* species. Medicinal and aromatic plants are among the high priority species for conservation action, as wild-crafting will certainly continue to play a significant role in their future trade. The sustainable commercial use of these biological resources may provide a financial instrument for nature conservation (Lange, 2004). The present study strengthens the need for phytochemical cataloging of all those plant species appropriate for conducting multidisciplinary research. It is especially relevant to the plant species that plunge into the category of being rare, threatened or endangered. *Scutellaria* is a perennial, herbaceous genus in the Lamiaceae family with 300 - 360 species (Bailey and Bailey, 1978; Poton, 1992; Joshee *et al.*, 2002; Cole *et al.*, 2007). This genus is well adapted to the North American climate which is currently represented by about 90 species of *Scutellaria.*

All over the world, the medicinal plants are gathered from the wild. With the progressive depletion of native populations, there is an urgent need for conservation and domestication of these plants. Currently, many pharmaceutical and nutraceutical companies are beginning to develop products based on the traditional medicine. *Scutellaria* species is used in alternative medicine as an anti-inflammatory, abortifacient, antispasmodic, emmenagogue, febrifuge, nervine, sedative and a strong tonic. Skullcap is well known among the Cherokee and other Native American tribes, as a strong emmenagogue and female medicinal herb (Moerman, 1998). Flavonoids from the roots of *S. baicalensis* have been reported to have inhibitory effect on some human viruses (Li *et al.*, 1993; Baylor *et al.*, 1992; and Konoshima *et al.*, 1992) whereas anti-inflammatory properties of *S. rivularis* extracts have been associated with three Flavonoids: baicalin, baicalein, and wogonin (Ching and Den, 1996). It has been suggested that baicalin may be therapeutically useful for mitigating the pathogenic effects of *Staphylococcal* exotoxins inhibiting the signaling pathway (Krakauer *et al.*, 2001). The price of skullcap has steadily increased during the past five years and is currently around US $ 8.80-17.60/kg. The demand for skullcap in the world market is predicted to grow steadily at an annual rate of 20-30 percent (Greenfield and Davis, 2004).

There is only scanty research information available on the phytochemicals of *Scutellaria angulosa* and *S. scandens*. The information that we could find in literature is related to the chemical compounds found in *S. scandens* as reported by Miyaichi *et al.* (1988a and b). In this study five new flavanones (I-V) were isolated from the roots of *Scutellaria scandens*, along with oroxylin A, dihydrooroxylin A, wogonin, norwogonin, chrysin, dihydrobaicalein, wogonin 7-O-glucuronide, dihydrobaicalin chrysin 7-O-glucuronide, baicalin and baicalein. There are few reports available on the micropropagation of *Scutellaria* species (Li *et al.*, 2000; Tascan *et al.*, 2007; Joshee et al., 2007; Sinha *et al.*, 1999, and Stojakowska *et al.*, 1999) but there is no such information on these two Himalayan species. We have successfully grown both of these two *Scutellaria* species in our specialty plants house which is located at Agricultural Research Station of the Fort Valley State University in Fort Valley, Georgia, USA (Fig. 1 A and C). In this communication we report a micropropagation protocol for *S. angulosa* to assist conservation. We also discuss the importance of screening bioactive contents of these species along with their therapeutic importance.

MATERIALS AND METHODS

Plant Materials

In this study explants were obtained from mature plants of *Scutellaria angulosa* growing in the specialty plant house at the Fort

Valley State University, Fort Valley, GA. Vigorously growing branches (containing 5-6 nodes from the tip) were harvested to obtain nodal explants and initiate cultures (Fig. 1 C).

Sterilization and Shoot Induction

Initial steps for plant material collection, washing in running water and sterilization were followed as described earlier (Joshee *et al.*, 2007). In brief, shoots cut into small pieces were washed under running tap water for 10 minutes and then transferred to distilled water with a drop of Tween-20 in a beaker continuously stirring for 30 minutes. After 30 min, water was decanted and explants were sterilized with 10 percent Clorox for 15 min and then washed three times with autoclaved distilled water (five minutes each) in a laminar flow hood. Following washing, plant material was trimmed to the final size to prepare explants. Approximately 1.0- to 1.5-cm-long nodes were selected as the final explants to initiate cultures. The plant culture medium used to study shoot induction consisted of MS salts (Murashige and Skoog, 1962) and B5 vitamins (Gamborg *et al.*, 1968). All media had 3 percent sucrose as the main carbon source. Media were gelled using 0.25 percent phytagel. Cultures were initiated in 100 x 15 mm petri dishes containing 25 mL culture medium. Shoot induction was studied in response to two cytokinins, BA at 0.44, 2.2, 4.4, and 8.8 μM, and kinetin at 0.46, 2.3, 4.6, and 9.2 μM concentrations. These four concentrations correspond to 0.1, 0.5, 1.0, and 2.0 mg/L of cytokinins used. The elongated adventitious buds / shoots that measured 5 mm or longer were evaluated after eight weeks in culture. These explants were incubated for three weeks in light at 24 - 26°C with 14 h photoperiod. Micropropagation experiments were repeated two times under the identical environmental conditions with at least ten explants used for each treatment.

Root Induction

Elongated shoots on basal MS medium measuring 4 to 5 cm in length were excised and transferred to MS-based two compositions of rooting media (MS + 4.9 μM IBA and MS + 9.8 μM IBA) under diffused light at 14 h photoperiod for three weeks. At the end of three weeks period, cultures were evaluated for root induction and growth. Once, three to four strong roots were visible, plants were gently washed under tap water to remove agar prior to acclimatization.

Phytochemical Screening, Cell Culture and Flow-Cytometry

Analysis of six Flavonoids isolated (apigenin, baicalein, baicalin, chrysin, scutellarein, and wogonin) was carried out using a Hewlett-Packard 1050 HPLC system utilizing an Inertsil ODS-2 5ì, 250 mm X 4.6 mm column (MetaChem Technologies, Inc., Torrance, CA), and monitored at ë 270 nm.

The flavonoids were quantitated from a calibration curve of the standards (Sigma Chemicals) with 6-hydroxyflavone as the internal standard.

Human glioma (U87-MG) cell lines were plated with $5x10^5$ cells/mL/well in the 12-well plates in the presence of *Scutellaria* extracts. After 72 h, cells were washed and induction of apoptosis/necrosis was analyzed by flow-cytometry. Cells harvested after 72 h of culture were stained with FITC-conjugated Annexin-V and Propidium Iodide (PI) (BD Pharmingen) and then analyzed in a flow-cytometer (Becton-Dickinson FACScan) as described earlier (Parajuli *et al.*, 2004). Cells were gated based on forward scatter (FS) and side scatter (SS). Data analysis was performed using CellQuest software package (Becton-Dickinson ImmunoCytometry Systems). Cells positive for PI stain alone (upper left quadrant) represent necrosis, while cells for Annexin-V alone or those positive for both Annexin-V and PI represent early and late apoptosis, respectively.

RESULTS

Both *Scutellaria* species, *S. angulosa* and *S. scandens,* profusely flowered and were frequented by pollinators (Fig. 1B), eventually setting seeds. At the end of two weeks of incubation period in the shoot induction medium, explants slightly swelled and started to show changes at the surface while at the end of four weeks small shoots were observed. At this point, explants with shoot cluster were transferred to MS basal medium with 14 hrs photoperiod. Two weeks culture in hormone-free medium, promoted shoot elongation and 2-11 shoots were visible on the nodal explants.

Response to Cytokinins and Auxin

In an experiment using BAP and NAA as growth regulators, shoot bud induction was achieved in all treatments having BAP. Nodal segments initially responded by axillary bud break, and 0.5 mgL^{-1} BAP and 0.5 mgL^{-1} BAP + 0.1 mgL^{-1} NAA were the most suitable treatments

(Fig. 1 D). Treatment containing NAA along with BAP induced callus at the cut end of the explant stuck in the medium. In experiments where BAP was replaced by kinetin, nodal segments showed response similar to that of BAP but the overall growth looked better in the case of BAP treatments.

Rooting

Shoots (3 to 4-cm-long) of *S. angulosa* were cut and inoculated into different rooting media as mentioned earlier. Microshoots were inoculated in the medium and were incubated at 25°C under diffused light for three weeks. Following three weeks of incubation all cultures were transferred to 12 hrs photoperiod. In two weeks time, more than 90 percent shoots produced roots. These roots were allowed to elongate for a month and then plants were transferred to a mist chamber for further acclimatization in pots containing autoclaved vermiculite. Prior to transfer, agar was gently removed from roots under running tap water. Well-acclimatized plantlets routinely showed a survival rate of 80 - 90 percent or above. To get fully established plantlets took 7-10 days.

Phytochemical Screening

In a separate study (Parajuli *et al.*, 2008, communicated), to examine whether the anti-proliferative activity is tumor-specific, two human malignant tumor cell lines (U87-MG glioma and MDA-MB-231 breast carcinoma) and the corresponding non-malignant cells (NHA, normal human astrocytes primary culture, and HMEC, non-malignant human mammary epithelial cell line) were cultured for 96 h in presence of selected *Scutellaria* extracts and isolated flavonoids. Baicalein as well as *S. angulosa* (leaf), and *S. scandens* (leaf) extracts significantly inhibited proliferation of the malignant glioma (U87-MG) cells. Similarly, *Scutellaria* extracts and six flavonoids significantly inhibited proliferation of breast cancer (MDA-MB-231) cells. On the other hand, proliferation of non-malignant cells, NHA and HMEC, were not affected by any of the extracts or flavonoids tested. This indicated specificity of bioactive compounds present in these two skullcap species. The flavonoids isolated from *Scutellaria* have demonstrated anticancer activities *via* inhibition of proliferation and induction of apoptosis (Middleton *et al.*, 2000, Zhang *et al.*, 2003, Kuwashima *et al.*, 2005).

DISCUSSIONS

In our experience, BAP is one of the most suitable cytokinins for shoot bud induction (Joshee *et al.*, 2007; Tascan *et al.*, 2007) though thidiazuron (TDZ) has also been used successfully (Li *et al.*, 2000). Our results show that kinetin can also be used with same efficacy. Callus produced at the cut ends of nodal segments was non-morphogenic since subsequent subculturing to basal or cytokinin-supplemented media did not result in any response. Micropropagation experiments with *S. scandens* are underway and preliminary results are quite encouraging. Phytochemical screening and animal cell studies indicate the presence of bioactive compounds in both, *S. angulosa* and *S. scandens, species* (Parajuli *et al.*, 2008). The more we use medicinal herbs on a commercial scale the more important it is to ensure that they come from sustainable sources, so that these plants will continue to exist in wild places. Our study clearly indicated the importance of wild plants and potential therapeutic compounds present therein. This calls for a long term plan for the conservation efforts along with thorough phytochemical screening and biomedical evaluation research.

ACKNOWLEDGMENTS

Financial assistance received from the USDA-CSREES, 1890 Institute Capacity Building Grants Research Project (GEOX-2002-02989) is thankfully acknowledged.

References

Bailey, L.H. and Bailey, E.Z. 1978. Hortus third: A concise dictionary of plants cultivated in the USA and Canada. Macmillan Publishing Co., Inc. New York, pp. 1920.

Baylor, N.W., Fu, T., Yan, Y.D. and Ruscetti, F. W. 1992. Inhibition of human T cell leukemia virus by the plant flavonoid baicalin (7-glucuronic acid,5,6-dihydroxyflavone). J. Infect. Diseases. 165, 433-437.

Ching, L.C. and Den, E.S. 1996. The anti-inflammatory activity of *Scutellaria rivularis* extracts and its active components, baicalin, baicalein, and wogonin. Amer. J. Chinese Medicine. 24, 31-36.

Clardy, J. and Walsh C. 2004. Lessons from natural molecules. Nature. 432, 829-837.

Cole, I.B., Saxena, P.K. and Murch, S.J. 2007. Medicinal biotechnology in the genus *Scutellaria*. In-vitro Cell Dev. Biol-Plant. 43, 318-327.

Duke, J.A. 1993. Medicinal plants and the pharmaceutical industry, pp. 664-669. In: Janick, J. and Simon, J.E. (Eds.), New Crops. Wiley, New York.

Gamborg, O.L., Miller, R.A. and Ojima, K. 1968. Nutrient requirements of suspension cultures of soybean root cells. Exp. Cell Res. 50, 151-158.

Greenfield, J. and Davis, J.M. 2004. Skullcap (*Scutellaria laterifolia* L.). Medicinal Herb Production Guide. http://www.naturalmedicinesofnc.org, July 12, 2006.

Joshee, N., Mentreddy, S.R. and Yadav, A.K. 2007. Mycorrhizal fungi and growth and development of micropropagated *Scutellaria integrifolia* plants. Industrial Crops and Products. 25, 169-177.

Joshee, N., Patrick, T.S., Mentreddy, S.R. and Yadav, A.K. 2002. Skullcap: Potential medicinal crop, pp. 22-30, in Janick, J. and Whipkey, A. (Eds). Trends in New Crops and New Uses, ASHS Press, Alexandria, VA.

Konoshima, T., Kokumai, M., Kozuka, M., Iinuma, M., Mizuno, M., Tanaka, T., Tokuda, H., Nishino H., and Iwashima, A. 1992. Studies on inhibitors of skin tumor promotion. XI. Inhibitory effects of flavonoids from *Scutellaria baicalensis* on Epstein-Barr virus activation and their anti-tumor-promoting activities. Chem. Pharm. Bull. 40, 531-533.

Krakauer, T., Li, B.Q. and Young, H.A. 2001. The flavonoid baicalin inhibits superantigen-induced inflammatory cytokines & chemokines. FEBS Letters. 500, 52-55.

Kuwashima, N., Nishimura, F., Eguchi, J., Sato, H., Hatano, M., Tsugawa, T., Sakaida, T., Dusak, J.E., Fellows-Mayle, W. K., Papworth, G. D., Watkins, S. C., Gambotto, A., Pollack, F., Storkus, W. J. and Okada, H. 2005. Delivery of dendritic cells engineered to secrete IFN-á into central nervous system tumors enhances the efficacy of peripheral tumor cell vaccines: dependence on apoptotic pathways. J. Immunol. 175(4), 2730-40.

Lange, D. 2004. Medicinal and aromatic plants: Trade, production, and management of botanical resources. Acta Horticulturae. 629, 177-197.

Li, B-Q., Fu, T., Yan, Y-D., Baylor N.W., Ruscetti, F.W. and Kung, H.F. 1993. Inhibition of HIV infection by baicalin- a flavonoid compound purified from Chinese herbal medicine. Cell Mol. Biol. Res. 39, 119-124.

Li, H., Murch, S.J. and Saxena, P.K. 2000. Thidiazuron-Induced *de novo* shoot organogenesis on seedlings, etiolated hypocotyls and stem segments of Huanq-qin, Plant Cell, Tissue and Organ Culture. 62, 169-173.

Li, J. and Khan, I. 2006. The advanced bioactivity studies of *Scutellaria baicalensis* Georgi and its phenolic compounds. Acta Horticulturae (ISHS) 720: 157-169.

Middleton, E. Jr., Kandaswami, C. and Theoharides, T.C. 2000.The effects of plant flavonoids on mammalian cells: implications for inflammation, heart disease, and cancer. Pharmacol. rev. 52(4), 673-751.

Miyaichi, Y., Imoto, Y., Kizu, H., and Tomimori, T. 1988a. Studies on the Nepalese crude drugs. X. On the flavonoid constituents of the root of *Scutellaria scadens* Buch.-Ham. D. Don. Chem. Pharm. Bull. 36, 2371-2376.

Miyaichi, Y., Imoto, Y., Kizu, H., and Tomimori, T. 1988b. Studies on the Nepalese crude drugs.X. On the flavonoid and stilbene constituents of the leaves of *Scutellaria scadens* Buch.-Ham. D. Don. Shoyakugaku Zasshi 42, 204-207.

Moerman, D. E. 1998. Native American Ethnobotany. Timber Press, Portland, OR;

Murashige, T. and Skoog, F. 1962. A revised medium for rapid growth and bioassays with tobacco tissue cultures. Physiologia Plantarum. 15, 473-497.

Parajuli, P., Mathupala, S. and Sloan, A.E., 2004. Systematic comparison of dendritic cell-based immunotherapeutic strategies for malignant gliomas: in vitro induction of cytolytic and natural killer-like T cells. Neurosurgery. 55(5), 1194-1204.

Parajuli, P., Joshee, N., Rimando, A., Mittal, S. and Yadav, A.K. 2008. Anti-tumor mechanisms of various *Scutellaria* extracts and constituent flavonoids. Planta Medica (communicated - under review).

Poton, A.1992. Kew Bull. 45, 400.

Sinha, S., Pokhrel, S., Vaidya, B.N. and Joshee, N. 1999. *In vitro* micropropagation and callus induction in *Scutellaria discolor* COLEBR.- A medicinally important plant of Nepal. Indian J. Plant Genetic Resources. 12(2), 219-223.

Stojakowska A., Malarz, J. and Kohlmuenzer, S. 1999. Micropropagation of *Scutellaria baicalensis* Georgi. Acta Soc. Bot. Pol. 68, 103-107.

Tascan A., Adelberg, J., Joshee, N., Yadav, A.K. and Tascan, M. 2007. Liquid culture systems for *Scutellaria* species. Acta Horticulturae (ISHS). 756, 163-170.

Zhang, D. Y., Wu, J., Ye, F., Xue, L., Jiang, S., Yi, J., Zhang, W., Wei, H., Sung, M., Wang, W. and Li, X. 2003. Inhibition of cancer cell proliferation and prostaglandin E2 synthesis by *Scutellaria baicalensis*. Cancer Research. 63, 4037-4043.

contd...

Fig. 1. Plants of *Scutellaria scandens* **(A)** and *Scutellaria angulosa* **(C)** growing in the specialty plants house at the Fort Valley State University, Fort Valley, GA, USA; a pollinator is visiting the *Scutellaria scandens* flower **(B)**; and the successful *in vitro* micropropagation of *Scutellaria angulosa* germplasm by means of single node explants on medium containing 0.5 mgL^{-1} BAP + 0.1 mgL^{-1} NAA **(D).**

□□□

28

Molecular Characterization of Indigenous Apricot (*Prunus armeniaca* L.) Accessions from Trans-himalayan Region using ISSR Markers

MEETUL KUMAR, GYAN P MISHRA, RAGHWENDRA SINGH, JITENDRA KUMAR, MANISH BHOYAR AND SHASHI BALA SINGH
Defence Institute of High Altitude Research, DRDO, Leh, Ladakh, J & K, India

ABSTRACT

The Phylogenetic relationships of thirty-six locally grown nondescript apricot (*Prunus armeniaca* L.) genotypes which are collected from nine sampling sites from two valleys *viz*. Nubra (9,600 ft) and Leh (11,500 ft) of trans-Himalayan region of Ladakh (Jammu & Kashmir, India) were analyzed using 20 ISSR markers. Of the 20 primers used in the present study, 11 primers resulted in amplifications and produced 58 bands across 36 genotypes, of which 56 were polymorphic, with an average of 5.09 polymorphic fragments per primer. The number of amplified fragment ranging from three (ISSR 3) to seven (ISSR 7 and ISSR 8) and which varied in size from 200 bp to 2,500 bp. Percentage of polymorphism ranged from 80% (ISSR 9) to a maximum of 100% (nine primers). The primers based on (CT)n produced maximum number of bands (nine) while, (AT)n and many other motifs gave no amplification at all with any of these thirty six genotypes. Na, Ne, h, I, Ht, Hs and Nm estimates for ISSR primers were 1.9483, 1.5683, 0.3310, 0.4948, 0.331, 0.1803 and 0.2992, respectively. The results of PCA analysis were comparable to the cluster analysis. These analyses, allowed us to identify the groups corresponding to the two *Apricot* collection sites. The results of the present study can be seen as a starting point for future researches on the population and evolutionary genetics of these genotypes from the cold arid desert of trans-Himalayas.

Keywords: Apricot, *Prunus armeniaca,* genetic diversity, ISSR, trans-Himalaya

INTRODUCTION

Most cultivated apricots of the world belong to one species, *Prunus armeniaca* L., which is native to Asia particularly the slopes of trans-Himalayan region and the Caucasus region. In India, apricot is dispersed over a vast area bordering the trans-Himalayan region encompassing the whole of northern India. Ladakh, which lies between 32° to 36° N latitude and 75° to 80° E longitude is home to most diverse flora and fauna. The apricot occupies 3200 ha with a total production of 5200MT (Dwivedi and Attrey, 2002) making it an important sources of livelihood and economy of this region. In addition, apricot prevails well in dry agro-climatic conditions of Ladakh which is characterized by rocky and sandy soils and has the capability to tolerate freezing temperatures during winter, when dormant. This perhaps makes it an ideal fruit crop for this region. However, the accelerated and uncontrolled use of this important fruit crop in recent times, has led to an imbalance in the ecosystem and consequently a gradual loss in genetic diversity. Therefore, appropriate measures need to be adopted for conserving the genetic makeup of natural population of *Prunus armeniaca* L. Classical methods for identification of cultivars based on morphological, phenotypic and agronomical traits were initially used to identify the cultivars. However, due to environmental influence, phenotypic traits in many cases, fail to serve as unambiguous markers for systematics and diversity analysis. In addition, the morphological and agronomical traits are limited and don't cover the whole genome. Molecular markers developed during the last two decades, have largely overcome the problems associated with phenotypic-based classifications. Initially isozymes and Restriction Fragment Length Polymorphisms (RFLPs) served as reliable markers for genetic analysis in plants, but PCR based techniques developed in early 1990s, like Random Amplified Polymorphic DNA (RAPD), Inter Simple Sequence Repeats (ISSR), Amplified Fragment Length Polymorphism (AFLPs) and Simple Sequence Repeats (SSRs) also known as microsatellites, provide DNA markers that are dispersed throughout the plant genome and are easy to reproduce and analyze. High levels of polymorphism and co-dominant nature have made SSRs as ideal markers for understanding genetic diversity in plants. However, the cost effectiveness of these SSR primers has limited their potential in plants diversity studies. ISSR markers, which have dominant inheritance, use SSR repeat-anchored primers and are useful in detecting polymorphism. ISSR markers detect genetic polymorphism among accessions by generating large number of markers that target multiple microsatellite loci distributed across the genome. In addition, they are simpler than SSR markers since no prior knowledge of the sequences are required. Limited studies done on European and Central Asian apricot accessions by SSR markers revealed them to be clustered into six major groups (Gómez *et al.*, 2005). The identification and estimation of genetic relationships of indigenous promising accessions of Apricot is essential since it can provide basic information for breeding programs and increase the repository of Apricot accessions. The aim of the present study was to provide a comphrensive genetic data for theoretical database for genetic improvement and property right protection of indigenous Apricot accessions of trans-Himalayan region.

MATERIALS AND METHODS

Plant Material and Genomic DNA Amplification

Thirty six apricot genotypes were collected from various villages located in Ladakh (Table1; Figure1). Genomic DNA was isolated from freshly harvested young leaves following the modified CTAB method (Saghai-Maroof *et al.*, 1984). The quality and quantity of DNA was estimated both spectrophotometrically, as well as visually, by ethidium bromide staining on agarose gel (Life Science Technologies, USA). DNA was diluted to a concentration of 20 ng/μl. For ISSR-PCR, 20 primers synthesized (Integrated DNA Technologies, Calif USA) were used for amplification, (Table 2). A single primer was used for PCR amplification reaction which compromised of 2 μl of genomic DNA (40ng), 2.5 μl of primer (0.2μM concentration), 2.5μl of 10X buffer (0.1M Tris pH 8.8; 0.5M KCl; 15mM $MgCl_2$; 0.1% gelatin), 2.5μl of 200μM dNTPs and 1.0U of Taq polymerase. The PCR reactions were performed on 96 well-plates on a gradient thermal cycler (Biometra) Germany; under the following conditions: a hot start of 94°C for 2 min. followed by 40 cycles of denaturation at 94°C for 40 s annealing for 45 s, extension at 72°C for 1min 30 s and final extension at 72°C for 7 min. The annealing temperature was specific for each primer

(Table 2). The molecular weight markers, 100 bp ladder (Bangalore Genei) were used for band sizing. Consistency of the bands was checked twice, and only the reproducible bands were scored in all the samples for each of 18 primers, separately. Bands were recorded as present (1) or absent (0). Numerical and taxonomical analyses were conducted using the NTSYS-pc software version 2.0 from Exeter software. A similarity coefficient matrix was constructed using SIMQUAL subroutine. This matrix was subjected to unweighted pairwise group method using arithmetic average analysis (UPGMA) to generate a dendogram using SAHN subroutine and Tree plot of NTSYS-pc. Polymorphism information content (PIC) and resolving power (Rp) values were calculated for each primer according to procedure of Prevost and Wilkinson (1999).

Data Collection and Analysis

The banding patterns obtained from ISSR were scored as present (1) or absent (0), each of which was treated as an independent character regardless of its intensity. Gel-Pro analyzer version 3-1 software was used to score ISSR profile. Jaccard's similarity coefficient (J) was used to calculate similarity between pairs of accessions. Where, [$J = n_{xy} / (n_t - n_z)$], n_{xy} is the number of bands common to accession A and accession B; n_t the total number of bands present in all samples and n_z the number of bands not present in A and B but found in other samples. The similarity matrix was subjected to cluster analysis by unweighted pair group method for arithmetic mean (UPGMA) and a dendrogram was generated using the program NTSYS-PC (Rohlf, 1992). Support for clusters was evaluated by bootstrapping analysis (Felsenstein, 1985). FreeTree software (Pavlicek, *et al.,* 1999) was used for construction of phylogenetic trees on the basis of distance data and for bootstrap analysis of the trees robustness. POPGENE software was used to calculate Nei's unbiased genetic distance among different species with all markers, including monomorphic markers. Nei's unbiased genetic distance is an accurate estimate of the number of gene differences per locus when populations are small. Within species diversity (Hs) and total genetic diversity (Ht) (Nei, 1978) were calculated within the species and within four major groups (as per their collection site) by POPGENE software. The ISSR data were subjected to a hierarchical analysis of molecular variance (AMOVA), as described by Excoffier *et al.* (1992), using three hierarchical levels; individual, population and their regions. The analysis was performed using GenALEx software (Peakall and Smouse, 2001). PCA analysis was also employed to evaluate the resolving power of the markers to differentiate between the genotypes. GenAlEx software was used to calculate a principal coordinates analysis (PCA) that plots the relationship between distance matrix elements based on their first two principal coordinates. According to Prevost and Wilkinson (1999) the resolving power (Rp) of a primer is: Rp = Ó *IB* where *IB* (band informativeness) takes the value of: 1–[2x (0.5–P)], P being the proportion of the 36 genotypes (*Apricot* genotypes analyzed) containing the band.

RESULTS

ISSR Band Patterns

Twenty ISSR oligonucleotides were used for amplification of all the thirty-six genotypes. These ISSR primers were obtained from Applied Biosciences and out of 20 only 11 primers give rise to reproducible amplification products, while 9 primers did not gave any amplification (Table 2). The sequences of these 20 primers seem to indicate that microsatellites more frequent in Apricot contain the repeated dinucleotides (AG)*n*, (GA)n, (TG)n, (CT)n, (AC)*n*, (GT)*n*, and trinucleotides (CCG)n, (GGC)n. The number of bands produced with different repeat nucleotide were more with the (GT)*n*, (GA)*n*, (CT)*n*, and (AC)*n* primers (ISSR8, 6, 7 and 9). The primers that were based on the (GA)*n*, poly (CT)*n* and (GT)*n* motif produced more polymorphism on average (7 bands per primer) than the primers based on any other motifs used in the present investigation. The primer sequences which did not amplified in the present investigation contain the dinucleotides repeat sequences as (AT)n, (GT)n, (TC)n, (TA)n, (GT)n, (CT)n with T at the 5′ end, the repeated trinucleotides (ACC)n, (TGC)n and the repeated tetranucleotides (CTAG)n and (TGCA)n (Table 2). The 11 primers on an average produced 58 bands across 36 genotypes, of which 56 bands were polymorphic, accounting for 96.55%. Number of bands varied from three (ISSR 3) to seven (ISSR 6, 7 & 8), and sizes ranged from 200 bp to 2,500 bp. Average numbers of bands and polymorphic bands per primer were 5.27 and 5.01 respectively. Percentage of polymorphism ranged from 80% (ISSR9) to 100% (nine primers), with an average polymorphism of 96.5% across all the genotypes. Total no of bands amplified ranged from 45 (ISSR3) to 174 (ISSR8). The resolving power (Rp) of the eleven ISSR primers ranged from 2.5 (ISSR3) to 9.67 (ISSR8). Primer amplification details as obtained for each ISSR primer are shown in Table 2.

Genetic Variability Details Across all the Populations Using ISSR Markers

The respective values of Na, Ne, H, I, Ht, NPL and PPL across all the nine populations were found maximum for all these parameters for *Stok* genotypes while minimum values are observed for *Khalsar* and *Hundar* genotypes (Table 3.1). When these parameters were analyzed for valley divisions (i.e. Leh and Nubra valley) then all these respective values were found higher for Leh valley genotypes indicating that there is more variability in Leh valley than in Nubra valley for the samples analyzed in the present investigation (Table 3.2). The respective values for overall genetic variability for Na, Ne, H and I for 36 genotypes were also calculated and were found as 1.9483, 1.5683, 0.331 and 0.4948 respectively. Value for Ht and Hs were found to be 0.331, 0.1803 respectively. Mean coefficient of gene differentiation (Gst) value 0.4552 indicated that 54.48 % of the genetic diversity resided within the population. While, the estimate of gene flow in the population was found nil (Table 3.3). Molecular variance among valley was found as 22%, 16% variance among population while, within the population this value was found to be 62% (Table 4) indicating that there is more variations with in the population. This is helpful in making strategy for germplasm collection and evaluation.

Dendrogram Analysis for Apricot as obtained with ISSR Markers

The PCR amplification using 11 ISSR primers gave rise to reproducible amplification products. The complete data was based on a total of 1367 bands. Jaccard's similarity coefficient ranged from 0.05 to 0.39. The 36 genotypes were clustered into five clusters (I, II, III, IV and V) with 6, 3, 13, 3 and 11 genotypes respectively. Cluster I, II and V represents all the genotypes from Leh valley while, cluster III, and IV contains all the genotypes from Nubra valley. A dendrogram based on UPGMA analysis with ISSR data is shown in Figure 2.

Cluster I comprised of all the four genotypes from Leh and two genotypes from *Stok* village. Cluster II consisted of two genotypes from *Stok* and one from *Hemis*. Cluster III consists of all the four genotypes from *Khardung*, *Hundar* and *Khalsar* along with one genotype from *Thoise* village. Cluster III is again sub-classified into three sub-clusters i.e. IIIa, IIIb and IIIc. Three genotypes from *Thoise* village are the part of cluster IV. Cluster V consisted of four genotypes each from *Stakna* and *Nimu* along with three genotypes from *Hemis* village. This cluster is again sub-classified into three sub-clusters i.e. Va, Vb and Vc. The results of PCA analysis were comparable to the cluster analysis. Two genotypes i.e. *Stok*3 and *Hemis*3 (i.e. 7 and 11) appear to be distinct from other genotypes in the PCA (Figure 3).

DISCUSSIONS

Polymorphism in a given population is often due to the existence of genetic variants represented by the number of alleles at a locus and their frequency of distribution in a population. Heterozygosity corresponds to a probability that two alleles taken at random from a population can be distinguished using the marker in question. Thus a convenient quantitative estimate of marker utility and the polymorphism detected can be given in terms of the Na, Ne, H, I, Ht, Hs, Gst and Nm (Zhao *et al.,* 2006). Accessions with the most distinct DNA profiles are likely to contain the greatest number of novel alleles. Our results indicate the presence of great genetic variability among local nondescript genotypes of *Apricot.* ISSR markers are very useful in the assessment of *Apricot* diversity, the detection of duplicate sample in germplasm collection, and the selection of a core collection to enhance the efficiency of germplasm management for use in *Apricot* breeding and conservation. Our results also indicate that these markers have great potential to discriminate these thirty-six genotypes. In fact the inter-simple sequence repeats are regions lying within the microsatellite repeats, have a high capacity to reveal polymorphism and offer great potential to determine intra-genomic and inter-genomic diversity (Zietkiewicz *et al.,* 1994). Such characteristics, which were confirmed in the present study, makes the both the markers as better potential tool to carry out future population genetics studies. Nagaoka and Oigihara (1997) in their studies found that ISSR primers, compared with RAPD primers, produce more reliable and reproducible bands. However, in our experiments, once the PCR conditions were well set up, we obtained a high reproducibility for both RAPD and ISSR markers; only very faint fragments were not reproducible and such fragments were discarded. As previously pointed out, during the ISSR screening we obtained good amplification products from primers based on (CT)*n* and (GT)*n* repeats while (AT)*n* and some other primers gave no amplification products (Table 2), despite the fact that (AT)*n* dinucleotide repeats are thought to be the most abundant motifs in plant species (Martín and Sánchez-Yélamo, 2000; Morgante and Olivieri, 1993). Similar results were obtained in grapevine (Moreno *et al,* 1998), rice (Blair *et al,* 1999) and wheat (Nagaoka and Ogihara, 1997). A possible explanation of these results is that ISSR primers based on AT motifs are self-annealing, due to sequence complementarily, and would form dimers during PCR amplification (Blair *et al,* 1999). Reason behind non amplification of other repeats may be their absence in the genome.

The results of the present study can be seen as a starting point for future researches aimed at defining the level of intra- and inter-specific genetic diversity and to detect hybrids among these species. For this purpose, a larger number of natural populations of the species collected from the whole distribution area should be analyzed and additional primers tested. Furthermore, in order to design new diagnostic primers more effective in genetic discrimination among genotypes, discriminate bands could be cloned and sequenced. These studies have given important clues in understanding genotype relationship, which may further assist in developing and planning breeding strategies.

ACKNOWLEDGEMENTS

The authors are thankful to DRDO HQ for support and funding the project.

References

Blair, M.W., Panaud, O., Mccouch, S.R. 1999. Inter-simple sequence repeat (ISSR) amplification for analysis of microsatellite motif frequency and fingerprinting in rice (*Oryza sativa* L.). Theor. Appl. Genet. 98, 780–792.

Dwivedi, S.K. and Atrey, D.P. 2002. Possibilities of apricot production in high altitude regions of Uttaranchal. In: National Seminar on Development of Horticulture in Uttaranchal. Directorate of Horticulture, Govt. of Uttaranchal, Dehradun. 3-4 September (2001) pp.17-20.

Excoffier, L., Smouse, P.E., Quattro, J.M. 1992. Analyses of molecular variance inferred from metric distances among DNA haplotypes: application to human mitochondrial DNA restriction data. Genetics. 131, 479-491.

Felsenstein, J. 1985. Confidence limits on phylogenies: an approach using the bootstrap. Evolution. 39, 783-791.

Martín, J.P., Sánchez-Yélamo, M.D. 2000. Genetic relationships among species of the genus *Diplotaxis* (*Brassicaceae*) using inter-simple sequence repeat markers. Theor. Appl. Genet. 101, 1234–1241.

Martínez-Gómez, P., Sánchez-Pérez, R., Rubio, M., Dicenta, F., Gradziel, T.M. and Sozzi, G.O. 2005. Application of Recent Biotechnologies to *Prunus* Tree Crop Genetic Improvement. Cien. Inv. Agr. *32(2), 73-96.*

Moreno, S., Martin, J.P., Ortiz, J.M. 1998. Inter-simple sequence repeat PCR for characterization of closely related grapevine germplasm. Euphytica. 101, 117–125.

Morgante, M., Olivieri, A.M. 1993. PCR-amplified microsatellite markers in plant genetics. Plant J. 3, 175–182.

Nagaoka, T., Ogihara, Y. 1997. Applicability of inter-simple sequence repeat polymorphisms in wheat for use as DNA markers in comparison to RFLP and RAPD markers. Theor. Appl. Genet. 94, 597–602.

Nei, M. 1978. Estimation of average heterozygosity and genetic distance from a small number of individual. Genetics. 89, 583– 590.

Pavlicek, A., Hrda, S., Flegr, J. 1999. Free Tree – freeware program for construction of phylogenetic trees on the basis of distance data and bootstrap/jackknife analysis of the tree robustness. Application in the RAPD analysis of genus *Frenkelia*. Folia Biol. (Praha). 45, 97–99.

Peakall, R., Smouse, P.E. 2001. GenAlEx V5: Genetic Analysis in Excel. Population genetic software for teaching and research. Australian National University, Canberra, Australia, (http://www.anu.edu.au/BoZo/GenAlEx/).

Prevost, A., Wilkinson, M.J. 1999. A new system of comparing PCR primers applied to ISSR fingerprinting of potato cultivars. Theor. Appl. Genet. 98, 107–112.

Rohlf, F.J. 1992. NTSYS-PC: Numerical taxonomy and multivariate analysis system version 2.0. State University of New York (Stony Brook, New York).

Saghai-Maroof, M.A., Soliman, K.M., Jorgensen, R.A. and Allard, R.W. 1984. Ribosomal spacer length in barley: Mendelian inheritance, Chromosomal location and population dynamics. Proc. Natl. Acad. Sci. 81, 8104-8118.

Zhao, W.G., Zhang, J.Q., Wangi, Y.H., Chen, T.T., Yin, Y.L., et al. 2006. Analysis of Genetic Diversity in Wild Populations of Mulberry from Western Part of Northeast China Determined by ISSR Markers. J. Genet. and Mol. Biol. 196-203.

Zietkiewicz, E., Rafalski, A., Labuda, D. 1994. Genome fingerprinting by simple sequence repeat (SSR)-anchored polymerase chain reaction amplification. Genomics. 20, 176–183.

contd...

Table 1: Nine populations of *Apricot* collected from two valleys covering 36 sampling sites and their altitude.

Name of Valley	Accession No.	Sampling Site	Altitude (m)
Leh valley	Leh 1-4	Leh village	3505
	Stok 5-8	Stok village	3414
	Hemis 9-12	Hemis village	5500
	Stakna 13-16	Stakna village	3371
	Nimu 17-20	Nimu village	3900
Nubra valley	Khardung 21-24	Khardung village	3960
	Khalsar 25-28	Khalsar village	3500
	Hundar 29-32	Hundar village	3017
	Thoise 33-36	Thoise village	2900

Table 2: List of primers used for ISSR amplification, sequence, GC content, total number of loci, the level of polymorphism, size range of fragments and resolving power. Where, (Y = C, T; R = A,G)

Primer	Primer Sequence (5′~3′)	Tm °C	GC (%)	Total number of loci	Number of polymorphic loci	% of polymorphic loci	Total number of fragments amplified	Resolving power
ISSR 1	$(AG)_8T$	47.0	47	4	4	100	111	6.17
ISSR 2	$(GA)_8T$	45.4	47	5	5	100	136	7.56
ISSR 3	$(AC)_8T$	51.4	47	**3**	**3**	100	**45**	**2.50**
ISSR 4	$(TG)_8A$	51.3	47	4	4	100	125	6.94
ISSR 5	$(AG)_8YT$	49.2	47.2	6	6	100	96	5.33
ISSR 6	$(GA)_8YT$	47.4	47.2	7	6	85.7	159	8.83
ISSR 7	$(CT)_8RA$	47.1	47.2	7	7	100	156	8.67
ISSR 8	$(GT)_8YC$	52.7	52.7	7	7	100	**174**	**9.67**
ISSR 9	$(ACC)_6$	60.6	66.6	5	4	**80**	154	8.56
ISSR 10	$(CCG)_6$	76.8	100	5	5	100	104	5.78
ISSR 11	$(GGC)_6$	77.3	100	5	5	100	107	5.94
ISSR 12	$(AT)_8T$	23.1	0	0	-	-	-	-
ISSR 13	$(TA)_8RT$	25.6	2.7	0	-	-	-	-
ISSR 14	$(AT)_8YA$	26.0	2.7	0	-	-	-	-
ISSR 15	$(CT)_8T$	45.7	47	0	-	-	-	-
ISSR 16	$(TC)_8A$	47.0	47	0	-	-	-	-
ISSR 17	$(GT)_8A$	49.4	47	0	-	-	-	-
ISSR 18	$(TGC)_6$	62.4	66.6	0	-	-	-	-
ISSR 19	$(TGCA)_4$	53.3	50	0	-	-	-	-
ISSR 20	$(CTAG)_4$	43.9	50	0	-	-	-	-
	Total	-	-	**58**	**56**	**96.5**	**1367**	-

Table 3.1: Summary of genetic variation statistics for all loci of ISSR among the nine *Apricot* populations

Populations	Sample Size	Na*	Ne	H	I	Ht	Number of polymorphic loci (NPL)	% of polymorphic loci (PPL)
Leh	4	1.3621 (0.4848)	1.2724 (0.3833)	0.1530 (0.2082)	0.2216 (0.2993)	0.1530 (0.0433)	21	36.21
Stok	4	**1.6207** (0.4895)	**1.4276** (0.3622)	**0.2500** (0.2014)	**0.3671** (0.2927)	**0.2500** (0.0406)	**36**	62.07
Hemis	4	1.5000 (0.5044)	1.3828 (0.4109)	0.2134 (0.2197)	0.3082 (0.3143)	0.2134 (0.0483)	29	50.00
Stakna	4	1.4310 (0.4995)	1.2931 (0.3558)	0.1724 (0.2025)	0.2537 (0.2960)	0.1724 (0.0410)	25	43.10
Nimu	4	1.4655 (0.5032)	1.3621 (0.4146)	0.2004 (0.2208)	0.2888 (0.3154)	0.2004 (0.0488)	27	46.55
Khardung	4	1.3621 (0.4848)	1.2379 (0.3297)	0.1422 (0.1923)	0.2104 (0.2830)	0.1422 (0.0370)	21	36.21
Khalsar	4	**1.3448** (0.4795)	**1.2345** (0.3395)	**0.1379** (0.1941)	**0.2029** (0.2839)	**0.1379** (0.0377)	**20**	**34.48**
Hundar	4	**1.3448** (0.4795)	**1.2345** (0.3395)	**0.1379** (0.1941)	**0.2029** (0.2839)	**0.1379** (0.0377)	**20**	**34.48**
Thoise	4	1.5517 (0.5017)	1.3586 (0.3408)	0.2155 (0.1984)	0.3193 (0.2921)	0.2155 (0.0394)	32	55.17
Mean		**1.4425**	**1.3115**	**0.1803**	**0.2639**	**0.1803**	**25.67**	

*Na = Observed number of alleles; Ne = Effective number of alleles; H = Nei's gene diversity; I = Shannon's Information index; Ht = Heterogeneity. The value in the bracket are standard deviation.

Table 3.2: Summary of genetic variation statistics for all loci of ISSR among the *Apricot* populations with respect to their distributions among two valleys

Valley divisions	Sample size	Na	Ne	H	I	Ht	Hs	Gst	Nm	Number of polymorphic loci	Percentage of ploymorphic loci (%)
Leh	20	1.9138 (0.2831)	1.5282 (0.3630)	0.3065 (0.1706)	0.4611 (0.2222)	0.3065 (0.0291)	0.00 (0.00)	1.00	0.0	53	91.38
Nubra	16	1.7414 (0.4417)	1.3970 (0.3715)	0.2355 (0.1895)	0.3587 (0.2629)	0.2355 (0.0359)	0.00 (0.00)	1.00	0.0	43	74.14
Mean		**1.828**	**1.463**	**0.271**	**0.409**	**0.271**					

Na = Observed number of alleles; Ne = Effective number of alleles; H = Nei's gene diversity; I = Shannon's Information index; Ht = Heterogeneity; Hs = Homogeneity; Gst = Gene differentiation; Nm = Gene flow

Table 3.3: Overall genetic variability across all the populations of *Apricot* based on ISSR analysis

Sample size	Observed number of alleles	Effective number of alleles	Nei's gene diversity	Shannon's information index	Ht	Hs	Gst	Nm	Number of polymorphic loci	Percentage of polymorphic loci (%)
36	1.9483 (0.2234)	1.5683 (0.3281)	0.3310 (0.1552)	0.4948 (0.2000)	0.3310 (0.0241)	0.1803 (0.0115)	0.4552	0.0	55	94.83

Na = Observed number of alleles; Ne = Effective number of alleles; H = Nei's gene diversity; I = Shannon's Information index; Ht = Heterogeneity; Hs = Homogeneity; Gst = Gene differentiation; Nm = Gene flow

Table 4: Summary of nested analysis of molecular variance (AMOVA) based on ISSR among the populations of *Apricot* (levels of significance are based on 1000 iteration steps, d.f.: degree of freedom; S.S.D.: sum of square deviation; P-value: probability of null distribution)

Source of variation	d.f.	S.S.D.	Variance component	Percentage	P-value
Among valley	1	58.528	2.499	22	< 0.001
Among populations/valley	7	98.750	1.784	16	< 0.001
Individual/within populations	27	188.250	6.972	62	< 0.001

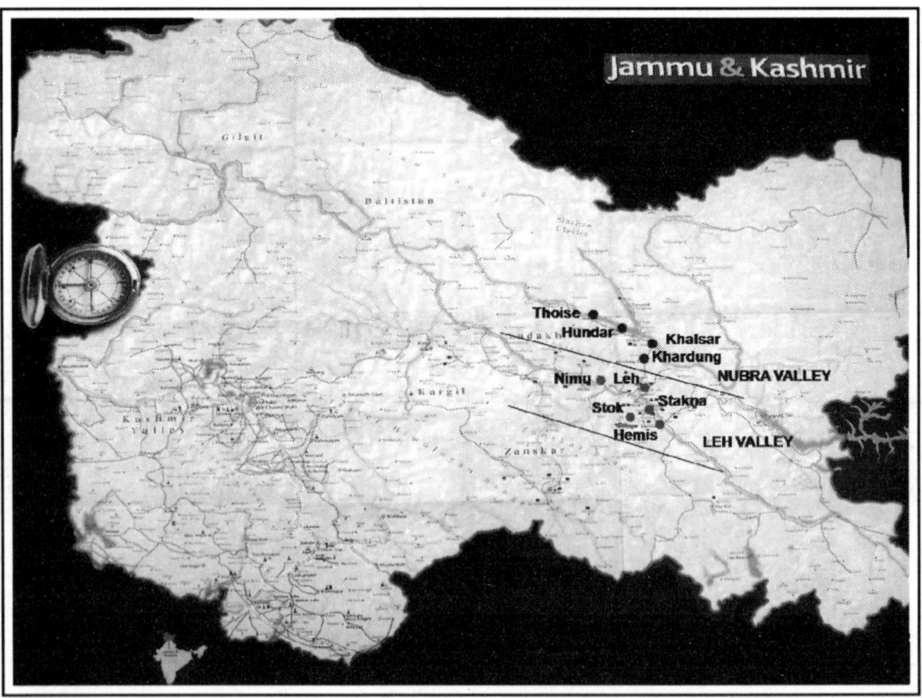

Fig. 1: Collection sites of thirty six apricot genotypes from two valleys (Leh and Nubra) and nine villages located in Ladakh (Jammu and Kashmir, India)

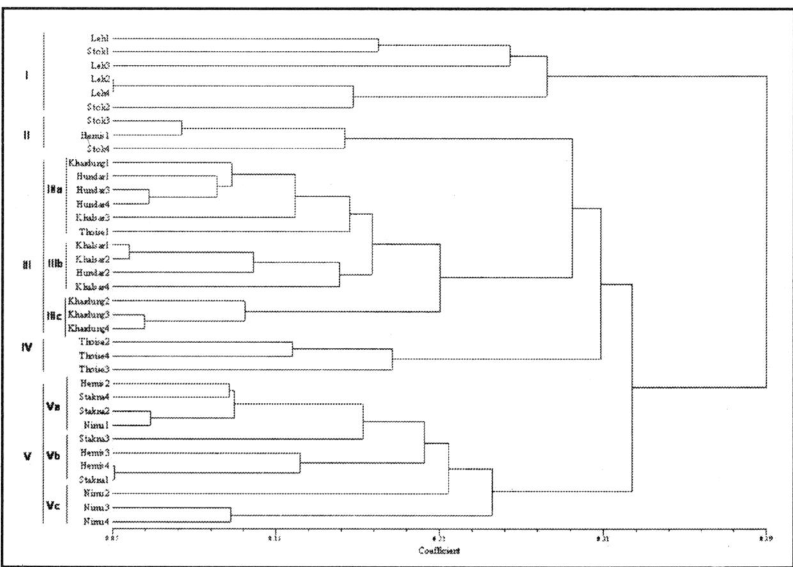

Fig. 2: Dendrogram generated using unweighted pair of group method with arithmetic average analysis, showing relationships between 36 *Apricot* genotypes, using ISSR data.

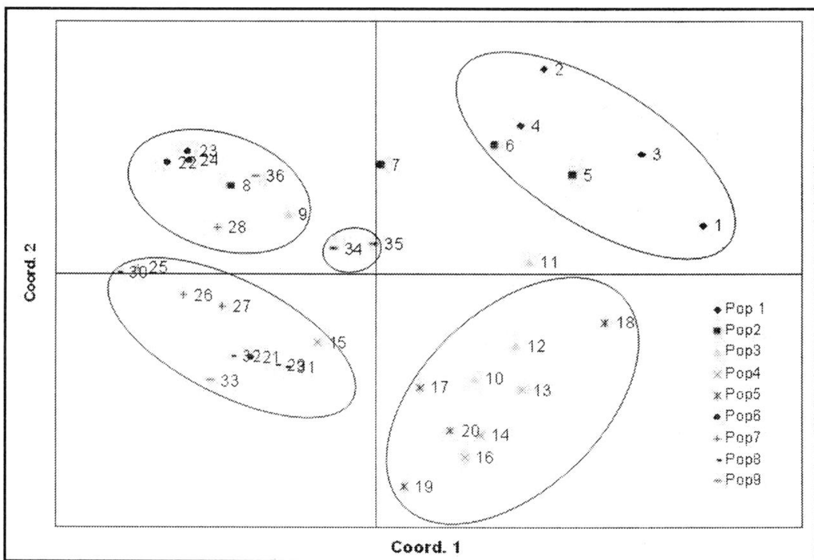

Fig. 3: Two-dimensional plot of principal component analysis of thirty-six local *Apricot* genotypes using ISSR analysis. The numbers plotted represents individual cultivars. (Where, Pop1=*Leh*, Pop 2=*Stok*, Pop 3=*Hemis*, Pop 4=*Stakna* , Pop5=*Nimu*, Pop6=*Khardung*, Pop7=*Khalsar*, Pop8=*Hundar*, Pop9=*Thoise*)

❑❑❑

29

Sustainable Fodder Production for Livestock in High Altitude Regions of India

K.A. SINGH, B. K. MISRI AND M. M. ROY
Indian Grassland & Fodder Research Institute, Jhansi, India

INTRODUCTION

The high altitude regions in northern India exhibit diverse climate, topography, vegetation, ecology and land use patterns. The annual average rainfall varies from 80 mm in Ladakh to over 200 cm in some parts of Himachal Pradesh and Uttar Pradesh. The vegetation varies from the scrub in the lower hills to arctic in the Greater Himalaya.

In this region, Ladakh is a unique land form encompassing distinct physical features, ecology, environment, ethnicity, agricultural systems and livestock rearing practices. This region experiences a very hostile climate and winter temperatures may go as low as -30°C. Summer temperature exceeds upto 40 °C, but the thin air makes the sunlight scorching hot during summer. The altitude ranges between 2750 m (Kargil) to 3505 m (Leh). The area receives very little precipitation and because of the Great Himalayan barriers, the monsoon does not reach here. The landscape can be classified into different land forms *viz.*; alluvial fans (rocky areas formed by washed up deposits), sandy plains (gentle and flat sandy areas generally about 4600 m), low slopes (areas below 4900 m and above 4600 m), high slopes (area above 4900 m) and marsh meadows (moist/wet areas around lakes).

In hilly regions, cultivation of fodder is restricted to the lower hills. At higher elevations, the available cultivable land is about 10 percent of the total geographical area and is used for production of cereals, pulses and oil seeds etc., and very little fodder. Rangelands and forests, comprising 50 to 70 per cent of the total geographical area are the major sources of forage for sheep, goats, rabbits, cattle and other animals. Nomadic graziers (*Khadwals, Gaddis, Gujjars, Khas*) use sub-alpine and alpine pastures for rearing their livestock. Due to ever-increasing demand for animal products, the livestock population has increased putting increased grazing pressure on these pastures. This has resulted in the deterioration of the grass cover as well as valuable forest species. As a result of this a significant area of this natural resource has been covered by noxious plant species like *Aconitum, Anemone, Adonis, Aquilegia, Cincifuga, Clematis, Lepidium, Artemisia, Ranunculus, Stipa, Sorbinia* and *Sambucus*. All these factors have led to the decrease in carrying capacity of these pastures.

However, high altitude regions, including Ladakh, have traditionally been surviving on the land base resources. The traditional agriculture is indeed a blend of excellent crop management and conservation of natural resources. Since independence, there has been a decline in dependence on land resources for agricultural production and greater emphasis on things like tourism in Ladakh. As livestock farming is a major source of economic activity in the high altitude regions of the country; an attempt has been made in this paper to describe and suggest suitable options for sustainable fodder production in these regions.

Livestock Rearing & Fodder Resources

Livestock is reared under three systems *viz.*, sedentary, semi-migratory and migratory. Sedentary rearing is confined only to the urban areas, whereas with the semi-migratory system, livestock are taken to pastures or ranges for grazing during summer and is brought back to their respective villages during winter. The migratory system involves year long transhumance in high slopes and is practiced by tribes like *Changpas*, the natives of *Changthang*. This is an elaborate system of year long transhumance which includes very efficient use of indigenous technical knowledge. Rearing of cattle, yak, sheep, horses, goats and crosses of yak x cattle is prevalent everywhere in this reason.

The region is essentially a mono-cropped area. During summer a single crop of wheat or barley is harvested, and vegetables and fruits are also grown. In the agricultural scenario crop production is as important as livestock rearing. Grazing, crop residues and alfalfa are the main source of forage base. The most productive pastures are confined to river valleys or lake peripheries. Most of the grazing takes place on sandy plains, low and high slopes which support only 5 percent vegetation. This vegetation contains some very nutritious plants like *Eurotia ceratoides* which contains more than 20 percent crude protein. Though most of the plants found in dry areas are noxious, plants like *Alyssum canescens, Carex* sps, *Carex melantha, Euphorbia tibetica, Eurotia ceratoides, Kobresia schenoides Leymus secatinus Stipa orientalis* are highly nutritious and contain 8-22 percent crude protein which substantially help in maintenance of the animal health. However, due to constant heavy grazing pressure, the population of these species has dwindled considerably in almost all of Ladakh, especially in the Changthang region. The yield levels have gone down from the potential of 60-70 to a mere 2-3 t green forage/ha. The crop residues are fed to animals during winter, but are also very low in quantity. Willow tree leaf fodder is also used as forage. However, except for alfalfa, which is still cultivated in over 4000 ha, other forage resources have deteriorated and a huge imbalance in demand and supply has occurred.

Alfalafa (*Medicago* complex) is the mainstay of livestock rearing. In Leh and Kargil, 8.40 percent and 31.37 percent of the total sown area is under alfalfa crop production which is much higher than the national figure of 4 percent area under fodder cultivation. However, the total availability of fodder does not suffice and there exists a huge gap in demand and supply. Currently 60 percent of the region's food requirement is met by means of the government sponsored public distribution systems besides co-operatives and commercial traders by importing it from other states. These imports include large quantities of livestock products like milk and mutton. It may not be commercially viable to import fodder from other states for the common man but the entire animal transport system of the army depends upon huge imports of hay from places as far away as Maharashtra.

Sustaining Fodder Production

The following points are considered important to tackle the issue of amelioration and sustaining of fodder production for livestock in high altitude regions, with special reference to Ladakh.

Build Up of Forage Germplasm

Ladakh is a repository of bio-diversity, unique plants and old land races. Diffused efforts have been made to collect, evaluate and introduce forage germplasm to the area. The Dras region supports excellent populations of temperate grass and legume species like *Agrostis, Phleum,Elymus, Agropyron, Dactylis, Festua, Trifolium* etc. 49 species of grass and legumes of forage value have been enumerated from Ladakh. *Prangos pabularia* is a unique forage plant of the Dras region which is not grazed / eaten by the animals when green. Its hay is highly prized and is a major source of animal rearing during winter However, the area under its growth is constant and no efforts have been made to domesticate this highly nutritious (CP up to 20 percent) forage plant. Old land races of various crops can still be found in the area. These land races have been saved from extinction as farmers by and large neither exchange nor buy seeds. They produce their own seed. Eleven collections of buckwheat from 11 houses in Kargil have been made. All these houses were in a cluster. When evaluated under uniform environmental conditions, all the collections were different from each other. This clearly exhibits the amount of bio-diversity available in the area.

The 14 accessions of wheat and 34 accessions of barley from Ladakh were evaluated at Manasbal (Kashmir) for their possible introduction as forage in Kashmir. Significant diversity was found in all these accessions for various traits. In no. of wheat days needed for 50 percent flowering ranged between 184-208; while green and dry forage yield ranged between 11.8-39.8 t/ha and 2.8-17.63 t/ha, respectively. Similarly, the number of barley days to 50 percent flowering ranged between 157-183 while green and dry forage yields ranged between 8.2-38.3 t/ha and 4.1-17.4 t/ha, respectively. The collections were made from the *Sanku, Kargil* and *Leh* regions and the diversity occurring can be exploited for development of suitable plants types for the area. The DIHAR may establish a bio-diversity group to collect and conserve this important gene pool. IGFRI can collaborate in this venture and offer mid-term conservation of material in its gene bank. Biotechnological interventions, if required, can also be provided by the IGFRI.

Use of Improved Fodder Varieties

Though alfalfa is the major cultivated crop in Ladakh region, there is scope for introducing other forage crops in the extant farming systems. The list of available varieties of clover, oats, forage sorghum and cowpea suited to high altitude regions is depicted in Table 1. It is considered that introduction of these species will increase the fodder availability in the area considerably.

Improvement of Alfalfa

Alfalfa complex is the main stay of livestock in the region of Ladakh. The native alfalfa is *Medicago falcata* and Ladakh is one of the centres of origin of this species. Traders travelling on the southern flank of the old silk route observed better performance and higher biomass production of *Medicago sativa* in Yarkand and introduced this species in Ladakh. Both these species have undergone so much hybridization that it is almost impossible to find a pure stand of any of the two species. Plants having any conceivable form are present. In a single field, erect, decumbent, prostrate, and climbing types of plants can be found. The variability also occurs in the height, leaf size, branch length etc. Detailed studies were made on pod characters of germplasm of alfalfa complex collected from 28 populations. Pod shape is one of the main diagnostic features for identification of various species of *Medicago*. The pod of *Medicago sativa* is double spiral or much curved, small, indehiscent, smooth or spiny and 1- few seeded. In case of *M. falcata* the pod is linear, sickle shaped and glabrous with 5-10 seeds. However, six different shapes of pods were found in the *Medicago* complex of Ladakh suggesting high level of diversity created by the natural hybridization.

Some effort has been made in past to select appropriate plant types from the *Medicago* complex but the information generated is diffused and scattered. DIHAR has made pioneer efforts in this direction. The credit for introducing multi-cut systems in alfalfa in the 1970's goes to them. Similarly, SKUST (K) after its establishment made some selections from the complex and has enumerated 12 local selections that yielded 52.85 - 98.73 t/acre dry forage. However, no variety has been released from the area.

In order to give an impetus to a holistic and multi- disciplinary approach to exploit the real potential of the diversity existing in *Medicago* complex the lucerne improvement group of IGFRI can collaborate with the on-going efforts of DIHAR to develop highly productive varieties.

Introduction of Short Duration Crops

Under the present cropping pattern after the harvest of wheat and barley the farmers sow a second crop of buckwheat or minor millets. In case of inclement weather these crops do not mature and farmers harvest the crop and use it as fodder. There is an urgent need for search of more suitable short duration forage crops. IGFRI's Srinagar Centre evaluated 19 collections of buckwheat, and out of this, six collections were made from Ladakh while others were procured from the Kashmir University. The Ladakh collections came to 50 percent flowering between 24-75 days. Two collections *viz.*, cv *Botansoba* and 86-20 came to 50 percent flowering within 24 days and produced 112 q/ha and 148.8 t/ha green fodder and dry fodder, respectively. Similar short duration materials can be identified out of fodder peas, minor millets, turnips, *kale* and other non-conventional forage crops.

Grassland / Pasture Management

Pasture improvement is indeed a very challenging task in a tough terrain like Ladakh. However, a well planned, long term, inter-disciplinary approach can accomplish the task. The populations of palatable species like *Eurotia, Euphorbia, Stipa* etc. have dwindled considerably. The rehabilitation of these species along with other suitable pasture species (Table 2 and 3) is of prime importance in grassland/pasture improvement and management in high altitude regions.

None of the natural grasslands in sub-temperate, temperate and sub-alpine regions are fertilized in any form except through dung and urine addition by the grazing animals. This leads to continuous depletion of plant nutrients, resulting in poor growth of plant species. Significantly higher herbage yields may be achieved with the application of nitrogen (60 kg/ha) and phosphorus (30 kg P_2O_5/ha).

Introduction of pasture legumes in the native vegetation not only helps in increasing total biomass production but also in improving the quality of the material produced. However, initial competition from the native vegetation must be checked to allow establishment of introduced species. In unstable terrains, disturbing the soil is not advisable and can lead to soil and nutrients losses. Use of sod killing herbicides, like paraquat may be useful.

Generally, uncontrolled and continuous grazing is prevalent in most of the grasslands throughout the hilly regions. Although, rotational systems of grazing give better results, it is not easy to adopt these systems on a large scale in hilly situations. The most desirable system of grazing would be that of periodic or deferred grazing by limiting the number of livestock on the basis of carrying capacity of grasslands. This system would give a better chance for self-seeding and stand recovery.

It is felt that in the Ladakh region, detailed phenology and reproductive biology on selected species found in the pasture is required. The growth pattern, seed dispersal mechanisms, seed bank of soil and related studies on these species can determine the optimal method and time for their sowing. Such studies can initially be confined to small areas and once the technology is perfected it could be tested in larger areas. Climatic changes in the area have increased the total precipitation manifold. In depth studies of the present rainfall pattern can help in the determination of suitable times to re-sow the pastures with promising species selected out of the existing rich flora or some improved materials. All the studies should incorporate active participation of farmers and other stake holders who can later assure better management of the improved pastures.

As far as river bed and valley basin pastures are concerned, the vegetation cover and species composition is almost adequate but all these pastures lack in nutrient management. Detailed studies need to be initiated on integrated nutrient management of pastures so that a package of practices is formulated and disseminated to the farmers and other stake holders.

Other Interventions

There is considerable potential pasture area under orchards in temperate regions. Inter spaces between fruit trees could be utilized for the production of fodder by growing perennial grasses and legumes depending upon the topography, age of the trees, spacing etc. The cultivation of legumes would be more desirable because of their nitrogen fixing ability.

Growing of improved grasses on terrace bunds and risers could be a very effective non-competitive land use system for forage production in the hills. There is an added advantage of producing forage without any fertilizer or manure inputs since it is available from the cropped terrace.

Post Harvest Technology

Winters are the biggest worry for farmers in Ladakh, and every effort is made to conserve all types of forage for winter feeding. For conservation of forage, traditional methods are used. The harvested grasses, weeds and alfalfa plants are stacked on the roof tops under some weight so that it is not blown away by strong winds. This saves space and provides insulation to the houses, but in the process the quality and quantity of the conserved forage dwindles considerably. In case of alfalfa significant losses occur due to leaf fall from the branches and consequent blowing off by winds. IGFRI has very successfully demonstrated in Himachal Pradesh the technology of silage making in Poly-bags, cemented and *Kacha* pits, and farmers have very eagerly adopted this technology. IGFRI can collaborate with DIHAR in demonstrating this technology in the area. Similarly, leaf meal making technologies have been perfected and the same could be used for alfalfa and *Hippophae* leaf meal.

CONCLUSIONS

Ladakh is undoubtedly a difficult terrain, having a very harsh climate and a dwindling natural resource base. Nevertheless, the local population still survives on the natural resources in this area, and there is an urgent need for various research and development interventions which could bring about a balance in demand and supply. Forage resources are no exception. The situation may be alarming but can be addressed with appropriate measures. Creation of a research and development base is required. Looking at the limitations of all the organizations based at Ladakh to work on natural resources, the situation can not be corrected by any individual agency in isolation. A wide and sustainable coordination needs to be created not only in Ladakh, but on national level, to save this marvel of nature (Ladakh) from further degradation.

IGFRI, with its micro-specialization in cultivated forage production, pasture improvement and management, soil water conservation, plant-animal relationship and agro-forestry, may play an important role in working with local organizations in this region.

References

AICRPFC. 2007. Varieties identified for different agro-climatic situations of India. All India Coordinated Research Project on Forage Crops. IGFRI, Jhansi.

Bailey, L.H. 1968. Manual of Cultivated Plants. Macmillan Co, NewYork.

De Candolle, A. 1967. Origin of Cultivated Plants, Hafuer Publishing Co., NewYork.

GJK. 2004. District Handbook. Government of Jammu and Kashmir, Srinagar.

Gohil, R.N. and Misri, B. 1986. Comparative performance of eleven local buckwheat germplasm under uniform environmental conditions. Buckwheat Research. 2, 23-28.

Hooker, J.D. 1897. Flora of British India (Vol. II). L. Reeves & Co., London.

Judy, B.S. Jesssup and Juantegny, E. 1992. The Silk Road from Xian to Kashgar. Odyssey, HongKong.

Koranne, K.D. and Singh, J.P. 1989. Technologies for increasing forage production of grasslands and non-agricultural lands in the hilly region. Paper presented in the workshop on *Pasture and Grasslands Improvement* at HPKV, Palampur

LAHDC. 2005. http://leh.nic.in/profile.htm. Leh Autonomous Hill Development Council.

LAHDC. 2005b. *Ladakh 2025 Vision - Document*. http://leh.nic.in/VISION-DOCUMENT.pdf.

Mir AA. 1986. Prospects of fodder production in Ladakh. Paper presented in the *All India conference on Cold Desert Development* at Leh.

Misri B. 1981. A preliminary survey of grasses and legumes of Ladakh. *Australian Plant Introduction Review*, 13(1):37-45

Misri B. 1988. Forage production in alpine and sub-alpine of North-West, Himalayas. *In: Pasture and Forage Crops Research* (ed. Panjab Singh). IGFRI, Jhansi: 43-55.

Misri B. 1989. Evaluation of wheat and barley germplasm from Ladakh as winter fodder in Kashmir. *Rachis* 8:20-21

Misri B. 1996. *Variability in Alfalfa of Ladklh*. http:\\www.fao.org/AGP/AGPC/doc/ bulletin/Ladakh.htm.

Misri B. and RK Bhatt. 2004. Alfalfa cultivation in Ladakh - a commercial enterprise. *Indian Farming*, 54: 9-10.

Misri, B., S Radotra and Inder Dev. 2007. *Research Initiatives for Fodder Resource Development in North- Western Himalaya*. IGFRI, Jhansi: 123 pp.

Sharma JR and KK Jindal. 1989. Introduction of superior grasses in orchards. Paper presented in workshop on *Pasture and Grasslands Improvement* at HPKV, Palampur during 12 -13 October.

Singh V. 1995. Technology for forage production in Hills of Kumaon. *In*: *New vistas in Forage Production* (eds Hazra, CR and Bimal Misri). IGFRI, Jhansi: 197-202.

Uniyal S., A. Awasthi and GR Rawat. 2005. Biomass availability and forage quality of *Eurotia ceratoides* Mey in the range leands of Changthang, eastern Ladakh. *Curr. Sci.* 89: 201-205.

Table 1: List of improved forage varieties suitable for high altitude regions of India

Crop	Name of Variety	GFY (t/ha)	Special Features
Berseem	BL-180	62-65	More CP and tolerant to stem and root rot
Oats	IGFRI-99-1	65-70	Single cut
	SABZAR	27-30	Moderately resistant to loose smut, aphids and armyworm
Forage Sorghum	PCH-109	75-80	Resistant to shoot fly and stem borer
Cowpea	UPC-8705	30-42	Resistant to dry root rot and collar rot
	UPC-622	25-30	Resistant to yellow mosaic virus, root rot and anthracnose
Maize	African Tall	30-40	High yielding, very tall, leaves long with dark green colour stem

Table 2: Suitable grass and leguminous species for high altitude regions of India

Zone	Species	
Temperate (1800-3200 m)	Grass	*Setaria anceps*, *Panicum maximum*, Napier bajra hybrids, *Pennisetum orientale*
	Legume	Siratro, *Desmodium* spp.
Alpine zone (above 3200 m)	Grass	*Dactylis glomerata*, *Bromus* spp., *Phleum pratense*
	Legume	*Trifolium repens*, *Trifolium pratense*
		Agropyron spp. Medicago spp.
		Poa bulbosum, *Agrostis stolonifer*, *Onobrychis vicifolia*

Table 3: List of improved pasture species for high altitude regions of India

Crop	Name of Variety	GFY (t/ha)	Special Features
Setaria grass	S-92	40-50	Drought and frost tolerant
Tall Fescue grass	Hima-4	30-40	High yielding cold tolerant
Red Clover	PRC-3	26-30	Root rot resistant and high yielding
Guinea grass	PGG-616	100-105	Multi-cut variety

❑❑❑

30

Strategies for Poultry Production in High Altitude Areas

R.P. SHARMA, M.V.L.N. RAJU AND M. NIRANJAN
Project Directorate on Poultry, ICAR, Rajendra Nagar, Hyderabad, India

INTRODUCTION

Chicken rearing has a history of over 5000 years in India. Red jungle fowl, the wild native chicken is believed to be the basic source of all the modern breeds reared through out the world. Poultry farming in India was confined to rearing of few chickens and ducks in the back yards till early 1960s. The R&D efforts initiated subsequently coupled with importation of high yielding breeding stocks changed the face of poultry activity in the country. Availability of high yielding layer (300-310 eggs) and broiler (2-2.2 kgs at 6 wks) varieties together with standardized package of practices on nutrition, housing and management and disease control have contributed to spectacular growth rates in egg production (5-7per cent per annum) and broiler production (10-12 per cent per annum) in India during the past 3-4 decades.

Currently, India ranks third in egg production (2492 thousand MT of egg mass) and fifth in broiler production (1900 thousand MT) in the world. The Poultry Industry is contributing about Rs.352 billion to the national GDP and providing employment to about 3 million people either directly or indirectly. The intensive large scale poultry industry is concentrated in certain pockets of the country, like in the states of Andhra Pradesh, West Bengal, Maharashtra, Tamil Nadu, Karnataka etc.

World over, the popularity of poultry meat is on the rise in the recent times. It is presently accounting for about 27 per cent of the total meat consumed and is the most popular meat from any single livestock species. The per capita availability of poultry meat in India was forecast as just 2.0 kg as against more than 7 kg in Asia and 12 kg in the world. Similarly, the availability of eggs was also much lower at 1.8 kg per year as against 9.0 kg eggs in Asia and 8.6 kg eggs in the World. Availability of eggs is highly non-uniform in different parts of the country primarily due to wide variation in the production levels. Ironically, in India 75 per cent of eggs produced are consumed by the 25 per cent population living in urban and semi-urban areas.

Poultry Production in High Altitude Areas - Present Situation and Future Strategies

The poultry production has been mainly concentrated in the plains of the country. The states of Andhra Pradesh, Tamil Nadu, West Bengal, Maharashtra and Karnataka have major share of the local poultry population. In the states having high altitude, viz. Arunachal Pradesh, Jammu & Kashmir, Himachal Pradesh, Uttaranchal and North Eastern Hill states, the poultry production is very limited and primarily confined to rural backyard poultry. Majority of the requirement is met from the imports of eggs from other states of the plains. Of the total 46,230 millions of eggs produced in the country, only 1,217 millions eggs are produced in the high altitude states, contributing to only 2.63 per cent of the country's egg production. Barring few states in the NEH region, chicken is the most popular poultry species in the high altitude areas of the country.

The poultry production in high altitude areas is yet to take a shape of comprehensive poultry industry. The contribution of desi chicken to the total egg production is much higher than the improved layers. It was observed that more than 90 per cent farmers procure the chicks from outside the Kashmir valley through chick suppliers and 95 per cent of farmers purchase feed imported from other areas. In central high lands of Ethiopia, about 7 to 10 chickens were reared per household in the backyard with inadequate housing, feeding and health care. Significant effect of season was observed on flock performance, particularly with regard to disease and hatching of chicks. In Bhutan, 5 to 10 chickens are raised per household in most parts of the country. Majority of the chickens are the local breeds which produced about 80 eggs per year, while improved chickens which can produce about 240 eggs are also reared in certain farms adopting intensive system of management.

The reasons for poor growth of poultry industry in these areas are the limitations associated with geographic location and extreme climatic conditions. The raw feed materials required for feeding poultry (like maize, sorghum, bajra, soybean meal, sunflower meal etc.) are not available making their importation unavoidable. Transportation of these materials from surrounding states involves considerable expenditure raising the cost of production. Another major bottle neck for expansion of poultry operations in these areas is extreme climatic conditions, where the environmental temperature fall much below the ideal temperature for chicken. The traditional open house system may not be suitable in this extreme cold weather, while environmentally controlled housing is cost prohibitive in the present marketing system.

Adverse Climate - The Issue of Concern in High Altitude Areas

a) Low Oxygen - Poor Hatchability

At higher altitudes, as atmospheric pressure reduces; air expands and there by results in reduced oxygen concentration. This results in poor hatchability. These problems become pronounced from an altitude of 1050 meters. Above this level, hatchability reduces by about 5 per cent for every 300 meters elevation. For managing the situation and avoiding hatchability problems at high altitudes, oxygen may be injected in to the incubating machines. This intervention is necessary in the artificial incubators. However, chickens living in high altitudes show certain adaptive changes in egg shell porosity and generally reproduce normally.

b) Low Ambient Temperature – Unfavorable for Young Chicks

The cool temperatures that prevail in the high altitude areas pose problems for raising chickens, particularly during the initial brooding phase. Chickens feel comfortable in the temperature zone of 12.8 to 23.8°C temperatures. Above this, birds experience stress and reduce performance. However, chickens covered with plumage generally tolerate low temperatures. Temperatures up to 0°C are generally not problematic, except for managemental issues in the shed like dampness of liter. But below 0°C, hens will be uncomfortable and at or below -6.7°C they fail to maintain the body temperature, subsequently leading to freezing of combs and death at -30°C. However, at low ambient temperatures the feed conversion efficiency reduces due to greater energy expenditure on body heat maintenance.

Chicks during brooding phase (up to 3 to 4 weeks of age) find it difficult to maintain their body temperature due to lack of insulating feather cover. If supplementary heat is not provided in such situations, increased mortality of chicks due to huddling/suffocation is noticed. For avoiding problems of adverse weather conditions at high altitudes, environmentally controlled poultry houses may be the choice, wherein the shed temperature, ventilation and lighting are controlled at optimum range through artificial means. Ventilation is to be given proper attention while attempting to conserve heat in the shed during cool conditions. Accumulation of toxics gases like CO_2, ammonia etc. will drastically affect the health and performance of poultry.

c) Metabolic Diseases

Though broilers grow faster throughout, much of their growth is manifested primarily in the first 4 weeks of life. The relative growth rate of a broiler decreases as the age advances (from 300 per cent in 1st week to 150 per cent in the 6th week). For the well being of an animal, synchronization of growth between supply organs (lungs, heart, liver etc) and demand organs (skeletal muscles, adipose tissue etc) is necessary. However, with greater breeding emphasis on muscle growth giving little or no attention to the growth of supply organs, this growth synchronization is likely to be disturbed. As a result, insufficient cardiopulmonary efficiency with concomitant development of metabolic disorders like ascites and sudden death syndrome is expected in the flock.

Ascites is typically a disease of high altitude areas (1500 m). Among the various known factors, hypoxia due to low-atmospheric oxygen levels and fast growth rate is considered the primary cause responsible for the greater incidence of the disease. High metabolic rate associated with fast growth rate is expected to increase the tissue oxygen demands substantially. Further, the inequality between the metabolic demands of rapid body growth and cardiopulmonary performance is also attributed for the development of the syndrome. Thus improved chicken varieties are more prone to the problem than the slow growing native chicken. In turkeys, low lung capacity per unit body weight in fast growing line was observed in comparison to an unselected one.

Therefore, delay of early growth with a modified concave shaped growth curve is expected to allow the somatic growth to remain in pace with the functional growth of organs in the cardiopulmonary and renal systems. Thus, birds may be able to enter the subsequent growth period with stronger cardio-pulmonary so as to cope-up with the physico-physiological demands of the subsequent fast growth rate. Further, slowing down somatic growth rate genetically during initial 3-4 weeks and allowing rapid growth rate later has been suggested for increasing their resistance to metabolic disorders. Diet dilution by inclusion (50 per cent) of rice hulls, skip-a-day feeding programme and restricted feeding may be adopted for reducing the incidence of these disorders. Similarly, restricted lighting programme in controlled houses may also be effective in minimising the losses due to mortality and culling rates.

d) Non-Availability of Inputs

Poultry feed is compounded using several feed ingredients like cereals, protein meals, minerals & vitamins supplements and several other additives. Production and availability of all these raw materials may not be possible in high altitude areas. The requirement may be met from limited imports from other areas and using locally available materials.

e) Strategic Distribution of Production Chain and Use of Environmentally Controlled Housing

Since hatching and brooding of chicks under cold climatic conditions is troublesome, breeding and hatchery facilities may be maintained in the adjacent region, where normal environmental conditions prevail. Thus chicks can be produced and then transported to the high altitude areas to propagate as family poultry operation. Also, broilers and table eggs can be produced under controlled environment. This process may perhaps be more economical than purchasing the poultry products from far of places. This will also ensure employment to the local population besides supplying the nutritious poultry products.

Back Yard Poultry – A Promising Strategy

The fruits of commercial poultry production in India have been limited to the urban and semi-urban areas and the rural poultry sector remained unchanged, more so in the remote high altitude areas. Egg production from native chicken contributes only 21% of total egg production of the country. In the hilly regions, the poultry products are either not available in required quantities or if available, sold at higher price than in plains. Further, the incidence of protein deficiency is prevalent among the susceptible groups like children, pregnant women, nourishing mothers and aged people in these areas. This can be alleviated by adopting small scale poultry farming in backyards of house holds. Besides, the backyard poultry production relies on minimal cost inputs in the form of kitchen waste, locally available grains, tender leaves, worms, insects and other material available for scavenging. Therefore, it is necessary to promote free range and backyard poultry farming in these high altitude areas of the country.

The production potential of the native chicken breeds is very low. Crosses like Vanaraja, Giriraja, Gramapriya, Girirani, Krishna J, Gramalakshmi, CARI Gold etc. have been evolved by ICAR and several state agricultural universities. These improved varieties need to be propagated extensively in the country. Going by the present international consumer market trends, eggs and meat produced under such free range/back yard system will have a great demand in the days to come. In view of its varied agroclimatic conditions and vast expanse of flora in most parts, India has a great potential for production of poultry products in the free range conditions and capturing a great share of the international market besides meeting the needs of the rural households.

The probable means for promotion of poultry production in high altitude areas like Kashmir have been suggested. The free range poultry farming can be adopted in these areas by developing suitable chicken germplasm that can sustain in the extreme cold conditions and perform well. This can be achieved by crossing / upgradation of native chicken germplasm (like Kashmiri Febrola in J & K) with the proven rural germplasm elsewhere in other parts of the country. An example of chicken germplasm that are surviving and performing well at a very high altitude are Vanaraja and Gramapriya. Vanaraja chicken is being successfully reared in four states of NEH region (Nagaland, Arunachal Pradesh, Mizoram and Manipur) and Jammu & Kashmir (Tables 1 & 2). Both varieties of germplasm proved well for their adaptability and production in the region under test conditions. Therefore, attempts need to be made to propagate these birds in the high altitude areas.

Table 1: Field Performance of Vanaraja Adult Females in Four States of NEH Region

Particulars	Manipur	Mizoram	Arunachal Pradesh	Nagaland
Body wt. (g) 6 wks	535	625	795	858
20 wks	2190	2200	2270	2175
40 wks	2800	3000	2540	2425
Age at maturity (d)	164-178	161-196	172-187	175-210
Egg prodn. (no)/wk	2.8	2.4	3.3	3.5

Source : Annual Report, PD on Poultry (2004-05)

Table 2: Performance of Vanaraja Under Backyard in Kashmir Valley

Trait	Male	Female	Combined
Body wt. (g)			
11 wks	664.5	557.5	611
12 wks	782.2	699.2	725.7
13 wks	1132	850	991
28 wks	3150	2550	2850
Mortality (%) between 5 -11 wks	--	--	3 - 7
Age at maturity (d)		177 - 182	

Source : Annual Report, PD on Poultry (2006-07)

In a study growth performance of Vanaraja chicks was evaluated under scavenging system of rearing in high altitude areas (5000 to 8000 ft. above msl) of Arunachal Pradesh. The average body weight at 6 weeks of age was observed to be 810g (ranging from 588.2g to 1000.0g). Comparatively higher mortality was recorded during this period, which was attributed to the period of the study (Oct-Nov.) and the cool weather as the average temperature was in the range of 12-24°C. The average egg production per bird per month was 13.8 nos and the average egg size varied from 46-55g. The egg and meat of Vanaraja were found very popular in Arunachal Pradesh. In another study on village chicken production system under hill agro-ecosystem using a dual purpose improved variety (Vanaraja), adult body weight at 18 months was reported to be 3.6 kg and the annual egg production was 176 eggs. The farmers used locally available materials such as bamboo and wood for housing the birds, particularly during night and village chickens served as an important income source for house hold expenses.

References

Banday, M .T, Anand, G.S. and Pampori. Z.A. 2005. Status of broiler farming in Kashmir valley. *Proc. XXIII Annual Conference and Symposium of Indian Poultry Science Association, 2-4 Feb., PDP and ANGRAU, Hyderabad vol.2: 429.*

Bhattacharya, M., Buragohain, R., Ahmed, F.A., Pathak, P.K. and Ghosh, M.K. 2005. Laying performance of Vanaraja birds in high altitude areas of Arunachal Pradesh under backyard system of rearing. *Proc. XXIII Annual Conference and Symposium of Indian Poultry Science Association, 2-4 Feb., PDP and ANGRAU, Hyderabad vol.2: 369.*

DAHD. 2008. Livestock Statistics. "www.dahd.nic.in". *Dept. of Animal Husbandry and Dairying, Govt. of India, New Delhi, accessed August 2008.*

Dessie T. and Ogle, B. 2001. Village Poultry Production Systems in the Central Highlands of Ethiopia. *Tropical Animal Health and Production, 33: 521-537.*

FAOSTAT. 2008. Statistical data base. "http:/www.fao.org". FAO.

Ghosh, M.K., Ahmed, F.A., Buragohain, R., Pathak, P.K. and Bhattacharya, M. 2005. Growth performance of Vanaraja birds in high altitude areas of Arunachal Pradesh under backyard system of management. *Proc. XXIII Annual Conference and Symposium of Indian Poultry Science Association, 2-4 Feb., PDP and ANGRAU, Hyderabad vol.2: 368.*

Kumaresan A., K., Bujarbaruah, M., Pathak, K. A. ,Bijoy Chhetri, Ahmed, S. K. and Santosh Haunshi. 2008. Analysis of a village chicken production system and performance of improved dual purpose chickens under a subtropical hill agro-ecosystem in India. *Tropical Animal Health and Production, 40: 395-402.*

Sharma, R.P., Raju, M.V.L.N. and Rama Rao, S.V. 2003. Intensive and free range poultry production in India and its applicability to Jammu and Kashmir. *Proc. National Symposium on Strategies for Sustainable Livestock Development, S.K. University of Agricultural Sciences and Technology, Shuhama, Srinagar, October 17-19 : 14-19*

Sherpa, D.L. 2008. Technology for poultry production in Bhutan. *www.vercon.bt/technology/poultryproduction.*

Watt executive guide to world poultry trends. 2007-08. Watt publishing co. www.wattpoultry.com.

□□□

31

Broiler Rabbit as A Source of Meat and Fur in High Altitudes

RAJEEV GULYANI
Rabbit Unit, Central Sheep and Wool Research Institute,
Avikanagar-304 501, Rajasthan India

INTRODUCTION

In recent years there has been rising awareness that due to rapidly increasing human population in developing countries like India, food shortages may be a major problem in the future. To increase food production, all reasonable options must be explored. Among these is the use of livestock which have not yet played a major role in animal agriculture. One such livestock species is the rabbit. Its importance as a source of meat for human consumption is widely recognized throughout the world. Rabbit meat production is currently, concentrated in a few countries, like Italy, France, Ukraine, China, Spain and Russia where large-scale commercial farming is practiced. In recent times attempts have been made to develop rabbit as a source of food in third world countries like India, Egypt, Nigeria, Ghana, Sri Lanka, Morocco, Indonesia, Algeria, Caribbean countries, South American countries and China.

Rabbit has a number of valuable attributes, including high rate of reproduction (Table 1), early maturity, rapid growth rate, high genetic selection potential, efficient feed and land space utilization, limited competition with man for similar foods and high quality nutritious meat (Table 2). In addition, its skin is soft and luxurious and can be used in the production of garments and toys in cottage industries. In developing countries where meat demand is higher, the potential of rabbit production is greatest. Small-scale rabbit projects have been gaining international attention as a feasible measure for poverty alleviation and for increasing self-reliance in food production, the two important elements of the World Food Summit Plan of Action. It is estimated that a family raising 5-6 does on locally available grain, forage and household waste can obtain 2 rabbits or 2.5 kg meat every week for domestic consumption or sale. Hence small scale backyard rabbit rearing can be a useful enterprise to improve the health and socio-economic condition of the tribal, rural and urban poor in our country. Commercial rabbit production also has great potential in and around metropolitan cities. Moreover, there are no religious inhibitions for the consumption of rabbit meat among the non-vegetarian population in India.

Table 1: Comparative Efficacy of Rabbit as Meat Producer

Type of animal	No. of Young/ year	Total carcass weight (kg) of offspring/year	In relation to dam weight
Cattle	0.9	173	0.35
Sheep	1.0	25	0.42
Goat	1.5	24	1.0-1.5
Pig	20	3200	16
Rabbit (intensive)	30-48	72-117	18-29

Table 2: Comparative Nutritive Value of Rabbit Meat

Type of animal/ meat	Protein %	Fat %	Moisture %	Calories/ 400 g (lb)	Cholesterol mg/kg
Rabbit	20.8	10.2	27.9	795	50
Chicken	20.0	11.2	67.0	810	60
Veal	18.8	14.0	66.0	910	100
Lamb	15.7	27.7	55.0	1420	-
Pork	11.9	45.0	42.0	2050	105

National Scenario

Taking in consideration, the scope and potential of rabbit rearing in our country, the Central Sheep and Wool Research Institute (CSWRI) of the Indian Council of Agricultural Research (ICAR) imported 60 New Zealand White rabbits (10 males+50 females) from U.K. in March 1978. Again during November 1979 another lot of 25 males and 100 females each of Soviet Chinchilla, Grey Giant and White Giant were received under Indo-USSR protocol for agricultural development as free gift. These rabbits were initially reared at NTRS Garsa and later shifted to Avikanagar and Mannavanur to study their adaptability and performance in different agro climatic conditions of India. Studies at temperate, sub-temperate and semi-arid regions have demonstrated that broiler rabbits can be raised successfully almost all over the country with some variation in managemental practices to suit the local climatic condition. As per the livestock census (2003), the total rabbit population of the country is estimated to be 406 000. Most of the states and union territories have a significant rabbit population. Kerala has the highest number (148 233) of rabbits followed by Nagaland (38 408), Bihar (29 380), Rajasthan (22 929) and Uttar Pradesh (22 191). Broiler rabbits are also concentrated in the states of Andhra Pradesh, Orissa, West Bengal, Maharashtra, Madhya Pradesh and other North Eastern States. Thus, in states which have a sizable tribal population and also have a sizable meat-eating population, which is desirous of healthy meat, rabbit meat production has great potential for development as a small backyard enterprise as well as for medium to large scale enterprises. In backyard and small scale systems one can expect higher outputs at lower inputs involving no off-farm purchases that would supply inexpensive meat for the family, improve farm productivity, and supplement income through the sale of surplus stock.

An important by-product obtained from broiler rabbits is the skin, which is covered with soft, luxurious fur. At present, most of the demand for fur pelts for making garments is met out of illegal trade in pelts from wild animals and through imports. Therefore, there is a good scope for obtaining high quality pelts at cheaper rates from broiler rabbits reared under farm conditions. Our country also has a large and thriving domestic fur industry, mainly centered in J&K. About 5 million fur skins are estimated to be imported every year, mainly from France, of which more than half are rabbit skins, as they are cheaper alternatives to high priced furs. A farm reared rabbits, particularly in high altitudes, can thus fulfill this domestic demand for rabbit skins and also relieve pressure on illegal hunting of animals and save valuable foreign exchange. Moreover, the fur produced in temperate regions is of superior quality as it is thick and has higher hair density and insulation capability.

At present, the major constraints to increasing rabbit production in high altitude areas of India is the shortage of germplasm. The available germplasm of broiler type rabbits at the CSWRI Avikanagar and its stations at Garsa and Mannavanur falls much short of the demand. Moreover, the breeders from far fetched areas have to bear high transportation cost and must in turn charge high prices for these rabbits to new breeders. Thus it is desirable that State Animal Husbandry Departments start rabbitries locally so that the new entrepreneurs may readily acquire breeding stock. The other major constraint is non-availability of a well-organized market for rabbit meat. Rabbit breeders in this reason either utilize the product himself or sells it in the local neighbourhood. Marketing of rabbit meat and fur skins deserves due attention from the development agencies in order to popularize broiler rabbit breeding in this country. There is also need to educate the law enforcement agencies like Forest Department and Police not to confuse domestic rabbit with wild rabbit and to allow free trade in rabbit and its products. Another problem is that people think of rabbit as pets and have to be educated about the rabbit as a producer of quality meat.

Broiler Rabbit Breeds in India

Exotic breeds of broiler rabbits, like the Soviet Chinchilla (SC), White Giant (WG) and Grey Giant (GG), were imported from the former USSR, while the New Zealand White (NZW) was introduced from the UK in the year 1978 by the Central Sheep and Wool Research Institute. These animals were kept at its North Temperate Regional Station (N.T.R.S.) at Garsa in the Kullu valley of Himachal Pradesh. Later, two new colors segregates viz., Black Brown (BB) and Dutch were developed from Grey Giant and Soviet Chinchilla breeds (Table 3). These segregates were selected strictly over the years for their color and type and are now breeding true to their characteristics. In 1983, a rabbit research unit was established at Avikanagar to study the performance of broiler rabbits as the semi-arid region constitutes about 50% of the geographical area of the country.

Table 3: Broiler Rabbit Breeds and their Characteristics

S.No	Breed	Fur color	Adult weight	Country of origin	Availability at
1.	Soviet Chinchilla	Chinchilla, steel grey	3.5 - 4.0 kg	Former USSR	CSWRI, NTRS, SRRS
2.	Grey Giant	Greyish brown	3.5 - 4.0 kg	-do-	CSWRI,NTRS
3.	White Giant	White	4.0 kg	-do-	CSWRI, NTRS, SRRS
4.	New Zealand White	White	3.0 - 3.5 kg	U.K.	NTRS
5.	Black Brown	Blackish Brown	3.5-4.0 kg	India	CSWRI, NTRS
6.	Dutch	Dutch pattern	3.0 kg	India	CSWRI,NTRS

CSWRI - Central Sheep and Wool Research Institute, Avikanagar, Via:Jaipur Rajasthan
NTRS - North Temperate Regional Station, Garsa, Kullu, H.P.
SRRS - Southern Regional Research Station, Mannavanur, Kodaikanal, T.N.

International Scenario

Broiler rabbit rearing for meat and fur in India is of fairly recent origin and, as such, is an imported concept. In the last few years though, rabbit rearing for meat has gained popularity and rabbit meat production is now estimated to be 7500 tons in India. At present, 82 per cent of the world's production of rabbit meat occurs in developed countries. It is estimated that 64 million rabbit does in the world and 709 million total rabbits. The distribution of rabbits in a few important countries are presented in Table 4.

Rabbit meat is consumed mainly in Western European and Latin countries (1.9 to 5.6kg/head/year) and secondarily in Eastern European countries (1.04kg/head /year).The average world rabbit meat consumption is 0.3 kg /head/year. Rabbit meat constitutes 5.6 per cent of the world meat production and is currently dominated by Italy, China, Hungary, France, Germany, Belgium, USA, Egypt and Indonesia. The European Union is the main consumer, producer and importer of rabbit meat and meat products in the world. In the European Union, Italy, Germany, France and Netherlands are the major producers, consumers and importers of rabbit meat (Table 4). The imports are mainly from China (47.05 per cent) and Hungary

Table 4: Rabbit Meat Production and Rabbit Population in a Few Countries of the World

S No.	Country	Production 000t	Export 000t	Import 000t	No. of females 000	Consum. kg/inhab	Value $/1000 $ of TGNP
1	Italy	300.0	0.65	23.0	7376	5.587	0.923
2	France	150.0	4.0	10.0	3916	2.756	0.423
3	Ukraine	150.0	.20	0.0	4958	2.886	3.252
4	China	120.0	40.0	0.0	6061	0.069	0.938
5	Spain	120.0	0.50	3.50	3864	3.152	0.803
6	Russia	100.0	0.00	0.00	4246	0.673	0.602
7	Nigeria	50.0	0.00	1.00	2107	0.455	50275
8	Indonesia	50.0	0.10	0.00	2698	0.273	1.415
9	Egypt	38.0	0.00	0.00	3235	0.696	3.996
10	USA	35.0	2.00	3.00	2422	0.142	0.020
11	Belarus	30.0	0.00	0.00	1341	2.913	2.683
12	Germany	30.0	0.00	6.50	1458	0.461	0.058
13	Poland	25.0	6.00	0.00	867	0.497	1.281
14	Belgium	25.0	10.30	11.00	553	2.612	0.434
15	Portugal	20..0	0.00	0.50	702	1.938	1.084
16	Czech Rep.	20.0	2.50	0.00	597	1.699	2.529
17	Morocco	20.0	0.00	0.00	992	0.779	2.465
18	Philippines	18.0	0.00	0.10	1609	0.288	1.352
19	Thailand	18.0	0.00	0.00	1028	0.310	0.660
20	Algeria	15.0	0.00	0.00	1043	0.583	1.047
21	Mexico	15.0	0.00	1.00	1018	0.182	0.190
22	Brazil	12.0	0.05	0.00	475	0.078	0.098
23	Hungary	12.0	12.0	0.70	275	0.067	1.303
24	Netherlands	10.0	3.75	3.70	225	0.661	0.119
25	Malaysia	9.0	0.00	0.10	494	0.501	0.626
26	Greece	9.0	0.00	0.10	203	0.899	0.457
27	Colombia	8.0	0.00	0.00	357	0.238	0.597
28	India	7.50	0.00	0.00	367	0.009	0.084
29	Sri Lanka	7.0	0.00	1.00	322	0.465	2.640
30	Ghana	7.0	0.00	0.00	310	0.451	3.969
31	Malta	3.20	0.00	0.00	110	8.889	4.310
32	Canada	3.0	0.00	3.00	121	0.227	0.017

(23.65 per cent). The six countries producing more than 100,000 tons/year are Italy, France, Ukraine, China, Spain and Russia (representing 58 per cent of world rabbit meat production) and eleven countries producing more than 20,000 tons/year are Indonesia, Nigeria, Egypt, USA, Germany, Belarus, Belgium, Poland, Morocco, Portugal and Czech Republic (representing 32 per cent of world rabbit meat production) (Table 4). The productivity of rabbit does is also the highest in Western Europe (36 kg rabbit meat / doe/year), followed by East Europe (27.5 kg), Far Eastern Asia (17.5 kg) and African continent (14.0 kg). For every thousand inhabitants in the world there are 12 rabbit does.

Rabbit Rearing Systems

Rabbits are reared mainly in three types of farms.

(A) *Traditional farms:* These farms have less than 8 does, are self consumption oriented and based on feeding raw ingredients. The productivity of these farms is low at 17.3 kg meat/year/doe but these types of farms contribute 40 percent of global production. Such farms are common in India.

(B) *Intermediate farms:* These farms have between 8-100 does, are simultaneously for commercial and self consumption and sale of rabbit meat and skins, and are based on complement and raw ingredient feeding. The productivity at such farms is 28.8 kg/year/doe and they contribute 33 percent of global production. The number of such farms is now increasing in our country.

(C) *Commercial farms:* These are usually big farms with more than 100 does, are commercial oriented and employ industrial feeding systems. The productivity of these farms is high at 54.3 kg/year/doe and they contribute 27 percent of the global production. Farms such as these have yet to be established in our country.

Performance of Broiler Rabbits in India

Studies at the Central Sheep & Wool Research Institute and its sub-stations in different parts of the country have shown that broiler rabbits have wide adaptability and have the potential to become an alternate source of meat in India.

Adaptability: Rabbits are highly susceptible to heat stress as they have few functional sweat glands and have a difficult heat output situation when environmental temperatures are high. In the semi-arid region, the rabbits show lower body weights from April to August and improvement in body weights as environment becomes more favourable (Table 5). Rabbits maintain a lower body weight in the semi-arid region as compared to the temperate and sub-temperature regions of the country indicating that during the course of adaptation to harsh environments there is a reduction in adult body weight.

Table 5: Average Adult Body Weight (kg) of Broiler Rabbits.

Breed	Jan	Feb	Mar	Apr	May	Jun	July	Aug	Sep	Oct	Nov	Dec
WG	3.36	3.33	3.40	3.36	3.19	3.05	3.07	3.21	3.24	3.34	3.34	3.51
GG	3.04	3.05	3.09	3.02	2.90	2.69	2.70	2.74	2.78	2.81	2.98	3.01
SC	3.20	3.25	3.22	3.25	3.07	2.79	2.84	2.80	2.90	3.12	3.17	3.27

Survivability: The survivability of rabbits varies among different age groups (Table 6). The mortality in kits (0-6 weeks) was highest in all the breeds, particularly in the first week of life as the kits are naked and depend totally on the mothering ability of the doe as well as the climatic conditions prevailing at that time. During winter kits generally die of hypothermia and in summer due to hyperthermia. The overall survivability was higher in GG and SC breeds and lower in the WG breed.

Table 6: Survivability (%) in Rabbits Among Different Age Groups

Age group	WG	NZW	GG	SC
0-6 Weeks	79.82	70.83	78.15	83.17
7-12 weeks	89.32	95.85	92.24	82.89
13-24 Weeks	92.56	95.48	93.70	97.38
Adults	82.44	85.73	88.25	88.13

Growth: Growth is an important aspect of broiler rabbit production. The growth performance of different broiler rabbit breeds is presented in table 7. The growth of the young during the suckling period, especially during the first three weeks, greatly affects their later performance. Apart from milk yield, litter size and doe's nursing ability also influence the pre-weaning growth. The 12 weeks weight ranged from 1711.11 g to 1853.00 g showing marginal breed differences. The growth is fastest during weeks 7-12, and the body weight at 12 weeks is of crucial significance as the broiler rabbit is usually marketed and slaughtered at this age. The higher the body weight at 12 weeks, the more the carcass yield and profitability. Under an experimental *ad lib* complete diet feeding system, average weight of 2.0 to 2.3 kg at 12 weeks has been achieved at this Institute.

Table7: Growth Performance of Broiler Rabbits.

Breed	Body Weights in grams			
	At birth	6 weeks	12 weeks	24 weeks
WG	54.05	806.41	1853.00	2645.30
SC	54.57	876.98	1788.46	2502.80
BB	46.08	769.76	1711.11	2448.60

Reproduction: Short generation time and high reproductive potential of rabbits makes this the ideal meat producing small livestock for our country. Rabbits are induced ovulators and can be bred within 24 hours of parturition. Thus it is theoretically possible to produce over 11 litters per year under an intensive production system. The average number of kindling/doe/year and average number of offspring/doe/year achieved in the semi-arid region under the semi-intensive production system was 3.5 and 23.9, respectively, while under an intensive system it was marginally higher at 4.0 litter/doe/year. However, in the temperate region it is possible to achieve an average of 5.16 litters/doe/year (range 4 to 9) or 35-40 offspring/doe/year under an intensive production system.

The reproductive performance of broiler rabbits is presented in table 8. The litter size at birth (LSB) ranged from 6.46 to 7.26 while the litter size at weaning (LSW) ranged from 4.92 to 6.17 in different breeds. High LSB and LSW were recorded in the Grey Giant (GG) and White Giant (WG) breeds. Similarly, high litter weights at birth (LWB) and weaning (LWW) were recorded in GG and WG breeds. The average LWB and LWW ranged from 325.34 to 390.82 g and 3023.80 to 3488.22 g, respectively.

Table 8: Reproductive Performance of Broiler Rabbits

Breed	Litter size at		Litter weight (g) at	
	Birth	Weaning (6 weeks)	Birth	Weaning (6Weeks)
GG	7.26	6.17	374.19	3488.22
WG	7.23	5.70	390.80	3681.09
NZW	6.46	5.18	345.99	3300.50
SC	6.65	5.26	258.00	3023.80
BB	7.06	4.92	325.34	3304.34

Milk production: Milk production by a rabbit doe is an important indicator of growth and health of young rabbits in their first month of life and is intimately related to the body weight achieved at the slaughter age of 12 weeks. The pre-weaning growth of kits is entirely dependent upon the does milk yield, particularly during the first 21 days of life. The milk yield and nursing ability of doe increases with parity order. The total milk yield of SC does was higher (4370.90 g) compared to WG (4326.43 g) and BB (4179 g) in a 35 days lactation period (Table 9). There was however, a sudden decline in milk production in all the three breeds studied from days 31.32 onwards if lactation is continued beyond 28 days. Therefore, in order to take a greater number of kindlings per doe per year, it is recommended to wean early, at 28 days as by this time the young start consuming some solid feed as well as roughage and the role of milk in their diet declines.

Table 9: Milk Production in Broiler Rabbit Does

Lactation Period	Milk yield/doe (g)		
	Soviet Chinchilla	White Giant	Black Brown
0-7 days	531.38	561.25	527.14
8-14 days	903.33	797.50	808.57
15-21 days	1055.83	923.75	892.50
22-28 days	941.11	975.45	915.83
28-35 days	700.00	680.00	725.00
0-28 days	3481.66	3278.18	3205.00
0-35 days	4370.90	4326.43	4179.00

Slaughter of Rabbits

The first step in slaughtering a rabbit is to render it unconscious. This is to be done quickly in order to prevent suffering and struggling. The two methods generally used are stunning or dislocating the neck. In the stunning method, the rabbit is first grasped across the loin and suspended head down. It is then stunned with a heavy blow at the base of the skull, between the ears with a stick or a small iron rod. In the dislocation method, the rabbit is held by its hind legs with one hand and the thumb of the other hand is placed on the neck and the four fingers extended under the chin. The animal is stretched by pushing down on the neck and with a quick movement the animal's head is raised to dislocate the neck. After making the animal unconscious by using either of the methods, the animal is suspended immediately on a hook and its head cut off in a quick movement to ensure proper bleeding so that the meat has a good colour.

Skinning and Preparation of Skins

For skinning, a 5-inch blade knife is a convenient size. The rabbit is suspended by inserting the hook between the tendon and bone of the right hind leg, just above the hock. First the tail is removed and the free hind leg cut off at the hock joint. The front feet are also cut off, and then the skin is slit just below the hock of the suspended leg. Now the skinning knife is inserted under the skin on the inside of the leg and opened up to the root of the tail. This is continued to the hock of the other leg. The edges of the skin are first carefully separated from the carcass and then the skin is pulled down over the animal. A knife is used to separate the fat completely from the skin so that the fat is left on the carcass. The skin removed by this method is known as a cased skin. Care should be taken while skinning as even small cuts lower the value of skin. While the skin is still warm, it is secured on a stretcher with the flesh side out. The stretcher is made of a No. 9 galvanized wire in varying sizes for small, medium and large skins. The ends of the skin are pulled down to remove all wrinkles and folds. Before hanging the skins to dry, sprinkle on these warm and fresh skins any insect repellent powder (mixed with saw dust in the ratio of 1:10) to protect it against flies and other insects. The skins should never be left to dry in the sun but always hung inside a room or shed. No salt or any other preservative should be used. After complete drying, the skins are removed from the stretcher, arranged in piles and stored in containers with naphthalene balls thrown in between. These dried skins can be sold as such to pelt dealers or processed at home for added value. The method recommended at our institute for processing dried skins is the chrome tanning method. The other methods are the Salt-Alum method and the Salt-Acid method. Rabbit skins, when properly processed and finished are very soft and luxurious. They fetch a good price and the garments made out of them are also priced high.

Carcass Characteristics

The first step in slaughtering rabbit is stunning before bleeding. In small scale operations, dislocation of neck is generally the method of choice, while on a commercial scale electrical stunning is mostly used. Feed is withheld at least 12 hours before slaughter to ensure an almost empty gut. After separation of head, the whole body is suspended by inserting the hook in the hock joint of the hind leg and skinning is carried out carefully to avoid any cut on the skin as it reduces the value of the skin, which is an important by-product of rabbit. This is followed by evisceration of all the visceral organs and separation from the carcass. The carcass is then cut into hind quarters, loin and fore quarters. Percentages of 55-60 per cent saleable meat yields are achievable in rabbit (Table-10).

Table 10: Carcass Traits of Broiler Rabbits

Traits	Soviet Chinchilla	Grey Giant	Pooled
Pre-slaughter weight (g)	2131.90	2206.60	2169.20
Blood weight (g)	80.40	80.60	80.50
Head weight (g)	186.40	191.90	189.20
Skin Weight (g)	224.00	238.40	231.20
Inedible offal weight (g)	477.10	516.20	496.60
Edible offal (g)	81.40	80.10	80.70
Head meat weight (g)	61.70	58.40	60.10
Leg cut weight (g)	423.40	422.50	422.90
Loin cut weight (g)	249.40	237.90	243.60
Rack, neck & BF cut Weight (g)	409.80	433.00	421.40
Total cut weight (g)	1082.60	1099.30	1090.90
Saleable meat yield (g)	1225.70	1237.80	1231.80
% PSW	57.45	55.68	56.57

Some 70 per cent of a rabbit carcass is composed of edible meat as compared to 50 per cent in chicken. Rabbit meat is also comparable to chicken in terms of low fat content (3.8 per cent) and high protein (20.7 per cent) and these values are more favourable than most other meats. Moreover the digestibility of rabbit meat is as high as 98.9 per cent. Rabbit meat is in fact appreciated for its low content of total fat and cholesterol and high proportion of polyunsaturated fatty acids (PUFA) and phospholipids, which make rabbit meat a healthy meat for humans.

Rabbit Diseases and their Control

Diseases and death are an integral part of any livestock operation. The same is true in rabbit production. Mortality usually averages between 20-25 per cent of all animals born in most of the rabbitries. This high rate of mortality in other livestock enterprises would be discouraging but is tolerable in rabbit raising because of its short generation interval and faster production. Since most diseases occur due to mis-management, it is possible to keep rabbit diseases at a minimum. Ventilation, sanitation and observation are the most important factors involved in disease control. Less respiratory problems are encountered in well-ventilated rabbitries as air dilution is a very effective method of reducing the number of certain disease causing organisms. Physical or chemical removal of such organisms, called sanitation, is another effective method of reducing the number of potential disease organisms. Observation means training oneself to look for signs of discomfort in animals. A sick animal is usually off-feed and water, is lethargic and has a dull appearance. If one misses these signs, it is generally too late to save the animal. However, in the initial stages, treatment may be helpful and may save the life of the animal.

The two most important specific rabbit diseases are pasteurellosis and enteritis. Pasteurellosis is caused by the organism *Pasteurella multocida* and this term covers a multitude of clinical conditions like snuffles, pneumonia, abscesses, weepy eyes, orchitis, metritis and wryneck. The first signs of *P. multocida* infection are snuffles or a cold. Strict isolation and treatment with antibiotics will control the problem. Ventilation is very important as both ammonia levels and humidity are involved in the transmission and development of snuffles. Treatment involves draining the pus properly and putting the rabbit on antibiotics. Another major cause of death in weaner/grower rabbits is the complex of diseases called enteritis. Enteritis complex includes enterotoxaemia, mucoid enteritis, coccidiosis and Tyzzer's disease. The clinical signs of enterotoxaemia are profuse diarrhea, dehydration and reduced feed intake. It is most commonly seen in weaners and the post-mortem reveals an enlarged and reddish caecum. Enterotoxaemia is more common when diets low in fiber and high in energy are fed to rabbits. Causing agents are generally *Clostridium perfringens* and *E. coli*. Oxytetracycline in water gives some relief but feeding of hay or straw is more helpful in prevention of this disease.

CONCLUSIONS

Rabbit production, though of recent origin in India, has great potential for growth as an agricultural enterprise. Due to a rapidly increasing human population, food shortage may become a major problem in future. Since all reasonable options have to be explored, rabbit can play a major role as an alternative source of meat and fur, particularly in the hilly regions where the climatic conditions are most suitable for rabbit rearing. Studies at the Central Sheep and wool Research Institute has shown that rabbit can be reared successfully in most regions of the country in either a small-scale backyard operation or as a large-scale commercial enterprise. Our experience of close interaction with rabbit farmers over the last 10-15 years has shown that rabbit rearers have genuine interest in this enterprise, which is encouraging, and should be utilized to increase meat production and thereby food production in the country. Development of infrastructure facilities in terms of germnplasm centers, housing, feeding, health care, marketing, training, testing and processing of raw materials, etc. should be taken up for further expansion of broiler rabbit production, particularly in the states where rabbit rearing enjoys popular support and where there is great potential for expansion like in the high altitude regions of the country.

□□□

32

Fisheries Potential in High Altitude Water Bodies of Leh-ladakh

PREM KUMAR[1], A. BARAT[1], G. K. SIVARAMAN[1], P. C. MAHANTA[1] & S. AYYAPPAN[2]
[1]NRC on Coldwater Fisheries, Bhimtal-263136 (Nainital) Uttarakhand
[2]Fisheries Division, Indian Council of Agricultural Research, Pusa, New Delhi, India

INTRODUCTION

Fisheries Resources

Lakes

In the past, Ladakh and its mountains were apparently under an ice-cap which extended from the North Pole down to Kashmir and which on melting formed many lakes and streams. Some of these lakes were formed as a result of dead ice masses acting as blocks, damming the melted waters into huge reservoirs of water. Gradually, as the region's climate underwent a change and Ladakh acquired desert like properties, most of the water of the lakes evaporated. The collapse of the moraine walls that acted as a dams led to a cataclysmic emptying of these lakes. Today only a few of these landlocked lakes remain and their water, because of the blockage of inlets and outlets became brackish, with the exception of *Quin-Tso,* which is a fresh water lake.

The *Pangong Tso* is the largest Lake in Ladakh. It stretches to a length of 160 km, 8 km at its widest and situated north-east of Leh, at an elevation of approximately 4,300 m.

The *Tsomo-Riri* is approximately 40 km long with an average width of six to seven kilometers which is situated at an approximate height of 4500 m in the erstwhile region of Rupshu, south east of Leh.

The *Tso-Kar* to the west of Tsomo-Riri and Quin-Tso to its east are much smaller lakes and *Hanle Tso,* which was once a, large lake, is but a pond today, surrounded by a wide marsh.

Rivers

The main rivers draining Ladakh are the Indus, Hunza, Shigar, Shyok, Saltro, Nubra, Zanskar, Hanle, and the Galiwan.

One of the greatest rivers, the Indus is the lifeline for the people of Ladakh's scanty population, The Indus rises from the Tibetan plateau near Mansarovar lake, to the north of Mount Kailash. Taking a north westerly course, it enters Ladakh from Demchok in the Changthang area in the southeast. It enters the Kashmir region near the confluence with the river Gurtang, at an elevation of about 4,200 m, it makes a sharp bend south of Pangong lake and cuts across the main Ladakh range. Flowing through a barren mountain-scape and plunging down narrow gorges, it finally leaves Ladakh in the west. The river Indus takes a sharp bend past Nanga-Parbat before flowing into the plains of Pakistan, towards Skardu in Baltistan. From there it transverses down the length of Pakistan before flowing into the Arabian Sea near Karachi. During its sinuous journey through Ladakh, many mountain streams join the Indus. Some are small and insignificant, but others are of considerable importance.

The most important tributaries to the Indus to the north are the Shyok and the Nubra rivers. These two tributaries bring down an incessant flow of waters from perpetually snow bound peaks and glaciers after originating from the Depsang plains at an elevation of 5000 m.

The River Shigar arising from the glaciers on the north face of the Karakoram flows towards south-east for a short distance before joining the river Indus at Skardu.

The River Zanskar originates in the northern slopes of the Great-Himalaya. The headwaters are known as the river Tsarapchi. This river cuts across the Zanskar range and flows for some length in the north east direction before being joined by the river Khurana, and is finally joined with the Indus river.

The *Hanle River* is a short left bank tributary of the Indus River. It originates in the northern slopes of the Zanskar range.

Wetlands

The Government of India initiated a national inventory of wetlands, entitled the All India Wetland Survey, in the late 1960's. A large number of sites have been listed. In the 1980's the Department of Environment established a wetland Working Group. Based on an analysis of the responses, the Directory of Indian Wetlands was produced in the 1990's by the Ministry of Environment and Forest some of the sites designated as wetlands include:

Thiksey to Shey: From Thiksey onwards to Shey, the .Indus divides into a numbers of channels which have led to the formation of several islands where small pockets of forests have developed. The road from Thicksey to Shey runs along the north bank of the Indus River

Shey marshes: A small, marshy tract along the Indus. Recently, a road has been constructed bisecting the marsh into two unequal parts, and on the southwest side of the marsh, is the Shey fish farm.

Chushul marshes: A complex of shallow ponds, marshes, and wet meadows in a broad, sandy valley to the east (10,000 Ha) ana west (1000 ha) of Chushul, 15 km south of Pangong Tso. The ponds and marshes are created by springs and streams flowing down into the valley from the Ladakh Range to the south-west. Some streams terminate on the sandy plains in stagnant pools, which become saline as they evaporate; others carry sufficient water to flow into the Pangong Tso and ultimately the Indus River.

Hanle river marshes: A complex of fast/flowing streams, stagnant pools, saline marshes, seasonally flooded marshes, and bogs along the Hanle river above its confluence with the Indus. The Hanle river flows through a wide, sandy plain with seasonally flooded marshes and grasslands.

Tsokar basin: In a very remote and sparsely populated area, the Tsokar Basin is a land locked basin of a former large freshwater lake (now in a state of desiccation) has contracted into two principal water bodies:

Startsapuk Tso a brackish lake of about 300 ha in the south, and Tsokar itself, a hyper saline lake of 2200 ha.

Present Status

Fish culture has spread all over the world. Methods are diversified and systems differ according to the species cultivated, the aims (fish for consumption or repopulation) and the techniques and methods employed. An important distinction between different types of fish cultivation is between cold water and warm water cultivation. In practice, the temperature separating the two types of cultivation is 20°C. The culture fisheries of Ladakh are still in their developing stages. The efforts of this department by setting up two fish farms and hatcheries in Leh-Ladakh have shown encouraging results, with successful breeding and rearing of trout and common carp. The two varieties of cultured trout are the rainbow trout (*Oncorhynchus mykiss*) and brown trout's (*Salmo trutta fario).* Among the common carp are the leather carp (*Cyprinus carpio* var *nudus*), mirror carp *(Cyprinus carpio* var *specularis),* and scale carp *(Cyprinus carpio* var *communis).* The common carp are the sturdiest fish for culture practices. All of the three varieties stated above have been found to thrive well even at high altitude. The common carp (scale carp), is found to be a fast growing fish and is widely cultured in fresh water ponds.

Fish Farm Facilities, State Fisheries Department

Shey Fish Farm of Leh District

A small fish farm with a total area of 1 ha and water area of 0.5 ha (five ponds) has been established in the marshes/wetland of Shey village. The farm is located at an altitude of 11500 ft above mean sea level. This farm is not only utilized for rearing exotic fish species but has played an important role in developing the fisheries in Ladakh. Experimental breeding of the exotic fish species like rainbow trout and carp has been successfully carried out under local conditions.

Trout Fish Farm at Sindhu Ghat

This farm is located near Sindhu Ghat and the farm consists of 10′ wide and 1000' long channel covered with the thick mesh wire, with nets in place to create a number of small ponds. The ponds are stocked with rainbow trout, which grow well to 80-100 g in 6 months.

Trout Fish Farm, Nubra

Situated in the valley of Diskit, Nubra, this farm is one of a kind in the Leh district, where for the first time, trout raceways (six receways of 26′ X 9′8″ size) were developed to rear rainbow trout brought in from Kashmir. The first batch of live trout fish seeds brought in from Srinagar has been successfully reared under local conditions and is now ready for sale.

Planktons Observed in Waters

Phytoplankton

Chlorophyceae

Ankistrodesmus convolutes, Ankistrodesmus falcatus, Chlamydomonas cingulata, C. globosa, C. Microspherea, Volvox aureus, Chlorella conglomerata, Chlorella vulgaris, C. Pyrenoidosa, Closterium acerosum, Closterium microporum, Closterium simansis, Closteriopsis longissima, Chlorococcum humicola, Desmidium spp., *Pandorina granulosa, Pediastrum duplex, Pediastrum simplex, Selenastrum gracile, Hydrodiction reticulam, Chaetophora elegans, Oedogonium spp., Cladophora spp.,* and *Spirigyra communis.*

Bacillariophyceae

Fragilaria cupucina, Fragilleria pinnata, Navicula sublinearis, N. Viridula, N. cryptocyphala, Pinnularia microstatum, Amphora ovalis, Asterionella formosa, Cymbella ltinceolata, Diatoma vulgare, Pinnularia nobilis, Rhopalodia gibba, Synedra ulna.

Cyanophyceae

Anabaena spiroides, A. Varibilis, Aphanocapsa, Gomphosphaeria, Merismopoedia, Microcystis auruginosa, Oscillatoria spp.

Dominant Species of Phytoplankton Recorded in the Littoral Shey Farm

Hydrodiction reticulam, Chaetophora elegans, Oedogonium spp., Cladophora spp., and *Spirigyra communis* are the chlorophyta, identified from the river water samples collected from the different locations.

Zooplankton

Rotifera

Asplanchna spp., Macrochaetus spp., Brachionus bidenlatas, Brachionus caudatus, Brachionus dedutata, Brachionus nilsoni, Cephalodella forftcula, Cephalodella gibba, Cephalodella intuta, Cephalodella auriculata, Colurella adriatica, Dicranophorus robustus, Horaella brehmi, Keratella cochlearis, Lepadella ovalis, Lecane bifastigata, Monostyla closterocerca, Mytilina ventralis, Philodinavus paradoxus, Philodina roseola, Rhinoglena frontalis, Rotaria tridentate.

Cladocera

Alona affinis, Alona rectangular,Alonella excise, Chydorus gibbus, Chydorus sphaericus, Diaphanosoma exisum, Leydigill acanthocercoides, Moina micrur, Simocephalus vetulus.

Copepoda

Mesocyclops leuckarti, Microcyclops varicans, Phyllodiaptomus bland, Tropocyclopspra sinus, Testudinella frontalis, Testudinella patina, Trichocerca cylindrical, Trichocerca ratus.

Fish Genomic Studies

Classifications of coldwater fish species are generally based on classical morphological and osteological methods. However, accurate identifications of available trout fish species are very difficult due to intraspecific morphological variability. The characters involved can be morphological, ecological, behavioral, or physiological; and are assumed to be genetically based. The cause of variation in the morphometric and meristic characters may range from genetic variability to the influence of environmental parameters. In genetic analysis, various types of genetic markers such as morphological, chromosomal, and biochemical and molecular markers are used. Morphological and protein markers usually show low degree of polymorphism and, hence, are not very useful genetic markers. Biochemical markers have been tried extensively, but have not been found encouraging, as they are often sex limited, age-dependent, and are significantly influenced by the environment. Whereas, the molecular markers, capable of detecting genetic variation at the DNA sequence level, have removed these limitations and allowed evaluation of the genetic basis for observed phenotypic variability with greater genetic resolution. Our objective will be to identify RAPD marker and mitochondrial genome variability in available fish species from natural waterbodies of the Leh-Ladakh region, to estimate genetic variability, and to construct a phylogenic tree among these fish species. So far, no such study has been carried out for this fish species. This will help in genome analysis and genetic improvement of the fish species in this region.

CONCLUSIONS

High altitude, lands in Ladakh have been used for procuring food, energy and minerals. Individuals dependent on fish as a means of livelihood in this reason are now facing difficulties due to a sharp decline in fish catch in upland areas. With a growing population and reduced per capita land resources in these areas. The fish resource base is of great relevance, and increasing this resouce should be a matter of national concern, which needs different technological approaches and support services. Such upland regions have the potential to be tapped for increased fish production to allow increased rural development.

It is only after recognition and description of existing fish production practices in Ladakh that a possibility of generating global prescriptions for wise management practices can emerge. It will require effort and time to analyze various activities in the uplands that are coordinated under diverse technological, social and cultural backgrounds. It will surely the laborious and time consuming to unravel the hidden unwise components of plans and programmes and initiate ways of countering their effects, but, it is a necessity and an investment for a safe and wise future. Research on coldwater fisheries in India was mainly initiated under the ICAR system especially through NRC, but Academic Universities in hill states have also contributed to generation of various basic information on many aspects of coldwater fish species and the aquatic ecosystems sustaining them. Because development of fisheries was a high-priority need in the region, an exploratory survey was conducted during December 2007, but this attempt was limited by the extremely cold climate when most of the water bodies were partially frozen. There is a need for further survey information in order to develop suitable models for fish culturing in the region.

References

Gazetter of Kashmir and Ladakh (1974) (Compiled under the direction of the Quarter Master General of India in Intelligence Branch), Vivek Publishing House, Delhi-7 India

Kaul, Shridhar & Kaul, H.N. (1992) Ladakh through the ages towards a new identity. Indus Publishing Company, FS-5, Tagore Garden, New Delhi.

Parvez Dewan's Jammu, Kashmir, Ladakh (2004) Manas Publications, New Delhi-India

Report (2004-2005) Fisheries Department, Leh.

Table 1: Physico-chemical parameters of water

S. N	Location Lat/Long	Place Name	Elevation (m)	Date	Temp.		pH	DO ppm	CO_2 ppm	Alkalinity ppm
					Air °C	Water °C				
1.	34°05′06″ N 77°36′30″E	Indus River (Sindhu Ghat)	3215	13.12.07	-6.0	0.5	8.8	12.0	nil	72
2.	34°05′6.5″ N 77°36′35″E	Shey Trout Farm (Sindhu Ghat)	3247	14.12.07	-2.0	2.0	8.5	9.2	nil	36
3.	34°05′11″ N 77°36′0.0″ E	Shey Fish Farm	3262	14.12.07	-1.5	1.7	8.7	9.6	nil	64
4.	34°09′53″ N 77°19′59″E	Sangam (Sindhu& Zanskar	3075	15.12.07	-4.6	-0.6	8.2	9.2	nil	70
5.	34°09′53″ N 77°19′59″E	Nimboo Bazgo (Alchi) HEP*	3070	17.12.07	-5.2	-0.4	8.1	9.6	nil	36

*HEP Hydro-Electric Project

❑❑❑

33

Effect of Feeding Seabuckthorn Leaves and Lucerne Hay Based Complete Diets on Growth and Blood Biochemical Profile of Sheep at High Altitudes

JADHAV SUNIL, TILAK RAJ, GURU CHARAN AND SHASHI BALA SINGH
Defence Institute of High Altitude Research, DRDO, Leh-Ladakh, J & K, India

ABSTRACT

A study was conducted at Defence Institute of High Altitude Research, DRDO, Leh, India (3500 m above mean sea level) to evaluate the effect of feeding locally available Seabuckthorn leaves and Lucerne hay based complete diets on growth and blood biochemical profile of sheep. Environmental temperature during experimental period ranged from -1.2°C to +10.7°C (Av. 4.7°C). Eight male lambs of average 12.0 months of age and 19.7±1.20 kg body weight were divided into two equal groups (Gr.I and Gr.II) of four lambs in each group. Two complete diets with roughage: concentrate ratio of 45:55 were formulated. Animals in Gr.I were fed a complete diet containing Seabuckthorn leaves and animals in Gr.II were fed a complete diet containing Lucerne hay as a roughage source for the period of 60 days. Average daily gain in body weight (g) in Gr.I (101.6±13.17) was lower ($P<0.05$) as compared to Gr.II (167.5±22.37). Dry matter intake (DMI)/day and DMI per cent body weight in Gr.I and Gr.II did not differ. Serum total protein, urea and blood urea nitrogen concentration in Gr.I was lower ($P<0.05$) in Seabuckthorn leaves fed group.

Serum calcium and inorganic phosphorus concentration was similar in both the groups. Serum creatinine concentration in Gr. I animals was lower ($P<0.01$) than Gr.II animals; however there was no change in serum cholesterol, triglycerides, sodium and potassium concentration among the groups. Feed cost in rupees per kilogram gain in body weight in Gr.I (63.17±8.59) and in Gr.II (59.66±6.50) was also similar. Beside the inferior effect of feeding Seabuckthorn leaves on growth performance of sheep as compared to Lucerne hay, cost benefit analysis revealed that Seabuckthorn leaves can be used economically as a local feed ingredient in complete feed of sheep at high altitude of Ladakh.

INTRODUCTION

Ladakh is situated in the Himalayas on the far northern frontiers of India. It is one of the coldest regions in the world. Ecologically Ladakh, a high altitude desert and mountainous region, is situated at 2500 to 5000 meter above mean sea level. The prevailing hostile climatic conditions create difficulties in agricultural and animal husbandry practices in Ladakh which includes two districts i.e. Leh and Kargil. There is an ever increasing demand of fresh food of animal origin for troops deployed and growing civil population in this region.

Out of total area reported in both the districts, 55.71 per cent land in Leh and 23.53 per cent land in Kargil district are barren and uncultivable. Moreover, out of net sown area in both the districts, 19.76 per cent land in Leh and 54.07 per cent land in Kargil district comes under fodder crops cultivation. The total animal population of Leh district is 5.54 lakhs (5.96 per cent cattle, 32.12 per cent sheep, 53.78 per cent goat and 8.14 per cent Yak, Dzo-Dzomos, horses, ponies, donkeys and camels) (Statistical Handbook of Leh, 2002-03) and Kargil district is 3.38 lakhs (13.50 per cent cattle, 48.25 per cent sheep, 27.23 per cent goat and 11.02 per cent other animals) (Statistical handbook of Kargil, 2002-03). However, shortage of locally available nutritious animal feed resources limits an efficient animal husbandry practices in the region.

Seabuckthorn (*Hippophae rhamnoides* L.) is a thorny, dioceous bush, which grows wildly on riversides and slopes in dry, temperate, high altitude region of Indian Himalayas. This plant has been found growing wild in wastelands of Ladakh, in about 11,500 hectare area (Dwivedi et al., 2006). The plant is also growing in other parts of the country like Lahaul-Spiti (HP), Badrinath (Uttaranchal), Sikkim etc. It also has been found growing in other countries viz. China, Pakistan, Nepal,

Canada, Germany, and Russia. Local tribal population of Ladakh region has been using the leaves of Seabuckthorn as feed/fodder for cattle, sheep, goat and double humped camel. High content of crude protein and other nutrients in Seabuckthorn leaves makes it a suitable fodder plant for the livestock in Lahaul valley of India (Singh et al., 1999). In north China the seeds, leaves and fruit residues of Seabuckthorn were found suitable as a livestock and poultry feed where the body weights of livestock and poultry were found to be increased after feeding with Seabuckthorn (Hu and Guo, 2006).

There are no reports on studies on effect feeding Seabuckthorn leaves to animals in high altitude cold desert conditions of Leh-Ladakh. In view of above, a study was conducted to evaluate the effect of dietary inclusion of Seabuckthorn leaves on growth and blood biochemical profile of sheep under high altitude cold desert conditions of Ladakh.

MATERIALS AND METHODS

Animals, Feeds and Feeding

Eight healthy male crossbred (Muzaffarnagari x Changthangi) lambs of average body weight 19.7±1.20 kg and 12.0 months of age were divided into two equal groups viz. Gr.I and Gr.II. Locally available Seabuckthorn leaves were collected from their natural habitat. Two complete diets were formulated using Seabuckthorn leaves and Lucerne hay as roughage ingredients both at the rate of 45per cent. Concentrate was included at the rate of 55 per cent which was common in both the diets. Percent composition of both the complete feeds is given in Table 1. Animals in Gr.I were fed Seabuckthorn leaves based complete feed where as animals in Gr.II were fed Lucerne hay based complete feed. All the animals were treated for ecto and endo parasite before the start of the experiment. During day time all the animals were kept in open paddock with the facility of individual feeding and watering and at night animals were housed in suitable underground shed. An adaptation period of 20 days was followed before the start of actual experiment of 60 days duration. Environmental temperature during the experimental period ranged from -1.2 °C to +10.7°C (Av. 4.7 °C). All the animals were fed individually weighed amount of feed daily at 0930 hrs. Fresh drinking water was made available to experimental animals daily at 1100 hrs in the morning and at 1600 hrs in the afternoon.

Data Recording, Sample Collection and Analysis

Daily feed residue left by individual animal was weighed and recorded to calculate the feed intake. Body weight of all animals was recorded at the start of experimental period and there after at weekly interval in the morning before feeding and watering. At the end of the experimental period blood was collected aseptically from all animals by jugular veni-puncture in clean dry test tubes. Serum was separated and stored at -20°C for further analysis of biochemical parameters.

Feed samples were analyzed for chemical composition as per the procedure of AOAC (2000). The blood biochemical profile viz. serum total protein, urea, blood urea nitrogen, creatinine, cholesterol, triglycerides, calcium, inorganic phosphorus, sodium and potassium were analyzed spectrophotmetrically by Diagnostic Kits manufactured by Span Diagnostics Ltd., Surat, India. Data generated in the study was analyzed for test of significance as per Snedecor and Cochran (1994).

RESULTS

Crude protein content (CPI) in Gr.I and Gr.II diet was 21.79 per cent and 23.39 per cent, respectively. As shown in Table 2, average daily gain (ADG) in body weight in Gr.I (101.6 g) was lower ($P<0.05$) as compared to Gr.II (167.5 g). It was observed that mean dry matter intake (DMI) did not differ significantly ($P>0.05$) among the groups (937.0 and 1079.0 g/d in Gr.I and Gr.II, respectively). Mean DMI per cent body weight (4.76 in Gr.I and 4.40 in Gr.II), DMI g/kg$W^{0.75}$ (95.81 in Gr.I and 93.66 in Gr.II) and crude protein intake (204.2 g/d in Gr.I and 252.3 g/d in Gr.II) were comparable among the groups. The feeding cost (in rupees) per kg gain in body weight was also comparable ($P>0.05$) at 63.17 and 59.66 in Gr.I and Gr.II, respectively.

Table 3, depicts the effect of feeding Seabuckthorn leaves vs. Lucerne hay based complete feeds on blood biochemical profile of sheep. Serum total protein (g/dl), urea (mg/dl) and blood urea nitrogen (mg/dl) concentration in Gr.I (7.80, 43.51 and 20.32, respectively) was significantly lower ($P<0.05$) as compared to Gr.II (10.0, 59.36 and 27.74, respectively), where as serum creatinine concentration in Gr.I (1.58 mg/dl) was significantly lower ($P<0.01$) as compared to Gr.II (2.10 mg/dl). There were no significant differences in serum cholesterol and triglyceride concentrations among the groups as it was at 136.64 and 158.90 mg/dl in Gr.I and 151.23 and 164.19 mg/dl in Gr.II, respectively.

As shown in Table 4, concentration of serum calcium in Gr.I and Gr.II animals (9.95 and 9.78 mg/dl, respectively) and serum inorganic phosphorus (5.19 and 5.38 mg/dl in Gr.I and Gr.II, respectively) were comparable. Similarly concentrations of serum electrolytes viz. sodium (mM/L) at 142.84 in Gr.I and 142.53 in Gr.II and potassium (mM/L) at 6.32 in Gr.I and 6.50 in Gr.II were also similar.

DISCUSSIONS

Perusal of Table 2, shows that daily DMI (g/d) and DMI of per cent body weight of the animals in present investigation (937 and 1079 and 4.76 and 4.40 in Gr.I and Gr.II, respectively) was higher as compared to ICAR (1998) which quote daily DMI of 680 to 830 g and DMI of 3.4 to 4.1 per cent body weight in 20 kg lamb growing at the rate of 100 to 150 g/d. Lower environmental temperature which ranged from –1.2 °C to +10.7°C (Av. 4.7 °C) during the trial period might have resulted in increased DMI in animals. Patrick (www.wisc.edu) has also reported 3 to 8per cent increase in dry matter intake of growing heifers as a result of prevailing environmental temperature of +5 to -5°C. Moreover ambient temperatures below the lower critical temperature (i.e. cold stress) result in increased energy requirement in animals (Young, 1983). Further perusal of Table 2. shows that the dry matter intake was 13per cent less and crude protein intake was 19 per cent less in the animals fed Seabuckthorn leaves based complete feed than the animals fed Lucerne hay based complete feed. Comparatively less intake of dry matter and crude protein may have contributed to some extent in lower body weight gain of animals in Gr.I than Gr.II. Cost benefit analysis in the present investigation revealed that feed cost (Rs.) per kg gain in body weight was similar (P>0.05) at 59.66 in Gr.I and 63.17 in Gr.II. Lucerne hay though highly nutritious is more costly as compared to easily available Seabuckthorn leaves. As stated earlier Seabuckthorn is growing wildly in Ladakh and hence can be available in plenty for feeding to the animals.

The biochemical profile of sheep in both the groups is given in Table 3. The values of different blood biochemical parameters observed in present investigation were in normal range. As shown in Table 2, feeding of Seabuckthorn leaves in the complete feed of sheep resulted in significantly lower values of serum total protein, urea and blood urea nitrogen concentration as compared to Lucerne hay. Though leaves of shrubs and trees are important in providing nutrients to animals, they usually contain certain anti-nutritional compounds. It is reported that Seabuckthorn leaves contain 11.0 per cent of total tannin phenolics and 1.0 per cent of condensed tannins (Singh *et al.*, 1999) as against 0.1 per cent condensed tannin in Lucerne (Hoskin *et al.*, 2000). Tannins may form a less digestible complex with dietary proteins and may bind and inhibit the endogenous protein such as digestive enzymes (Kumar and Singh, 1984). The presence of the higher concentration of the tannins in the Seabuckthorn leaves might have hampered the availability of the protein to the animals resulting in lower concentration of serum total protein as compared to the animals fed Lucerne hay based complete feed. Serum creatinine concentration in Gr.II was significantly ($P<0.01$) higher than Gr.I. The absolute muscle mass and level of physical activity may influence the rate of creatinine production and serum creatinine concentration (Carlson, 2002). More muscle mass gives higher serum creatinine concentration and vice versa. Higher gain in body weight and resultant increased muscular mass and activity in Gr.II animals may have resulted in higher serum creatinine concentration as compared to Gr.I animals.

Similar concentration of the serum calcium and phosphorus in the animals fed Seabuckthorn leaves and Lucerne hay based complete feed shows that availability and utilization of dietary calcium and phosphorus from Seabuckthorn leaves is as good as Lucerne hay.

CONCLUSIONS

There is marked shortage of nutritious animal feed resources in high altitude desert conditions of Ladakh. Use of unconventional feed resources in animal diet reduces feeding cost. Beside the inferior effect of feeding Seabuckthorn leaves on growth performance of sheep and certain blood biochemical parameters as compared to Lucerne hay, cost benefit analysis revealed that Seabuckthorn leaves can be used economically as a local feed ingredient in complete feed of sheep at high altitude of Ladakh. Moreover, data generated in the present investigation will be of greater use in formulation of animal ration in high altitude cold conditions. Further studies towards identification and amelioration of anti-nutritional factor present, if any, in Seabuckthorn leaves and rate of its inclusion in animal ration will be important to exploit their use in animal ration.

References

AOAC. 2000. Official Methods of Analysis. Association of Official Analytical Chemists. (Edn16), USDA, Washington, DC.

Carlson, G.P. 2002. Clinical chemistry tests. In: Large animal internal medicine, Third Edn., Molsby publication, USA, pp. 408.

Dwivedi, S.K., Singh, R. and Ahmed, Z. 2006. The Seabuckthorn. Field Research Laboratory, Leh, Ladakh.

Hoskin, S.O., Wilson, P.R., Barry, T.N., Charleston, W.A. and Waghorn, G.C. 2000. Effect of forage legumes containing condensed tannins on lungworm (*Dictyocaulus sp.)* and gastrointestinal parasitism in young red deer (*Cervus elaphus*). Res. Vet. Sci. 68 (3), 223-230.

Hu, Jian-zhong and Guo, Xiao-feng. 2006. Evaluation of nutrient value of seabuckthorn in north China. http—scholarilib_cn_Abstract_aspxA=zglx-e200601010.mht

ICAR. 1998. Nutrient Requirement of Livestock and Poultry. Indian Council of Agricultural Research, New Delhi.

Kumar, R. and Singh, M. 1984. Tannins: Their adverse role in ruminant nutrition. J. Agri. Food. Chem. 32, 447-456.

Patrick, C.H. Optimum Growth Rates for Holstein Replacement Heifers. www.wisc.edu/dysci/uwex/heifmgmt/papers/Optimum %20 Growth % 20Rates %20for %20Holstein %20Replacement %20 Heifers. pdf

Singh, V.R., Singh, K. and Singh, Bhupinder. 1999. Fodder values of the foliage of Seabuckthorn (*Hippophae rhamnoides* ssp. Turkestanica) growing in Lahul valley, a dry temperate region of Himalaya, Shaanxi. Forestry Science and Technology. 83-87.

Snedecor, G.W. and Cochran, W.G. 1994. Statistical Methods, 8th Ed. IOWA State University Press USA.

Stastistical hand book of Leh, 2002-03. Ladakh Autonomous Hill Development Council, Govt. of Jammu and Kashmir, Series 24, pp. 24-39.

Stastistical hand book of Kargil. 2002-03. Ladakh Autonomous Hill Development Council, Govt. of Jammu and Kashmir, pp. 23-37.

Young, B.A. 1983. Ruminant cold stress: Effect on production. J. Anim. Sci. 1601-1607.

contd...

Table 1: Percent composition of complete feeds

Ingredient	Gr. I	Gr. II
Seabuckthorn leaves	45.0	—
Lucerne Hay	—	45.0
Crushed barley	22.50	22.50
Wheat bran	29.50	29.50
Mustard oil cake	02.0	02.0
Mineral/Vitamin mix*	0.50	0.50
Salt	0.50	0.50

* Each 100 g of Mineral/Vitamin mixture composed of Calcium – 30 g, Manganese – 1.10 g, Iodine – 0.04 g, Iron – 0.30 g, Zinc – 0.60 g, Copper – 0.08 g, Cobalt – 0.018 g, Vitamin A – 200000 IU, vitamin D3 – 40080 IU, vitamin E – 30 Units, vitamin K – 0.04 g, Calcium Pentothenate – 0.10 g, Nicotinamide – 04 g, vitamin B12 – 0.24 g and Choline chloride – 6.0 g

Table 2: Effect of feeding Seabuckthorn leaves vs. Lucerne hay based complete feed on growth, feed intake and feeding cost in sheep (Mean ± S.E; N=4)

Particulars	Gr.I	Gr.II
ADG (g/d)*	101.6±13.17	167.5±22.37
DMI (g/d)	937±78.39	1079±69.20
DMI % body weight	4.76±0.26	4.40±0.61
DMI (g/Kg $W^{0.75}$)	95.81±4.79	93.66±16.19
CPI (g/d)	204.2±17.08	252.3±16.19
Cost/kg gain (Rs.)	63.17±8.59	59.66±6.50

* Significant at (P<0.05)

Table 3: Effect of feeding Seabuckthorn leaves vs. Lucerne hay based complete feed on blood biochemical profile in sheep (Mean ± S.E; N=4)

Parameters studied	Gr. I	Gr.II
Serum total protein (g/dl)*	7.80±0.29	10.00±0.53
Serum urea (mg/dl)*	43.51±4.87	59.36±1.66
Blood urea nitrogen (mg/dl)*	20.32±2.27	27.74±0.78
Serum creatinine (mg/dl)**	1.28±0.04	2.10±0.13
Serum cholesterol (mg/dl)	136.64±2.05	151.23±12.33
Serum triglyceride (mg/dl)	158.90±1.68	164.19±3.09

* (P<0.05), ** (P<0.01)

Table 4: Effect of feeding seabuckthorn leaves vs. Lucerne hay based complete feed on serum calcium, phosphorus and electrolytes in sheep (Mean ± S.E; N=4)

Parameters studied	Gr. I	Gr.II
Serum calcium (mg/dl)	9.95±0.47	9.78±0.26
Serum inorganic phosphorus (mg/dl)	5.19±0.33	5.38±0.27
Sodium (mM/L)	142.84±4.12	142.53±6.07
Potassium (mM/L)	6.32±0.64	6.50±0.22

□□□

34

Bio-Medical Systems for High Altitude Applications

V.C. PADAKI
Defence Bioengineering and Electromedical Laboratory, DRDO, Bangalore, India.

ABSTRACT

Defence Bioengineering and Electromedical Laboratory (DEBEL) is engaged in R&D activities in the areas of Life Support Systems, Protective equipment and Bio-medical systems and devices that are specific to Service combatants who operate in physiologically hostile environments and difficult terrains. An effort is made in this paper to highlight the usefulness of some of Bio-medical systems developed at DEBEL for high altitude and extreme cold condition environments.

Key word : Biomedical system, High altitude, Hypoxia, Hyperbaric.

INTRODUCTION

Strategic compulsions necessitate the deployment of soldier at high altitude of the Himalayan region exposing them to reduced atmospheric pressure and extreme cold conditions. Emergency health management that include diagnosis, treatment, providing expert medical opinion at inaccessible location and some time the necessity of remotely monitoring health of a soldier are important among many other logistics that are essential for high altitude operations. Some of the biomedical systems developed at DEBEL come in handy while addressing these problems.

The Mil grade 'Portable Biomedical Data Acquisition System' with telemedicine capability serves in monitoring vital physiological parameters and obtain expert opinion from command hospital. Some times dedicated 'Light weight vital parameter monitoring system' without high end interface of communication capability are also very useful. HAPO treatment system helps in immediately reversing the pulmonary oedema of those affected due to high altitude and cold conditions. 'Portable automated resuscitator' comes extremely handy in reviving unconscious and patients with acute respiratory problem. 'Remotely monitoring vital physiological parameters' is important in many critical missions undertaken. The biomedical systems developed keeping these issues in mind are discussed in this paper.

Portable Telemedicine System for Armed Forces

The portable telemedicine system was designed to cater to the following specific requirements :

- Withstand harsh environmental conditions in which it will be operated.
- Portable light weight system which can be set up quickly in a make shift hospital situation and be capable of establishing telemedicine link to command hospital for expert opinion.

The general picture of medical assistance required in a forward area is very clearly discussed in their paper *by Lt Col KM Rai.* The paper cites that lack of trained medical staff, or lack of proper expert advice at the site of injury as one of the reasons for the increased mortality rate. Current approaches to battlefield trauma care focus on timely and effective evacuation of the soldier in a stable condition to a center with definitive treatment facilities. In these situations, if a system is equipped with portable medical diagnostic devices it will aid the field medic with accurate & fast triage along with preoperative diagnosis. Conveying the vital diagnostic information pertaining to the physiological condition of the injured to the specialist is essential. Of late battlefield hospitals are equipped with portable ultrasound device, compact and portable Computed Radiography devices wherein digital medical images are sent electronically to radiologists to help the on-site medical personnel.

Keeping in view the specific requirements of Armed Forces, DEBEL has developed the telemedicine system which consist of the following.

(i) Biodas Hardware

Rugged Biomedical data acquisition System [BIODAS] hardware equipped with required sensors / probes to acquire vital Physiological parameters of a patient such as Electro Cardio Graph (ECG), SaO2 (Oxygen saturation), Blood Pressure (BP), temperature, Heart Rate (HR), Respiratory Rate (RR).

(ii) Biodas Software

The customized software operates in "store and forward" mode and "continuous monitoring" mode. The software consists of two modules one at Telemedicine Consultation Center (TCC) and specialist /expert doctor's end Telemedicine Specialists Center (TSC) respectively.

At TCC, the user can acquire the following data, store the data into the database, establish communication with TSC and transmit the data.

- Patient demography and case history.
- Biomedical data /Vital parameters
- 12 lead ECG [10 seconds], Blood Pressure, Body temperature, SaO_2 [Oxygen saturation], Respiration Rate , Heart Rate, Mean arterial Pressure
- Three minute audio.
- Image [with annotations] acquired through camera or stored image at TSC the user can receive case data sent from TCC.
- Store the data in the database
- View received data and generate report (text or audio)
- Transmit report to TCC

The features offered by the software are versatile and enables the system to be used both in field areas where communication modality such as S-band SATCOM terminal / VSAT terminals can be utilized and in hospitals where PSTN / ISDN facility are available. The dual operation mode of "store and forward" and "continuous monitoring" provides option to the user to take advantage of available communication bandwidths and the usage context. Portable medical imaging device which has USB interface and provides digital image [JPG] output can be integrated with the system. Store & forward mode operation is more advantageous in situation of temporary snag in communication media. The System facilitates point to point, exchange of data using communication modalities such as ISDN, C-band VSAT, INMARSAT & PSTN, S-band briefcase SATCOM terminal. The user can switch between communication options with ease as long as the communication interface devices are in place.

The hardware for acquiring the biomedical data (vital parameters) has been designed and certified for its intended use in harsh environmental conditions. It has successfully undergone testing for military environmental conditions as per JSS 55555 & EMI / EMC for electronic ground equipment for army. It has also undergone Patient safety tests as per IEC 60601-1. The BIODAS has also successfully passed the EMI / EMC conducted /radiated emission and immunity tests. It has successfully passed the following environmental tests viz., Vibration, Low and High operational and storage temperature, Bump, Drop, High altitude and Relative humidity.

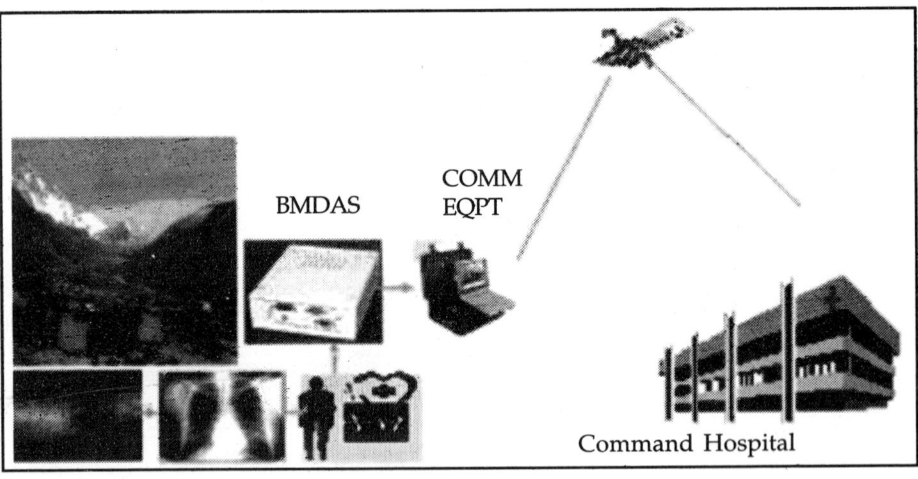

Fig: Portable telemedicine system for armed forces

The system stands apart from some of the similar systems that are available commercially by way of

- Supporting communication of data on the secure S-band briefcase Satellite terminal currently used by Indian Army.

- Providing various communication options in the single application.
- Providing store and forward and continuous mode of operation.
- Incorporating rugged BIODAS hardware for harsh environmental conditions.

Handheld Portable Vital Parameter Monitoring System

Troops are often deployed in inhospitable and far - flung posts during wartime and peacetime defence operations. One such situation is the deployment of the soldiers in the high altitude and snowbound areas for short and long term operations. In emergencies soldiers may have to be dropped via parachutes into the high altitude environments giving them very less time for acclimatization. Survival at high altitude conditions is dictated by the person's ability to adjust to less oxygen levels and pressure variations. The physiological effects may be decreased stamina, fatigue, acute mountain sickness, which can cause headaches, dizziness, nausea, and make it difficult to fall asleep. Acute mountain sickness symptoms will often dissipate once a soldier's body adjusts to the high-altitude environment, but sometimes the sickness can intensify into pulmonary edema, which is caused by a build up of fluid in the lungs and can lead to shortness of breath and heavy coughing. Acute mountain sickness can also transform into cerebral edema, which is caused by an increased blood flow to the brain. Cerebral edema can cause swelling, disorientation, hallucinations and can impact physical coordination. It can be deadly if left untreated. On such highly dispersed and fluid environments soldiers find themselves increasingly deficient in basic life support.

Vital parameter monitoring like oxygen saturation and blood pressure, Electrocardiogram (ECG) is a fundamental requirement to monitor the personnel deployed in high altitude areas. This will enable early detection of the physiological effects of high altitudes, in turn; help to increase the performance and reduce the mortality rate. The drawbacks of the conventional physiological monitoring equipments are:

- Bulky in Size, not easily portable
- Not designed for performing at extreme temperature and pressures as that at high altitude environments.
- Reduced battery life and other power constraints.

- Single, integrated, handheld multi parameter monitoring system is not available.
- Not rapidly deployable.

Rapid and portable vital parameter monitoring and treatment of casualties undoubtedly improves survival of soldiers. This clearly brings out a need for development of handheld portable vital parameter monitoring system for soldiers deployed at high altitudes. The handheld portable vital parameter monitoring system is portable, multiple physiological parameter monitoring capabilities, light weight, has extended battery life, easy to deploy and operate. The unit consists of the following parts. i) sensors, ii) data acquisition and processing hardware, iii) remote analysis system. Figure 1 gives the overall block diagram of the handheld portable vital parameter monitoring system.

i) Sensors

The physiological parameters that are being monitored at high altitudes are: 12 - lead ECG, oxygen saturation, temperature, blood pressure, and respiration rate.

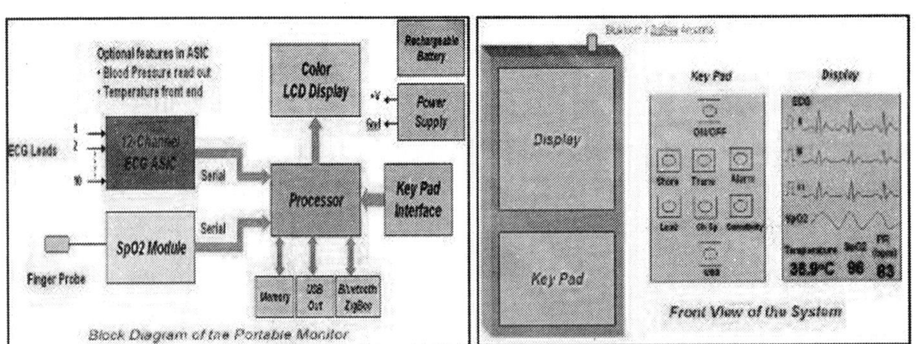

Fig. 1. **Fig. 2.**

The sensors for the same will be incorporated. Heart rate, respiration rate will be computed form ECG data thereby avoiding the use of additional sensors.

ii) Data Acquisition and Processing Hardware

The data acquisition and processing and hardware acquires the physiological data from the sensors, processes the data. To reduce the size and produce lightweight and portable system latest acquisition practices have been targeted such as such as 12 - channel ECG ASIC with Blood Pressure and temperature acquisition feature. The ECG ASIC is used to reduce the size of the data acquisition system. The unit is designed for giving output via USB. Additional memory is provided to store events for future reference and playback of data. Provision for Bluetooth/ZigBee wireless modules has also been provided for wireless transmission of data in cases of requirement. The system will be made compatible as per JSS55555 standards. The block diagram of the display unit is shown in Figure 2. The display unit could be a PDA /Handheld PC. It captures the data acquired from the data acquisition and processing hardware via USB or wirelessly and displays on the monitor.

iii) Remote Analysis System

Significant events will stored in the memory of the data acquisition and processing hardware and can be accessed through USB port or wireless modules. The Remote analysis system receives the data form the data acquisition and processing hardware and analysis of the acquired data can be performed manually or by using software.

The handheld portable vital parameter monitoring system has been conceptualized with the ultimate aim to provide integrated multiple physiological parameter monitoring capabilities for high altitudes and forward areas. The goal of the research and development effort is: to improve the physiological monitoring capabilities thereby increase the survival rate and minimize morbidity at high altitudes.

Portable Hyperbaric Chamber

Strategic compulsions necessitate the regular deployment of soldiers at high altitudes of the Himalayan regions thereby exposing them to reduced atmospheric pressures. The reduced atmosphere and the harsh mountain environment at high altitude, being alien to low Landers, evoke a series of physiological responses in human system. There is a wide variation in both the speed of onset and severity of physiological responses and also the height at which they develop. Sudden exposure to high altitude is detrimental to the performance, both physical and mental, and in some cases may lead to dreaded condition like High Altitude Pulmonary Oedema (HAPO). HAPO is a condition in which fluid accumulates in lungs causing severe illness and is a serious consequence of failure to acclimatize to hypobaric hypoxic atmosphere of altitude greater than 2700 m above mean sea level. This condition is characterized by breathlessness, abnormally rapid heart beat, mild cough, dry cough to blood stained sputum, and raised pulmonary artery pressure. In the Indian context, the military relevance of HAPO is evident by the experience of the Indian Army during the Sino–Indian border conflict in the 1960s, when the incidence of HAPO was noted to be as high as 15.5% in a group of "fresh inductees" who were rapidly transported to altitudes between 3,355 and 5,490 m.

Condition of HAPO can be reversed by increasing the alveolar and arterial oxygenation either by providing supplemental oxygen at higher pressure or by descent to lower altitudes. Supplemental oxygen may not be readily available in a combat situation; while rapid descent, the treatment of choice, is often not possible either due to the topography, adverse weather or non-availability of transport. In such a scenario, portable hyperbaric chambers present an inexhaustible, light weight source of additional oxygen pressure which leads to rapid relief of symptoms of high altitude illness. Portable hyperbaric chambers allow the environmental pressure around the subject to be increased equivalent to a descent of up to 8000 to 10000 ft.

Portable chambers are near cylindrical in shape and are made out of medium to heavy weight, impermeable coated fabric with an air-proof zipper to accommodate a patient in supine position. Such chambers are integrated with various valves to facilitate the ease of operation and safety. The inflation of the chamber and the maintenance of the pressure are facilitated by a foot-pump alone or in combination with power operated compressor and a pressure sensing unit. Internationally, several bags viz., Gamow bag, CERTEC chamber, and PAC bags are available for the treatment of HAPO. In the Indian scene, DEBEL is in the fore-front of the development of the HAPO bag. Years of R&D effort coupled with the evaluation has culminated in the development of an automated bag, a unique feature vis-à-vis the imported bags. The operating pressure inside the developed bag is 2.5psi vis-à-vis 2psi of the above mentioned internationally available bags thereby enhancing the virtual descent of the patient and hence the recovery. The developed bag is 2.1m long, 0.65m dia at the head end and 0.5m dia at the foot end, with a volume of approximately 600ltr to accommodate the HAPO patient even with the multiple layers of cold-clothing. The bag is provided with a pre-set leak control valve to bleed continuously at a rate of 20lpm in order to avoid the build-up of carbon dioxide inside the bag and a safety pressure relief valve in order to prevent the excessive pressure inside the bag. Several user friendly features have been incorporated in the developed bag. All the materials have been selected to withstand sub-zero temperatures encountered in the usage. The developed bag is shown in the figure below.

Automated Pneumatically Driven Resuscitator

The portable automated resuscitating unit designed for field application integrated with a twin cylinder cassette. Carbon composite cylinders of 2 liter water capacity are connected to a metal manifold block. These light weight cylinders can be used up to operating pressures/filling pressures of 300 bar. These bottles fitted in the cassette of two cylinders each has a charging port and pressure regulator fitted on the metal manifold. This version has provision to be run with gas bottle as auxiliary source. It is provided with a one meter long corrugated hose fitted with patient mask. The patient mask is provided with elastic band to secure it on the face /head of the patient. The patient mask and out casing are made of NBC resistant materials to with standing mild to moderate contamination for periods longer than 23 hours. The unit could be fitted with NBC canister for operation in contaminated environment. The outer casing is made of FRP (3mm thickness)chop strand matt the front controls(Breaths per minute, minute volume and safety pressure) of the unit are accessible by lifting the front window. The window flap is secured with screws. The pneumatic connections are made with quick disconnect couplings for ease of operation. The cylinders in the cassette are charged up to 300 bar with air/oxygen and the connecting manifold has a pressure regulator which reduces the output pressure to 4.5bar. The charging is done through a side window where a charging port is provided. The total unit weighs 14.5kgs. With one cassette having two cylinders charged to 300 bar. This can resuscitate a person for a period of 2 hours at 16 breaths per minute and 10 liters per minute flow. This unit can also be run using oxygen concentrators with suitable external boosting pump. Provision is also made to connect external 100% oxygen source either in ambulance or otherwise, with a provision to select 60% O_2 mode to reduce O_2 concentration with ambient air through canister and to economize the source to extend the endurance. The advantage of this design and important design specifications are given below.

Advantages of this Version are

1. No need of power, battery and compressor.
2. Works on compressed air source/bottled oxygen/ oxygen concentrator.
3. Light weight carbon composite cylinders.
4. Alarms to indicate low/high input pressures.
5. Alarms to indicate tongue blockage.
6. Front panel settable controls for breaths per minute and minute volume
7. Patient mask with 1m length NBC compatible corrugate hose.
8. Fitted with NBC canister for contaminant free breathable gas supply.

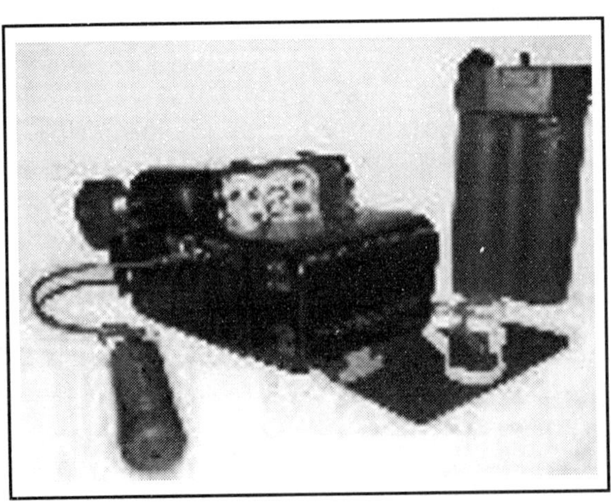

Design Specifications

The resuscitator is a manual/ automatic gas driven, time cycled volume preset pressure limited, flow generator and it neither requires AC mains nor battery nor runs without internal compressor.

Type	: Pneumatic
Power Sources	: Stand alone system driven by a compressed dry, oil free filtered gas (air//oxygen) delivered at the inlet of the resuscitator with 3.5 bar to 6 bar pressure and at 45 lpm.
Frequency Range	: 8-40 bpm
Minute volume range	: 14 litres
Flow range	: 6-42 lpm
I: E ratio(fixed)	: 1:2
Relief pressure (fixed valve)	: 60 cm H_2O with audible alarms
Relief pressure range (Adjustable value)	: 20to 60 cm H_2O with audible alarm
Inflation Pressure Monitor	: Indicates from -10 to 100 cm H_2O
Aspiratory Breathing Resistance	: -6cm HO_2 @60 LPM
Exhaulation Breathing Resistance	: 6cm H_2O @60 LPM
Pressure gauge accuracy	: ± 2% FSD
Air Mix	: 100% or 60% selectable reduces gas Consumption by 40% with O2 or air as gas source.
Gas Consumption	: 100% O_2 : VDEL +20 cc/cycle 60%O_2 : 60%VDEL+20 cc/cycle
Weight	: 14.5 kg

Gas Supply status indicator, input Pressure alarm, Airway pressure alarm, BSI alarm provided.

Smart Vest for Remote Physiological Monitoring

There are certain situations in the day-to-day life wherein the personnel are highly mobile and are vulnerable also. These may include a soldier fighting the enemy, fire fighters involved in fighting fire, law enforcement personnel, personnel involved in mining, deep sea divers and astronauts in space. Considering the mobility and vulnerability, it is very important to monitor the health status and geo-location of the personnel to ensure the safety and effective completion of the assigned job. Hence, there has been significant shift in the development of the clothing by way of introduction of the sensors into the clothing worn by the personnel or soldiers. Wearable physiological monitoring systems uses an array of sensors integrated into the fabric of the wearer to continuously acquire and transmit the physiological data to a remote monitoring station. The data acquired at the remote monitoring station is correlated to study the overall health status of the wearer. The wearable monitoring systems allow an individual to monitor his/her vital signs remotely and receive feedback to maintain a good health status. These systems alert medical personnel when abnormalities are detected.

The conventional sensors and medical instruments cannot be used for wearable physiological monitoring applications due to the following reasons.

- The conventional physiological monitoring systems are bulky to be used for wearable monitoring.
- The gels used in the electrodes dry out when used over a period of time, which lead to increase in the contact resistance and thereby degrading the signal quality.
- The gels used in the electrodes cause irritations and rashes when used for longer durations.
- There are number of hampering wires from the sensors to the data acquisition system.
- The signals acquired are affected with motion artifact and baseline wander as the electrodes float on the layer of gel.
- The sensors used in conventional monitoring systems are bulky and are not comfortable to wear for longer durations.

To overcome the above problems associated with the conventional physiological monitoring there is a need to develop sensors for wearable monitoring and integrate them into the fabric of wearer and continuously monitor the physiological parameters. A wearable data acquisition, processing and transmission hardware, which is portable, comfortable to wear for longer durations, and having sustainable battery power and a remote monitoring station is to be developed. The wearable physiological monitoring systems consists of three systems namely (a) vest with the sensors integrated (b) wearable data acquisition and processing hardware and (c) remote monitoring station. In the vest sensors for acquiring the physiological parameters are integrated. The sensors outputs and power cables are interconnected to the data acquisition and processing hardware by means of wires routed through the wires woven in the vest. In the wearable data acquisition and processing hardware, the circuits for amplification, filtering and digitization are housed. The digitized and processed data is transmitted wireless to a remote monitoring station. In the remote monitoring station the data is received and displayed in a form suitable for diagnosis. Figure illustrates the overall architecture of the wearable physiological monitoring system, consisting of a vest with sensors integrated, wearable data acquisition and processing hardware and a remote monitoring station.

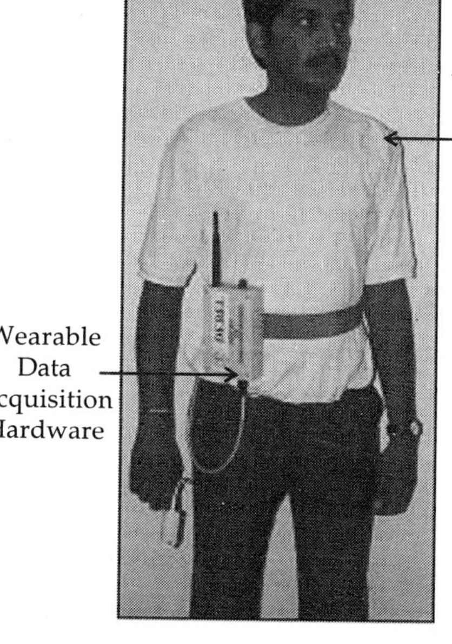

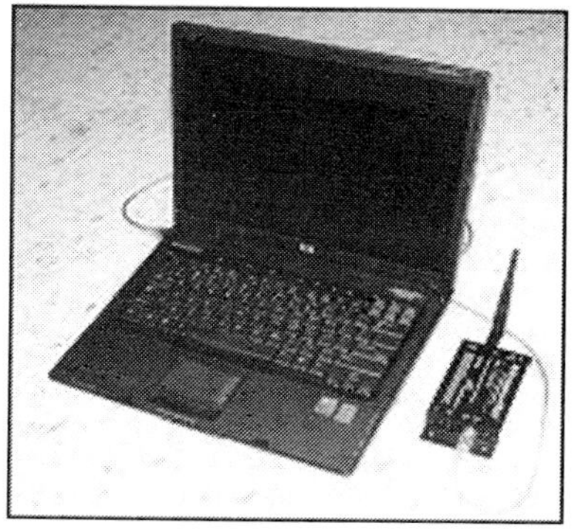

Remote Monitoring Station

Figure illustrates the overall architecture of a wearable physiological monitoring system. A number of wearable physiological monitoring systems have been developed worldwide to monitor the health status of the individual wearer. DEBEL had taken a project on the development of wearable physiological monitoring system named 'Smart Vest' to monitor various physiological parameters such as ECG, PPG, heart rate, blood pressure, body temperature and GSR. The acquired physiological parameters are transmitted wireless to a remote monitoring station along with the geo-location of the wearer to a distance of 200 meters.

The developed smart vest system is an initial effort to build a technology for remote physiological monitoring. The smart vest prototypes have varying degrees of success in achieving medical accuracy on all measurement combinations during the stationary and mobile conditions of the subjects. Although we have made progress in addressing many of the issues, there is considerable scope for improvement. Recent advances in sensor technologies, low power microelectronics and miniaturization and wireless networking will enable wireless sensor networks for human health monitoring. A number of tiny wireless sensors, strategically placed on the human body create a wireless body area network that can monitor various vital signs, providing real-time feedback to the user and medical personnel is going to revolutionalize health monitoring in the near future.

ACKNOWLEDGEMENT

All scientists of DEBEL belonging to various design teams involved in development of these products and providing the necessary information for the preparation of this article are hereby sincerely acknowledged.

References

Answering the Call. 2004. Medical Imaging,19(7), pp. 34–37.

Ari, T. and Adler, A. 2000. Cost effective portable telemedicine Kit for use in developing countries, M.S. Thesis, Massachusetts Institute of Technology.

Bledsoe, B. et al. 2003. Essentials of paramedic care. Upper Saddle River, New Jersey: Prentice Hall.

Chaoxina, Sima. et al. Vital signs services for secure Telemedicine applications, Concurrent engineering research centre, West Virginia University.

David, Hailey. et al, Assessments of Telemedicine applications – an update, Alberta heritage foundation for medical research.

GCS 7500A Portable Telemedicine System, www.globalcoms.com.

Gopalsamy, C., Park, S. Rajamanickam, R. and Jayaraman, S. 2005. "The Wearable Motherboard™: The first generation of adaptive and responsive textile structures (ARTS) for medical applications", Virtual Reality, pp. 152-168.

Halin, N., Junnila, M., Loula, P. and Aarnio, P. 2005. "The LifeShirt system for wireless patient monitoring in the operating room", J. of Telemedicine and Telecare, 11, pp. 41-43.

Halstead, D. 2001. Progress in pulse oximetry—a powerful tool for EMS providers. JEMS, 55-66.

Hoyt RW, Reifman J, Coster TS, Buller MJ. 2002. Combat medical infomatics: present and future. Proc AMIA Symp, 335-339.

http://www.high-altitude-medicine.com/hyperbaric.html

IEC 60601-1 Patient safety standards for biomedical electronic equipment

JSS 55555 : Joint services specifications for environmental tests for electrical & electronic equipment.

Kyriaco, E. et al, 2003. Multipurpose Healthcare Telemedicine Systemswith Mobile communication Link Support, Biomedical Engineering online, 2.

Majumdar, A.K. et al. 2004. TELEMEDIK. A Telemedicine system for providing quality health care in rural areas, Souvenir of seminar on Telemedicine- A Vision turned Mission, 19 March.

Menon, N.D. 1965. High-altitude pulmonary edema: A clinical study. N Engl J Med. 273, 66–73.

Michael P. Ward, James S. Milledge and John, B. 1989. High Altitude Medicine and Physiology, Chapman and Hall Medical, London, pp. 385.

MIL 461E: Military standard for EMC requirements & measurement methodology.

Milenkovic, A., Otto, C. and Jovanov, E. 2006. "Wireless sensor network for personal health monitoring: issues and an implementation", Computer Communications, 29, pp. 2521-2533.

Mode 1, GATS report, India: Telemedicine Initiatives in India.

Mundt, K. N. Montgomery, U. E. Udoh, V. N. Barker, G. C. Thonier, A. M. Tellier, R. D. Ricks, R. B. Darling, Y. D. Cagle, N. A. Cabrol, S. J. Ruoss, J. L Swain, J. W. Hines and G. T. A. Kovacs, 2005. "A Multiparameter Wearable Physiologic Monitoring System for Space and Terrestrial Applications", IEEE Trans Info. Tech. in Biomedicine, 9.

Negoslav, Daja et al. Telemedicine in cardiology-ECG Transmission by mobile phone –Annals of the academy of studenica, 2001.

Rai, K.M. et al, 2004. "Treatment of causalities in a forward hospital of the Indian Army : Nine year experience." MJFAI, 60,1.

Raskovic, D., Martin, T., and Jovanov, E. 2004. "Medical monitoring applications for wearable computing", The Computer Journal, 47(4), 495-504.

Rienzo, M. Di, Rizzo, F., Parati, G., Brambilla, G., Ferratini, M. and Castiglioni, P. 2005. "Magic System: a New Textile-Based Wearable Device for Biological Signal Monitoring. Applicability in Daily Life and Clinical Setting", Proc. IEEE-EMBS, pp. 7167-7169.

S.K.Mishra, Telmedicine at SGPGI, Lucknow- current programmes & future perspectives.

Satyamurthy, L.S., 2004. Telemedicine- A healing touch through space for speciality health care delivery to remote & rural population. Souvenir of seminar on Telemedicine- A Vision turned Mission.

Selvamurthy, W. 1983. Review on physiological and medical problems related to climates ranging from altitude to subtropics. World Meteorological Organization, Geneva, Document No. WCP- 78.

Singh I, Kapila CC, Khanna PK, Nanda RB, Rao BDP. 1965. High-altitude pulmonary oedema. Lancet. I, 229–231.

Singh I, Khanna PK, Srivastava MC, et al. 1969. Acute mountain sickness. N Engl J Med. 280, 175-184.

Technical working group on Telemedicine standardization, Recommended guidelines & standards for practice of Telemedicine in India. May 2003.

Telemedicine reference architecture project – Enabling the next generation of telemedicine Devices, Sandia Research Laboratory.

Vikram chatwal, Apollo hospitals group & Telemedicine. Souvenir of seminar on Telemedicine- A Vision turned Mission, 19 March 2004.

Young, A.J., Castellani, J.W., et al. 1998. Exertional fatigue, sleep loss, and negative energy balance increase susceptibility to hypothermia. J Appl Physiol. 85, 1210-1217.

□□□

35

Endothelin-1 and High Altitude Pulmonary Hypertension

**BAKTYBEK K. KOJONAZAROV[1], MAHABAT M. KOCHKOROVA[1]
JAINAGUL T. ISAKOVA[1], ABDYRASHID M. MARIPOV[2]
TALANTBEK M. SOORONBAEV[2], NICHOLAS MORRELL[3]
MARTIN R. WILKINS[3], ALMAZ A. ALDASHEV[1]**
[1]Institute of Molecular Biology and Medicine, Bishkek, Kyrgyzstan
[2]National Center of Cardiology and Internal Medicine, Bishkek, Kyrgyzstan
[3]Imperial College of Medicine, London, United Kingdom

ABSTRACT

It is known that endothelin-1 (ET-1) play a critical role in hypoxic pulmonary vasoconstriction. The role of endothelin system in genesis of high altitude pulmonary hypertension (HAPH) in highlanders is still to be elucidated. The aim of the study was to investigate the role of endothelin-1 in genesis of HAPH in highlanders. The pulmonary arterial pressure (PAP) was measured by ECHO in permanent residence of high altitude (3200-4000 m above sea level). The effect of 125 mg of bosentan on PAP was studied. The plasma ET-1 level was measured by ELISA assay. Genomic DNA was extracted from peripheral blood samples and -3a/-4a polymorphism of ET-1 gene was determined by PCR. The mean PAP in highlanders with HAPH was higher than in healthy highlanders (34.2±13.9 vs 20.1±2.2 mm Hg, p=0.001). The plasma ET-1 in HAPH highlanders was significantly higher than in healthy subjects (7.05±2.35 vs 4.65±1.65 pg/ml, p=0.002). After 125 mg of bosentan intake the mean PAP decreased from 32±2.2 to 26.5±3 mm Hg (p=0.001) and systolic PAP from 46.4±6.2 to 37.0±7.2 mm Hg (p=0.004). Moreover the effects of oxygen and bosentan on PAP were similar. -3a/-4a polymorphism of ET-1 gene associated with HAPH. We have found that the frequency of mutant allele

-4a was significantly higher in HAPH group compared to healthy highlanders (0.43 vs 0.3, $c^2 = 4.3$, p=0.03). Thus, hypoxia related increased level of ET-1 play an important role in development of HAPH in highlanders and the high level of ET-1 in HAPH subjects has inheritance trait.

Key words : High altitude pulmonary hypertension, endothelin-1, bosentan, gene polymorphism.

INTRODUCTION

It was shown that 4–6% of Kyrgyz highlanders of Tien-Shan and Pamir Mountains living at altitudes 3000-3500 m above sea level develop high altitude pulmonary hypertension with right ventricular failure (Mirrakhimov and Winslow, 1998). HAPH is characterized by increased pulmonary arterial pressure (PAP) and pulmonary vascular remodeling involving all elements of the vessel wall, including endothelial and smooth muscle cells and fibroblasts (Aldashev, 2000; Leon-Velarde *et al.,* 2005).

It has been proposed that an imbalance of endogenous vasodilator and vasoconstrictor might contribute to pulmonary vascular vasoconstriction and remodeling of the pulmonary vessels (Humbert *et al.,* 2004). The vasoactive factors implicated in the pathogenesis of pulmonary hypertension are endothelin-1 (ET-1), angiotensin converting enzyme (ACE), angiothensin II, serotonin, nitric oxide (NO), bradykinin, prostaglandins and other.

ET-1 is a one of the most potential vasoconstrictor produced by the vascular wall cells (Yanagisawa and Masaki, 1989). There have been several studies which investigated the role of ET-1 in pathogenesis of pulmonary arterial hypertension. Increased levels of plasma ET-1 and high level of ET-1 peptide in the lung were measured in rats exposed to short- and long-term hypoxia (Shirakami *et al.,* 1991; Yang *et al.,* 1997). The exposure to acute hypoxia leads to significant over expression in the lung of ET-1 peptide its gene and ETA receptor gene (Li *et al.,* 1998). The plasma ET-1 is elevated in patients with primary and secondary pulmonary hypertension and correlated with high pulmonary vascular resistance (Stewart *et al.,* 1991). Expression of ET-1 gene is increased in vascular endothelial cells of the pulmonary artery in various types of pulmonary hypertension (Giaid *et al.,* 1993) and in plasma and lung of patients with primary pulmonary hypertension (MacLean, 1998). Moreover, the acute increase of pulmonary arterial pressure in mountaineers was found to be closely related to an elevated plasma level of ET-1 (Goerre *et al.*, 1995; Maggiorini *et al.*, 2001). In the recent ACME-1study it was revealed that ET-1 play an important role in pulmonary vascular response during adaptation to high altitude and dual endothelin receptors antagonist and bosentan effectively attenuated hypoxic pulmonary vasoconstriction in healthy subjects (Modesti *et al.*, 2006). It is known that the effect of ET-1 on pulmonary vasculature mediated by two type receptors ETA and ETB and it is confirmed by several studies that bosentan effectively decreased pulmonary arterial pressure and reverse the morphological changes in monocrotaline- and hypoxia-induced pulmonary hypertension in animals (Miyauchi *et al.*, 1993; DiCarlo *et al.*, 1995; Chen *et al.*, 1995, 1997; Hill *et al.*, 1997; Houbloue *et al.*, 2003). The effectiveness of bosentan in the treatment of primary pulmonary hypertension is undoubted (Channick *et al.*, 2001; Rubin *et al.*, 2002; Humbert *et al.*, 2004), but the effects of bosentan in patients with HAPH still needs elucidation. Thus, bosentan could be a powerful medicine in treatment of high altitude related sickness, especially HAPH.

Due to hypoxic pulmonary vasoconstriction playing crucial role in pathogenesis of HAPH, the aim of our study was to investigate the role of ET-1 in pathogenesis of HAPH in permanent residents of high altitude.

METHODS

Patients: We have studied 138 (88 with HAPH and 50 healthy) Kyrgyz subjects, the permanent residents of altitude 2500-3800 m above sea level. All subjects underwent health screening by history, physical examination, ECG, spirometry, and blood pressure measurement. The pulmonary arterial pressure measured by Doppler-ECHO.

Doppler-Echocardiography

Doppler echocardiography was performed using a portable Doppler echocardiograph (SpectraMax, SonoSite, USA) and a transducer array of 2.5 MHz. Images were obtained from standard parasternal views (long axis and short axis) and an apical four-chamber view. The subjects were placed in the left lateral decubitus position for the parasternal views and in a supine position for the apical four-chamber view. Right ventricle end-systolic and end-diastolic diameters were measured by M-mode echocardiography from the left parasternal short- and long-axis views. Continuous Doppler tricuspid regurgitant jets were recorded and analysed for the estimation of systolic PAP (Yock and Popp, 1980). A pulsed Doppler pulmonary blood flow velocity signal was sampled in the right ventricular outflow tract for the estimation of mean PAP. For this purpose, an acceleration time was defined as the time interval from the onset of forward flow in the pulmonary artery to the peak velocity of this flow (Kitabatake *et al.*, 1983).

Acute Bosentan Study Protocol

The 15 highlanders with HAPH (4 women and 11 man), mean age 49.2±12.6 year, mean PAP 32.0±2.0 mm Hg, received bosentan in dose 125 mg. The PAP was measured by Doppler-ECHO before and in 3 hours of 125 mg bosentan intake. The heart rates (HR), SaO_2 and BP were measured at the baseline and in 3 hour of bosentan.

In 10 (3 women and 7 men) of 15 highlanders Doppler-ECHO measurement was performed at the baseline and in 30 minutes of oxygen breathing. After that all highlanders received 125 mg of bosentan and after 3 hours Doppler-ECHO was repeated before and in 30 minutes of oxygen breathing. The satisfactory pulmonary blood flow velocity was identified in all 15 highlanders, at the same time in 2 of 15 subjects the sufficient quality of tricuspid regurgitation signal was unequivocal reading. All measurements were performed at altitude of residency of highlanders.

Exhaled Nitric Oxide Measurement

Exhaled nitric oxide was measured by standard methods by NIOXMINO (Aerocrine, Sweden) in 15 highlanders before and 3 hours after 125 mg bosentan intake.

Plasma Endothelin-1

Blood samples were drawn from 32 highlanders (16 with HAPH and 16 Healthy) subjects in the supine position from an antecubital vein. The samples were immediately put on ice and then centrifuged at 3000 rpm for 10 minutes. Plasma samples were then snap-frozen in liquid nitrogen until their return to the laboratory, where the ET-1 concentrations measured by ELISA assay.

-3a/-4a Polymorphism of ET-1 Gene

DNA was extracted from blood samples of 133 highlanders by standard technique. The -3a/-4a polymorphism of ET-1 gene was determined by polymerase chain reaction (PCR) and the following primers were used 5′-TTGTCTGGGGCTGGAATAAAGTGG-3′ N 5′-CGAGCGTGCCTAÑÑTGTTTCTG-3′. The products of PCR were restricted by Bse II enzyme.

Criteria of HAPH

Diagnosis of HAPH was confirmed when the mean PAP = 30 mm Hg and/or systolic PAP = 50 mm Hg measured at altitude of residency (Leon-Velarde *et al.,* 2004).

Statistical Analysis

Statistical analysis was performed with SPSS statistical software (SPSS Inc). The data are presented as mean (SD). P=0.05 was taken as evidence of significance. The paired t-test was used for comparison effects of bosentan on PAP within group. An unpaired t-test was used for the comparisons between groups. Frequency of genotype or allele of *ET-1* was compared by use of the contingency table, and significance of difference was examined by χ^2 analyses.

The ethics committee of the National Center of Cardiology and Internal Medicine (Bishkek, Kyrgyzstan) approved the study and all subjects gave informed consent.

RESULTS

Highlanders

138 highlanders (88 HAPH and 50 Healthy) were included in this study, mean age 49.6±12.8 year, 111 male and 22 female. The groups of highlanders did not have significant differences in age, and altitude of residence. PAP mean in HAPH group was 32.4±9.8 mm Hg and in healthy 20.5±3.2 mm Hg (Table 1).

Plasma Endothelin-1

The level of plasma ET-1 in highlanders with HAPH was 7.07±2.26 pg/ml whereas in healthy highlanders it was 4.56±1.66 pg/ml accordingly (p=0.001, Table 1).

Acute Effects of Bosentan on Pulmonary Arterial Pressure

The PAP was measured by Doppler-ECHO in 15 highlanders before and in 3 hours after bosentan intake. Bosentan significantly reduced mean PAP from 32.0±2.2 to 26.5±3.0 mm Hg (p=0.001) and systolic PAP from 46.4±6.2 to 37.0±7.2 mm Hg (p=0.004; Figure 1). The systemic blood pressure, heart rate and SaO_2 did not significantly change after the bosentan intake (Table 2).

Effects of Bosentan and Oxygen

Doppler-ECHO study was performed in 10 of 15 highlanders four times: 1) baseline; 2) baseline +oxygen; 3) bosentan 4) bosentan+oxygen. After 30 minutes of oxygen breathing PAP mean significantly decreased compared with baseline by Δ=-5.5±3.2 mm Hg (95% CI -7.8 to -3.2 mm Hg, p=0.001) and PAP systolic by Δ=-7.6±4.2 mm Hg (95% CI -11.2 to -4.0 mm Hg, p=0.001). After 3 hours of bosentan intake both the systolic and mean PAP significantly reduced compared with baseline, PAP mean decreased by Δ=-5.7±2.8 mm Hg (95% CI -7.7 to – 3.7 mm Hg, p=0.001) and systolic PAP reduced by Δ=-10.6±4.9 mm Hg (95% CI -14.7 to -6.5 mm Hg, p=0.001). Interestingly, that bosentan as compared to oxygen more effectively decreased systolic PAP (Δ=-3.0±2.6 mm Hg, 95% CI -5.2 to -0.8, p=0.01). Moreover, combination of bosentan intake and oxygen breathing decreased mean PAP more ast compared to bosentan alone (Δ=-3.3±2.7 mm Hg, 95% CI -5.2 to -1.4 mm Hg, p=0.004), but in the case of systolic PAP it was not significantly changed compared to bosentan (Δ=-2.0±3.1, 95% CI -4.6 to 0.6, p=0.17; Table 2).

The individual analysis showed, that in highlanders there was no significant change of systolic PAP with oxygen breathing as compared to baseline (n=3) by Δ=-3.3±1.5 mm Hg (95% Cl -7.12 to 0.46 mm Hg, p=0.06), bosentan also did not significantly decrease systolic PAP (Δ=- 5.6±3.0 mm Hg, 95% Cl -13.2 to 1.9 mm Hg, p=0.08). From another side in highlanders with significant response to oxygen inhalation (n=5) (Δ=-10.2±2.9 mm Hg, 95% Cl -13.8 to -6.53 mm Hg, p=0.002), bosentan also significantly decreased systolic PAP (Δ=-13.6±2.8 mm Hg, 95% CI -17.2 to -10.0 mm Hg, p=0.001).

Association of -3a/-4a Polymorphism of ET-1 Gene with Development of HAPH

The -3a/-4a polymorphism of ET-1 gene was also analyzed. We have found that frequency of homozygous genotype was higher in healthy highlanders compared to HAPH subjects and frequency of heterozygous genotype -3a/-4a was higher in highlanders with HAPH (χ^2=11.15; p=0.0008). Accordingly the frequency of mutant allele -4à being significantly higher in highlanders with HAPH compared to healthy control (χ^2=4.3; p=0.03, Table 3), it was suggested that HAPH is associated with this mutation in the population of highlanders. This is why we did not find the mutant homozygous genotype -4a/-4a among the highlanders.

DISCUSSIONS

The aim of our study was to investigate the role of ET-1 in development of HAPH. One hundred thirty three (88 with HAPH and 50 Healthy) permanent residence of altitude 2500-3800 m above sea level were included in this study. The PAP mean in HAPH group was 32.4±9.8 mm Hg and in healthy 20.5±3.2 mm Hg (p=0.001).

The concentration of plasma ET-1 was measured in 32 highlanders. We have found that the level of plasma ET-1 was significantly higher in highlanders with HAPH compared with healthy subjects. So in highlanders with HAPH level of ET-1 was 7.07±2.26 pg/ml and in healthy subjects it was 4.56±1.66 pg/ml (p=0.001). ET-1 is a potent endogenous vasoconstrictor and smooth-muscle mitogen (Rubin *et al.*, 2002). Concentration of plasma ET-1 and expression of ET-1 protein is increased in experimental animals with pulmonary hypertension (Shirakami *et al.*, 1991; Yang *et al.*, 1997; Li *et al.*, 1998) and in humans with pulmonary hypertension (Stewart *et al.*, 1991; Giaid *et al.*, 1993; MacLean, 1998). Interesting finding was observed in secondary hypoxic pulmonary hypertension. It was shown that the level of ET-1 increased and correlated with level of PAP in secondary pulmonary arterial hypertension due to COPD (Ferri *et al.*, 1995; Fujii *et al.*, 1999) and interstitial lung disease (Yamakami *et al.*, 1997). The plasma ET-1 also elevated in patients with severe COPD (Channick *et al.*, 2004), moreover expression of ETA and ETB receptors increased in pulmonary artery of COPD patients (Davie *et al.*, 2002).

The level of ET-1 increased in subjects who are susceptible to development of high altitude pulmonary edema (HAPE) as compared with HAPE-resistance subjects and correlated with level of PAP (Sartori *et al.*, 1999). Several studies showed that the increasing of PAP during exposure to high altitude is related with high level of ET-1 concentration (Goerre *et al.*, 1995; Maggiorini *et al.*, 2001). In the present ACME-1 study, Modesti *et al.*, 2006, showed that the ET-1 play a significant role in pulmonary vascular vasoconstriction during adaptation to high altitude. The level of ET-1 increased in 1st day of adaptation to high altitude and then decreased in later days which correlated with the level of systolic PAP. Moreover authors in placebo-controlled study showed that mixed ETA and ETB receptors antagonist bosentan effectively controls the hypoxia-induced increase in pulmonary blood pressure but also causes a reduction in free water clearance (Modesti *et al.*, 2006). Thus, it will be logically correct to block the ET receptors which could improve the pulmonary hemodynamics and decrease the PAP in highlanders with HAPH.

In our study 125 mg of bosentan significantly decreased mean and systolic PAP. Mean PAP decreased from 32.0±2.2 to 26.5±3.0 mm Hg (p=0.001) and systolic PAP from 46.4±6.2 to 37.0±7.2 mm Hg (p=0.004). At the present time the effectiveness of bosentan in primary pulmonary hypertension is undoubtedly good. It was shown that the bosentan effectively reduce PAP, improve exercise capacity and quality of life in patients with pulmonary arterial hypertension (Channick *et al.*, 2001, Rubin *et al.*, 2002), as well as combination of bosentan with epoprostenol (Humbert *et al.*, 2004). In the recent placebo-controlled study showed that the bosentan compare with placebo not only failed to improve exercise capacity but also deteriorated hypoxemia and functional status in severe COPD patients without severe pulmonary hypertension at rest (Stolz *et al.*, 2008). In another study bosentan after 12 month of therapy compared to placebo also did not improve 6MWD, but showed a trend in delayed time to death or disease progression, and improvement in quality of life of patients with COPD (King *et al.*, 2008).

As the key role in pathogenesis of HAPH is high altitude hypoxia, it will be more reasonable to use oxygen in treatment of HAPH. Results of several studies showed that the oxygen breathing lead to decreasing PAP in highlanders and attenuated increasing of PAP during exercise (Grover, 1966, Penaloza, 1971, Mirrakhimov and Winslow, 1998). In our study we compared the effectiveness of 30 minutes oxygen breathing and bosentan on PAP in patients with HAPH. We have showed that the oxygen as well as bosentan significantly decreased PAP and it is interestingly that the level of decrease of PAP was the same. At the same time combination of bosentan and oxygen breathing led to significantly decreasing of mean PAP compare with bosentan along.

In 3 patients with HAPH PAP was not significantly decreased compared to baseline, and it indicates the morphological changes of the pulmonary vessels and the irreversibility of the pulmonary hypertension. On another hand in highlanders with significant response to oxygen breathing bosentan also effectively reduced PAP. Thus, bosentan lead to decrease of the PAP by excluding of vasospastic component. It is suggested that in patients who showed mild response to oxygen breathing should be tested for long-term effects of bosentan on PAP and blockade effects of ET-1 on smooth-muscle cells proliferation and migration.

It is well known that the ET-1 have opposite vascular effects mediated through different receptors. ETA receptors, located on vascular smooth muscle cells, mediate ET-1 vasoconstriction. ETB receptors are found on vascular smooth muscle cells, where they mediate vasoconstriction and on endothelial cells, where their stimulation produces vasodilation, presumably through the endothelial synthesis of NO and prostacyclin (Michael and Markewitz, 1996, Shetty *et al.,* 1993).

It has been shown that the infusion of ET-1 increases exhaled NO in guinea pigs (Malmstrom *et al.,* 2003) and the blockage of ET receptors by bosentan decreased exhaled NO in dogs (Hubloue *et al.,* 2003). On the contrary infusion of selective ETA receptor antagonist normalizes the decreased level of NO in animal with hypoxic pulmonary hypertension (Perreault *et al.,* 2001). It was also found that in Tibetans the level of exhaled NO is 2 times higher than in lowlanders (Beall *et al.,* 2001) and this phenomenon could explain why Tibetans have the low level of PAP (Groves *et al.,* 1993).

In our study the level of exhaled NO was not changed in patients with HAPH after 3 hour of bosentan intake compare with baseline. Girgis *et al.,* 2005, found that after the 3 month of bosentan therapy the level of exhaled NO and urinary nitrites significantly increased in patients with primary PAH, however level of endogenous plasma NO did not change significantly. In our study acute blockage of ET receptors led to decrease of PAP without change in the level of exhaled NO and it may be due to blockage of ETB receptors in smooth muscle cells and endothelium could block the release of NO and prostacyclin. From these findings it would be interesting to investigate the long-term effects of bosentan or selective antagonist of ET receptors on PAP and NO on large population of highlanders with HAPH.

It is well known that HAPH is a genetic trait, and it was confirmed by interindividual response of pulmonary vessels to hypoxiñ stimuli (Westcott *et al.,* 1951). In the previous study we have found that II genotype and I allele of ACE gene associated with development of HAPH (Morrell *et al.,* 1999, Aldashev *et al.,* 2002). It was very logically to investigate the genetic predisposition to increased levels of ET-1 in lungs.

One of the gene candidates is ET-1 gene. It was found that -3a/-4a polymorphism of ET-1 gene is functional and associated with level of ET-1. It was shown that the 4a allele associated with high level of plasma ET-1 compare with gomozygous -3a/-3a genotype in patients with essential hypertension (Tanaka *et al.,* 2004) and chronic heart failure (Vascu *et al.,* 2002). In the recent study Rajput *et al.,* have found that -3a/-3a genotype was associated with lower level of ET-1 compare with highlanders with -3a/-4a genotype (Rajput *et al.,* 2006).

We have shown that -3a/-4a polymorphism of ET-1 gene was associated with development of HAPH. We have found that the frequency of mutant allele -4a of ET-1 gene polymorphism as well as heterozygous genotype -3a/-4a significantly higher in highlanders with HAPH compared with healthy control. We could speculate that in each of -4a/-4a homozygous genotype in the population of the studied highlanders could be an evidence of a selection pressure against this genotype which could be associated with a very severe pulmonary vasoconstriction response to hypoxia and the early onset of pulmonary hypertensive phenotype.

Thus, we concluded that hypoxia related increased level of ET-1 play an important role in development of HAPH in permanent residence of high altitude and the high level of ET-1 in HAPH subjects has inheritance trait.

References

Aldashev, A.A. 2000. High altitude pulmonary hypertension and signal transduction in the cardiovascular system. J. Recept. Signal Transduct Res. 20, 255–278.

Aldashev, A.A., Sarybaev, A.S., Sydykov, A.S., et al. 2002. Characterization of High-Altitude Pulmonary Hypertension in the Kyrgyz: Association with Angiotensin-Converting Enzyme Genotype. Am. J. Respir. Crit. Care Med. 166, 1396-1402.

Bartsch P., Swenson, E.R., Maggiorini, M. 2001. Update: high altitude pulmonary edema. Adv. Exp. Med. Biol. 502, 89 –106.

Beall, C.M., Laskowski, D., Strohl, K.P., Soria, R., Villena, M., Vargas, E., Alarcon, A.M., Gonzales, C., Erzurum, S.C. 2001. Pulmonary nitric oxide in mountain dwellers. Nature. 414, 411-412.

Channick, R.N., Simonneau, G., Sitbon, O., Robbins, I.M., Frost, A., Tapson, V.F. et al. 2001. Effects of the dual endothelin-receptor antagonist bosentan in patients with pulmonary hypertension: a randomised placebo-controlled study. Lancet. 358, 1119–1123.

Channick, R.N., Sitbon, O., Barst, R.J., Manes, A., Rubin, L.J. 2004. Endothelin receptor antagonists in pulmonary arterial hypertension. J. Am .Coll. Cardiol. 43, 62S–67S.

Chen, S.J., Chen, Y.F., Meng, Q.C., Durand, J., DiCarlo, V.S., Oparil, S. 1995. Endothelin-receptor antagonist bosentan prevents and reverses hypoxic pulmonary hypertension in rats. J. Appl. Physiol. 79 (6), 2122–2131.

Chen, S.J., Chen, Y.F., Opgenorth, T.J., Wessale, J.L., Meng, Q.C., Durand, J., DiCarlo, V.S., Oparil, S. 1997. The orally active nonpeptide endothelin A-receptor antagonist A-127722 prevents and reverses hypoxia-induced pulmonary hypertension and pulmonary vascular remodeling in Sprague–Dawley rats. J. Cardiovasc. Pharmacol. 29, 713–725.

Davie, N., Haleen, S.J., Upton, P.D., Polak, J.M., Yacoub, M.H., Morrell, N.W., Wharton, J. 2002. ETA and ETB receptors modulate the proliferation of human pulmonary artery smooth muscle cells. Am. J. Respir. Crit. Care Med. 165, 389–405.

Ferri, C., Bellini, C., De Angelis C., De Siati L., Perrone, A., Properzi, G., Santucci, A. 1995. Circulating endothelin- 1 concentrations in patients with chronic hypoxia. I. Clin. Pathol. 48, 519-524.

Fujii, T., Otsuka, T., Tanaka, S., Kanazawa, H., Hirata, K., Kohno, M., Kurihara, N., Yoshikawa, J. 1999. Plasma Endothelin-1 Level in Chronic Obstructive Pulmonary Disease: Relationship with Natriuretic Peptide. Respiration. 66, 212–219.

Giaid, A., Yanagisawa, M., Langleben, D., Michel, R.P., Levy, R., Shennib, H., Kimura, S., Masaki, T., Duguid, W.P., Stewart, D.J. 1993. Expression of endothelin-1 in the lungs of patients with pulmonary hypertension. The New England journal of medicine. 328(24), 1732-1739.

Girgis, R.E., Champion, H.C., Diette, G.B., Johns, R.A., Permutt, S., Sylvester, J.T. 2005. Decreased exhaled nitric oxide in pulmonary arterial hypertension. Response to bosentan therapy. Am. J. Respir. Crit. Care Med. 172, 352–357.

Goerre, S., Wenk, M., Bartsch, P., et al. 1995. Endothelin-1 in pulmonary hypertension associated with high-altitude exposure. Circulation. 91, 359 –364.

Grover, R.F. 1966. Comparative physiology of hypoxic pulmonary hypertension. Cardiovscular and respiratory effects of hypoxia. Bsel: Karger, ...

Groves, B.M., Droma, T., Sutton, J.R., McCullough, R.G., McCullough, R.E., Zhuang, J., Rapmund, G., Sun, S., Janes, C., Moore, L.G. 1993. Minimal hypoxic pulmonary hypertension in normal Tibetans at 3,658 m. J. Appl. Phys. 74(1), 312-318.

Hill, N.S., Warburton, R.R., Pietras, L., Klinger, J.R., 1997. Nonspecific endothelin-receptor antagonist blunts monocrotaline-induced pulmonary hypertension in rats. J. Appl. Physiol. 83, 1209–1215.

Hubloue, I., Biarent, D., Abdel, K.S., Bejjani, G., Kerbaul, F., Naeije, R., Leeman, M. 2003. Endogenous endothelins and nitric oxide in hypoxic pulmonary vasoconstriction. Eur. Respir. J. 21, 19–24.

Humbert, M., Barst, R.J., Robbins, I.M., et al. 2004. Combination of bosentan with epoprostenol in pulmonary arterial hypertension: Breathe-2. Eur. Respir. J. 24, 353–359.

Humbert,, M., Morrell, N.W., Archer, S.L., Stenmark, K.R., MacLean, M.R., Lang, I.M., Christman, B.W., Weir, E.K., Eickelberg, O., Voelkel, N.F., Rabinovitch, M. 2004. Cellular and molecular pathobiology of pulmonary arterial hypertension. J. Am. Coll. Cardiol. 43, 13S–24S.

King, T.E. Jr., Behr, J., Brown, K.K., du Bois, R.M., Lancaster, L., de Andrade, J.A., Stähler, G., Leconte, I., Roux, S., Raghu, G. 2008. BUILD-1: a randomized placebo-controlled trial of bosentan in idiopathic pulmonary fibrosis. Am. J. Respir. Crit. Care Med. 177(1), 75-81.

Kitabatake, A., Inoue, M., Asao, M., et al. 1983. Noninvasive evaluation of pulmonary hypertension by a pulsed Doppler technique. Circulation. 68, 302–309.

León-Velarde, F., Maggiorini, M., Reeves, J.T., et al. 2005. Consensus statement on chronic and subacute high altitude diseases. High Alt. Med. Biol. 6(2), 147-157.

Li , H., Elton, T.S., Chen, Y.F., Oparil, S. 1994. Increased endothelin receptor gene expression in hypoxic rat lung. Am. J. Physiol. 266, 553–560.

MacLean, M.R. 1998. Endothelin-1: a mediator of pulmonary hypertension? Pulm. Pharmacol. Ther. 11, 125-132.

Maggiorini, M., Melot, C., Pierre, S., et al. 2001. High-altitude pulmonary edema is initially caused by an increase in capillary pressure. Circulation. 103, 2078 –2083.

Malmstrom, R.E., Tornberg, D.C., Settergren, G., Liska, J., Angdin, M., Lundberg, J.O., Weitzberg, E. 2003. Endogenous nitric oxide release by vasoactive drugs monitored in exhaled air. Am. J. Respir. Crit. Care Med. 168, 114–120.

Michael, J.R., Markewitz, B.A. 1996. Endothelins and the lung. Am. J. Respir. Crit. Care Med. 154, 555–581.

Mirrakhimov, M., Winslow, R. 1996. The cardiovascular system at high altitude. In: Environmental physiology / Edited by Fregley M, Blatteal C. Oxford: Oxford University Press, pp. 1241-58.

Miyauchi, T., Yorikane, R., Sakai, S., Sakurai, T., Okada, M., Nishikibe, M., Yano, M., Yamaguchi, I., Sugishita, Y., Goto, K. 1993. Contribution of endogenous endothelin-1 to the progression of cardiopulmonary alterations in rats with monocrotaline-induced pulmonary hypertension. Circ. Res. 73, 887–897.

Modesti, P.A., Vanni, S., Morabito, M., et al. 2006. Role of endothelin-1 in exposure to high altitude acute mountain sickness and endothelin-1 (ACME-1) study. Circulation. 114, 1410-1416.

Morrell, N.W., Sarybaev, A.S., Alikhan, A., Mirrakhimov, M.M., Aldashev, A. 1999. ACE genotype and risk of high altitude pulmonary hypertension in Kyrghyz highlanders. Lancet. 353, 814.

Penaloza, D. 1971. Discussion. High Altitude Physiology: Cardiac and Respiratory Aspects / Edited by Porter, R. - Churchill, Livingstone, Edinburg. pp.169.

Perreault, T., Berkenbosch, J.W., Barrington, K.J., Decker, E.R., Wu, C., Brock, T.A., Baribeau, J. 2001. TBC3711, an ET(A) receptor antagonist, reduces neonatal hypoxia-induced pulmonary hypertension in piglets. Pediatr. Res. 50, 374–383.

Rajput, C., Najib, S., Norboo, T., Afrin, T., Pasha, Q. 2006. Endothelin-1 gene variants and levels associate with adaptation to hypobaric hypoxia in high-altitude natives. Biochemical and Biophysical Research Communications. 341, 1218–1224

Rubin, L.J. 2006. Endothelin-1 and the Pulmonary Vascular Response to Altitude: A New Therapeutic Target? Circulation. 114, 1350-1351.

Rubin, L.J., Badesch, D.B., Barst, R.J., et al. 2002. Bosentan therapy for pulmonary arterial hypertension. N. Engl. J. Med. 346, 896–903.

Shetty, S.S., Okada, T., Webb, R.L., DelGrande, D., Lappe, R.W. 1993 .Functionally distinct endothelin B receptors in vascular endothelium and smooth muscle. Biochem. Biophys. Res. Commun. 191, 459–464.

Shirakami, G., Nakao, K., Saito, Y., Magaribuchi, T., Jougasaki, M., Mukoyama, M., Arai, H., Hosoda, K., Suga, S., Ogawa, Y., et al. 1991. Acute pulmonary alveolar hypoxia increases lung and plasma endothelin-1 levels in conscious rats. Life Sci. 48, 969–976.

Stewart, D.J., Levy, R.D., Cernacek, P., et al. 1991. Increased plasma endothelin-1 in pulmonary hypertension: marker or mediator of disease. Ann. Intern. Med. 114, 464–469.

Stolz, D., Rasch, H., Linka, A., Valentino, M.D., Meyer, A., Brutsche, M., Tamm, M. 2008. A randomized controlled trial of bosentan in severe COPD. Eur. Respir. J. [in press].

Tanaka, C., Kamide, K., Takiuchi, S., Kawano, Y., Miyata, T. 2004. Evaluation of the Lys198Asn and -3a/-4a (-134delA) Genetic Polymorphisms of the endothelin-1 Gene. Hypertens. Res. 27(5), 367-371.

Vasku, A., Spinarova, L., Goldbergova, M., Muzik, J., Spinar, J., Vitovec, J., Toman, J., Vacha, J. 2002. The Double Heterozygote of Two Endothelin-1 Gene Polymorphisms (G8002A and -3A/-4A) Is Related to Big Endothelin Levels in Chronic Heart Failure. Experimental and Molecular Pathology. 73, 230–233.

Voelkel, N.F. and Tuder, R.M. 1997. Cellular and molecular mechanisms smooth muscle cells in pulmonary hypertension. Pulmonary Pharmacology and Therapeutics. 10, 231-241.

Weir, E.K., Tucker, A., Reeves, J.T., Will, D.H., Grover, R.F. 1974. The genetic factor influencing pulmonary hypertension in cattle at high altitude. Cardiovasc. Res. 8, 745 - 749.

Westcott, R.N., Fowler, N.O., Scott, R.C., Hauenstein, V.D., McGuire, J. 1951. Anoxia and pulmonary vascular response. J. Clin. Invest. 30(9), 957-970.

Yamakami, T., Taguchi, O., Gabazza, E.C., Yoshida, M., Kobayashi, T., Kobayashi, H., Yasui, H., Ibata, H., Adachi, Y. 1997. Arterial endothelin-1 level in pulmonary emphysema and interstitial lung disease. Relation with pulmonary hypertension during exercise. Eur. Respir. J. 10, 2055–2060.

Yanagisawa, M., Masaki, T. 1989. Endothelin, a novel endothelium-derived peptide : Pharmacological activities, regulation and possible roles in cardiovascular control. Biochem. Pharmacol. 38, 1877–1883.

Yang, X., Chen, W., Chen, J. 1997. Change of level and expression of endothelin-1 in the lungs of rats with hypoxic pulmonary hypertension. Clin. Med. J. 110 (2), 104-108.

Yock, P.G., Popp, R.L. 1984. Noninvasive estimation of right ventricular systolic pressure by Doppler ultrasound in patients with tricuspid regurgitation. Circulation. 70, 657–662.

contd...

Table 1: Characteristics of highlanders

	HAPH (n=88)	Healthy (n=50)
Age, year	49.7±13.6	48.8±11.2
Gender (m, f)	69 m 19 f	47 m 3 f
Altitude of residency, m above sea level	2500-3800	2500-3800
PAP mean, mm Hg	32.4±9.8	20.5±3.2*
FEV_1, % predicted	93.0±7.0	92.0±8.0
Endothelin-1, pg/ml	7.07±2.26	4.56±1.66*

PAP-pulmonary arterial pressure; FEV_1 - forced expiratory volume in 1 second; *p=0.001 - HAPH vs Healthy.

Table 2: Acute effects of bosentan and oxygen on PAP in highlanders

Parameters	Baseline	Baseline+ Oxygen	Bosentan	Bosentan + oxygen
PAPmean, mm Hg (n=10)	31.2±1.5	25.7±3.3*	25.5±2.5*	22.2±2.0*†
PAPsyst, mm Hg (n=8)	45.5±7.1	37.9±7.7*	34.9±7.1*#	32.9±5.7*#
BP syst, mm Hg	125.0±10.8	124.5±10.6	123.0±10.8	122.8±10.3
BP diast, mm Hg	80.0±5.8	78.0±5.8	76.8±4.7	77.2±5.6
HR, beats per min	81.6+13.0	77.0±12.2	81.1±10.5	78.8±11.7
SaO_2, %	88.5±2.5	97.4±1.4*	89.3±1.8	97.1±1.0*
NO, ppb	9.4±3.0	-	10.4±3.9	-

*p=0.001 - compare with baseline; #p=0.01 - compare with baseline + oxygen; †p=0.004 - compare with bosentan.

Table 3: -3a/-4a polymorphism of ET-1 gene and HAPH

	Frequency of genotypes (number of highlanders)			Frequency of alleles (number of alleles)	
	-3a/-3a	-3a/-4a	-4a/-4a	-3a	-4a
HAPH (n=88)	15%(13)	85% (75)*	-	59% (101)	41%(70)**
Healthy (n=50)	40% (20)	60% (30)	-	70% (70)	30%(30)

*χ^2=11.15; p=0.0008; **χ^2=4.3; p=0.03

Figure 1: Mean and systolic PAP response to 125 mg of bosentan

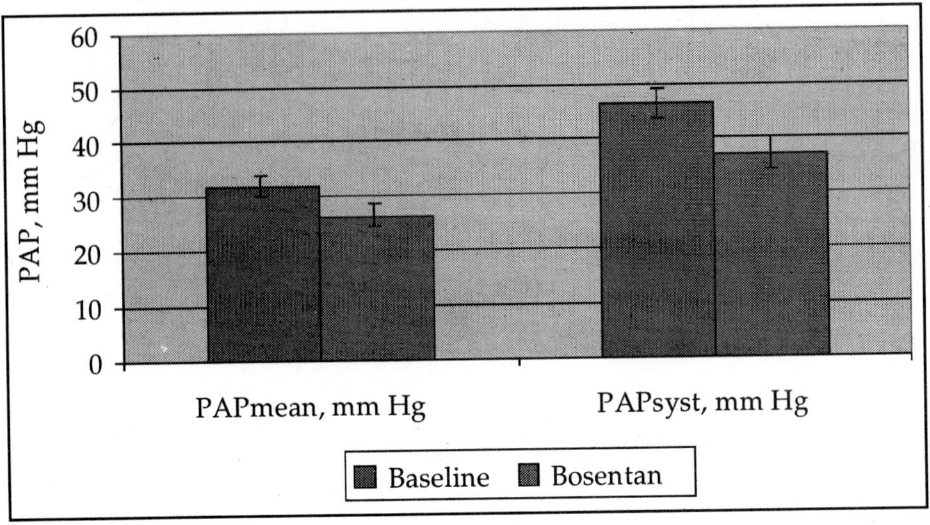

*p=0,001, **p=0,004 – compare with baseline

□□□

36

Hypoxic Trainings and Essential Hypertension

A.S. DZHUMAGULOVA
National Center of Cardiology and Internal Medicine
Togolok Moldo Str. 3, Bishkek, Kyrgyzstan

ABSTRACT

The aim of the study was to investigate the short-term and long-term effects of hypoxic training in patients with essential hypertension (EH). In the phase I hypoxic trainings were performed in 56 patients with EH (I-II stage of hypertension). During phase II of the study 26 patients with high amino oxidase activity underwent adaptation for 30 days in high altitude (Tuya-Ashu). Phase III consisted of prospective monitoring in 39 patients who underwent hypoxic training in barocamera (every 4-5 months, 22 patients) or course on highland climatotherapy (17 patients). Physical examination, ECG, Doppler echocardiography, bicycle test, serum amino oxidase, monoamio oxidase of platelets (MAO), serum rennin and aldosterone were the methods used for study. The study showed that blood pressure (BP) was decreased in most of the patients (62.5 percent - "responsiveness") and it also improved the clinical and functional symptoms of the disease under the hypoxic training. The group of "responsiveness" in the baseline conditions was characterized by increased amino oxidase and MAO activity as compared with a control group (healthy). On the contrary, the "nonresponsiveness" group in the baseline conditions had lower amino oxidase compared with healthy control group. Thirty day adaptation to the highlands conditions led to substantial improvement in myocardial relaxation. Results obtained at the end of phase III had shown normal levels of BP, myocardial function and physical activity. In contrast, in patients not receiving the hypoxic training revealed the progression of EH during the duration of study. Thus, our study indicates the usefulness of hypoxic training in a large number of patients with EH for treatment and secondary prevention.

Key words : Essential hypertension, Hypoxic trainings.

INTRODUCTION

Essential hypertension (EH) is a leading problem of medicine because the long-existing increase in blood pressure (BP) leads to injury of target-organs: heart, kidneys, vessels and eyes; it promotes the development of myocardial infarction, ischemic stroke, congestive heart failure, renal insufficiency, retinal hemorrhages (JNC 7 Report 2003).

Epidemiological studies had shown that over the past 20 years the prevalence of EH in Kyrgyzstan has increased to 40-50 percent, which in turn causes increase cardiovascular diseases (CVD). CVD are the main cause of mortality in the Kyrgyz Republic, with more than 45 percent of all deaths and a dramatic progressive increase. Mortality from cardiovascular diseases showed an increase by 23 percent from 261.8 events in 1991 to 313.5 events in 2006 among 100,000 individuals (Dzhumagulova *et al.*, 1997 and Young *et al.*, 2005).

On the other hand the effectiveness of treatment of EH remains quite low even in developed countries, despite of the large arsenal of antihypertensive drugs and non-pharmacological methods for its treatment. Thus, in the USA this figure is 25-27 percent, in Europe it varies from 9 percent to 30 percent. In Kyrgyzstan, BP is effectively controlled by only in 2.9 percent patients with EH (Young *et al.*, 2005). Therefore, the search of effective and safe alternative methods of BP correction seems to be very important.

One such method is using exogenous and endogenous hypoxia. In the last century epidemiological studies from different countries showed a lower frequency of EH among highland residents compared with the inhabitants of the plains (Ruiz *et al.*, 1997, Eckes, L. 1976 and Mirrakhimov *et al.*, 1985). Literature to date shows a complex and not fully explores mechanisms for therapeutic effect of exogenous hypoxia. Nevertheless, the results of experimental work suggest that the therapeutic effect of hypoxic training in patients with EH resulted in normalization of sympathetic nervous system and decreased activity of rennin-angiotensin system.

The aim of the study was to investigate the short-term and long-term effects of hypoxic trainings in patients with essential hypertension (EH).

PATIENTS AND METHODS

Hypoxic trainings was carried out in 56 patients with EH (I-II stage of hypertension) aged 38.4±1.3 years during the first phase of the study in the chamber with the volume of 5.6 m^3. Therapeutic trainings were conducted step by step, starting from the height of 1000 meters and gradually increasing up to 1500 m, 2000 m and 3200 m. Usually the height of 3200 meters was achieved after 2-3 day of hypoxic training. Daily patient's exposition in barocamera at "height" of 3200 m (pO_2- 112 mmHg) lasted for 70 minutes and course of treatment continued for 15 days. The positive and negative effects of hypobarotherapy method of treatment were estimated.

In the second phase of the study, 26 patients with I-II stage of EH aged 43.2±1.6 years were underwent adaptation for 30 days in a highlands, Tuya-Ashu (3200 m above sea level). The impact of high-altitude hypoxia on the structural and functional parameters of left heart, tolerance to physical activity, as well as the changes in neurohumoral system activity and lipid profile was assessed. Phase III consisted of prospective monitoring of 39 patients, received hypoxic training in barocamera (every 4-5 months, 22 patients) or course of highland climatotherapy (17 patients).

Activity of serum amino oxidase (AO) and monoamino oxidase (MAO) of platelets were determined by selecting benzaldehyde, from benzilamine. Contents of rennin and aldosterone were determined by radio immunological method with standard reactive from firm "Cis" (France) for aldosterone and firm "Par-Sorin-Biomedica" (Italy) for rennin. Serum cholesterol concentration was determined on analzator AA-II "Tehnikon".

The echo-CG was performed using the Sequioa-512 (Acuson, USA). Both M- and B-scanning were used for evaluation of the cardiac structure. The end-systolic and end-diastolic dimensions of the LV, volumes of the LV, thickness of the LV posterior wall and interventricular wall, LV ejection fraction were determined using standard technique. Myocardial mass index was calculated as suggested by Devereux and Reichek (Devereux *et al.*, 1977). Diastolic left ventricular function was estimated by transmitral blood flow on Sequioa-512.

The bicycle test was performed on the veloergometry "Elema" (Sweden), mainly in the morning, not less than two hours after eating and step-by-step increase in subtotal physical activity was used. Initial power load was 300 kGm/min, every 4 minutes it was increased by 150 kGm/min until of subtotal level of heart rate was achieved.

Clinical data was compared between groups using two-sample Student 't' test in case of normal distribution and non-parametric Manna-Whitney and Wilcoxon criteria in case of asymmetric distribution was applied for finding out the statistical significance between the groups.

RESULTS

In the first phase of the study hypoxic trainings were performed in 56 patients with EH. In 34 patients (60.7 percent) in baseline conditions had the I stage of EH, and 22 (29.3 percent) patients had the II stage of EH.

The study revealed different reactions in EH patients to hypoxic training despite the lack of distinction between the patients in baseline conditions on the basis of age, length of illness, clinical manifestations of disease, as well as the level of BP. BP was decreased in most of the patients (62.5percent - "responsiveness") and there was improvement of the clinical and functional current of the disease by means of the hypoxic training. In some cases (the smaller group: 37.5 percent - "nonresponsiveness") there were negative manifestations in 5 patients (increased heart rate, heartache, supraventricular arrhythmia).

In order to reveal the possible causes of different response to hypoxia in examined patient's baseline activity of serum amino oxidase and monoamino oxidase (MAO) of platelets was studied, as well as changes in these indicators during the course of hypoxic training.

Activity of AO serum and MAO in platelets in "responsiveness" group was 0.58±0.04 nmol/mg protein and 42.3±2.1 nmol/mg protein respectively in baseline conditions and it significantly exceeded the value in healthy people (0.47±0.03 nmol/mg protein and 35.7±2.3 nmol/mg protein respectively, $p<0.05$), indicating an increased activity of the sympathetic nervous system in this category of patients. The activity of monoaminergic system was normalized during hypoxic training that was shown as decreasing of AO activity in control group (0.41±0.01 nmol/mg protein, against 0.58±0.04 nmol/mg protein before treatment, $p<0.05$) and MAO activity (34.5±0.9 nmol/mg protein, against 42.3±2.1 nmol/mg protein before treatment, $p<0.05$).

"Nonresponsiveness" group in the baseline conditions, in contrast, were characterized by significant decrease in activity as compared to the normal. The MAO values in nonresponsive group was 0.36±0.02 nmol/mg protein, and normal group was 0.47±0.03 nmol/mg protein ($p<0.01$) and showed tendency towards decreasing of MAO activity (31.2±1.9 nmol/mg protein, against 35.7±2.3 nmol/mg protein in normal ($p>0.05$). The barocamera hypoxic training patients from this group had no significant changes in AO activity, and at the same time, it registered a further decline in MAO activity (24.7±2.0 nmol/mg protein, against 31.2±1.9 nmol/mg protein before treatment, $p<0.05$).

At the second phase of the study 26 patients aged 43.2±1.6 years under went adaptation for 30 days in highlands, Tuya-Ashu. For high climatotherapy patients with AO activity more than 0.47 nmol/mg proteins without contradictions for hypoxic training were selected.

The study had shown that patients with initial stage of EH and increased activity of monoaminergic system in baseline had good response on treatment with decrease of BP during 30-days adaptation to the highlands conditions (Table 1). The BP reduction was mainly due to decrease of the total vascular peripheral resistance (TVPR) during highlands adaptation, indicating decrease tone of resistive vessels under the influence of exogenous hypoxia. It was combined with the relative reduction of heart rate.

Echo-CG had shown significant improvement of left ventricle myocardial relaxation after 30-day of adaptation to hypoxia. Thus E/A ratio reached by the end of the course of therapy were 1.37±0.03 and in healthy people it was 1.42±0.02 units. By the end of the course of highland climatotherapy there was a significant reduction of left atrial size in EH patients. Besides that, ejection fraction of left atrial was reduced by 77.5±1.0 percent at the end of treatment and the values were not different from the values in healthy people (75.6±0.6 percent, $p>0.05$). At the same time, substantial dynamics of the wall thickness and left ventricular myocardial mass was not registered ($p>0.05$).

Improvement of myocardial function activity by means of highland climatotherapy was accompanied by the significant increase of physical activity (512.8±17, kGm/min/m^2 after treatment against 442.3±19.8 kGm/min/m^2 before treatment, $p<0.05$).

We have also examined the impact of 30-day adaptation to high-altitude hypoxia on neuro-hurmoral system and lipid profile. This study showed that staying in the mountains of EH patients contributed to the normalization of monoaminergic system. The AO activity decreased (0.45±0.02 nmol/mg protein, against 0.62±0.03 nmol/mg protein before treatment, $p<0.05$) and MAO activity (36.7±1.0 nmol/mg protein, against 42.6±2.2 nmol/mg protein, $p<0.05$). Besides that, patients with high initial activity of renin and aldosterone showed decreased activity by the end of course of treatment. The value of renin fell to 1.48±0.31 ng/ml/h against 2.53±0.36 ng/ml/h before treatment, $p<0.05$), and the aldosterone level dropped to 153.0±20.7 pg/ml against 235.8±32.2 pg/ml before treatment, $p<0.05$).

We found decrease in total cholesterol by 18 percent, low density lipoprotein (LDL) by 29 percent and increase in high density lipoprotein (HDL) by 15 percent after 30-days of stay at the Tuya-Ashu.

In order to study the long-term effects of hypoxia the patients with EH were provided with prospective monitoring. The 39 patients underwent hypoxic training in barocamera (every 4-5 months, 22 patients) and course of highland climatotherapy (17 patients) during 1.5 years. Control group consisted of 12 patients with EH, adjusted by age, clinical and functional symptoms of disease, and were refused therapy because of various reasons.

At the end of 1.5 years monitoring, results showed that patients with EH treated by regular hypoxic training had normal levels of blood pressure during all period of monitoring. It was accompanied by positive clinical changes, including favorable effect on the diastolic left ventricular function and physical activity.

A control group of patients refused hypoxic trainings through 1.5 years of observation revealed signs of disease progression. Thus, at the end of prospective observation of this group of patients a significant increase in blood pressure was recorded accompanied by increase in TPVR (1920±42 din*s*cm^{-5} in a 1.5 year, against 1567±27 din*s*cm^{-5} before observation, $p<0.01$). The echocardiography parameters revealed significant progression of myocardial remodeling by the end of prospective monitoring, this was the effect of physical activity in this group of patients (411.6±15.5 kGm/min/m^2 in 1.5 years, against 464.3+17.2 kGm/min/m^2, p <0.05).

DISCUSSIONS

Even 60-70 years ago it was revealed that prevalence of EH, as well as level of BP in highlands residents was lower than among the inhabitants of the plains (Ruiz *et al.*, 1997, Eckes, L. 1976 and Mirrakhimov *et al.*, 1985). Experimental and clinical studies have shown that hypotensive effect of exogenous hypoxia is associated with decrease of sympathetic nervous system (SNS) tone and development of relative hyperactivity of parasympathetic nervous system on the one hand, and reduction of rennin-angiotensin system (RAS) activity on the other hand (Khitrov *et al.*,1975). Possible EH prevention in highlands residents may be caused by moderate physiological hypofunction of thyroid gland (Khmel'nitskii *et al.*, 2006). Based on the results of earlier studies, raises hope for treating EH by high-altitude climate. Indeed, work by several groups (Potievskia *et al.*,1992, Vorobev *et al.*, 1993, Berezovskyi *et al.*, 2000) have shown positive therapeutic effect of hypoxia in hypertensive conditions. But accumulated data indicate different reaction of BP, TPVR and activity of cardiovascular system in highland. Therefore, it is important to find the predictors of efficiency of hypoxia in the treatment of EH.

Our study revealed different response of cardiovascular system on hypoxia in EH patients. BP decreased in most of the patients and there was improvement in the clinical and functional current of disease. But in some cases the response of hypoxic training was not good and on the contrary, there were some negative manifestations (heart rate increases, heartache, supraventricular arrhythmia).

According to Meerson *et al.*, 1978, one of the main mechanisms of antihypertensive action of exogenous hypoxia is by reducing the SNS activity by normalization of catecholamine exchange. According to a study of Baumanis 1981, one of the main ways of catecholamine exchange is oxidative deamination catalyzed by amino oxidase, and, consequently, their activity reflects on the activity of SNS (Baumanis *et al* 1981). In this context, to determine possible causes of different response to hypoxia among examined patients, we examined baseline AO and MAO activity, and also assessed changes in the biochemical markers under the course of hypoxic training.

The "responsiveness" group in the baseline conditions was characterized by increased AO and MAO activity which was normalized during the treatment. On the contrary, the "nonresponsiveness" group in the baseline conditions had low values of AO and MAO enzyme activity.

Thus, we assumed that "responsiveness" and "nonresponsiveness" groups of patient showed different pathogenic variants of diseases that differ not only by features of regulatory mechanisms, but also identity by several of hemodynamic responses to the treatment of exogenous hypoxia. Hypoxic training lead to improvement of clinical course of disease, providing a stable normalization of blood pressure and decreasing of TPVR in patients with EH and high baseline activity of monoaminoergic system. In patients with reduced baseline activity of monoaminergic system hypoxic training were ineffective and accompanied by the development of side effects. Consequently, AO activity can be used as a predictor of hypoxic training efficiency in patients with initial stage of EH.

This assumption finds confirmation in the results of 30 days adaptation to highlands in patients with high baseline AO activity. Hypoxia in such patients leads not only to reduce blood pressure and expressed TPVR, but also improve the myocardial relaxation, increase in physical activity, normalization of lipids, and these positive effects persisted during a long period (up to 1.5 years) of prospective monitoring.

Besides that, highland climatotherapy contributed to normalization of monoaminoergic system, as well as rennin and aldosterone level.

Thus, our study recognizes the usefulness of hypoxic training for a large part of patients with EH for treatment and secondary prevention of disease. It is necessary to take into account the degree of activity and reactivity of MAO as predictor of hypoxic trainings effectiveness in selecting patients for therapy while viewing the reaction of patients to exogenous hypoxia.

References

Baumanis, E.A., Kalninia, I.E., Iriste, A.A., Gorkin, V.Z. 1981. Modification of mitochondrial monoamine oxidase activity by certain amine inhibitors. Vopr. Med. Khim. 27(1), 72-77.

Berezovs'kyĭ, V., Levashov, M.I. 2000. The correction of arterial hypertension by oxygen deprivation. Fiziol Zh. 46(2), 82-90. (in Russian)

Devereux, R.B., Reichek, N. 1977. Echocardiographic determination of left ventricular mass in man. Anatomic validation of the method. Circulation. 55(4), 613-618.

Dzumagulova, A.S., Mirrakhimov, E.M. 1997. Primary and secondary prevention of arterial hypertension and dislipidemia in Kyrgyz Republic. Central Asian Medical Journal. 1, 35-39. (in Russian)

Eckes, L. 1976 Altitude adaptation. Morbidity and mortality. Gegenbaurs Morphol Jahrb. 122(6), 826-863. (in German)

Khitrov, N.K., Svistukhin A.I. 1975. Relationship between certain indicators of the activity of sympathico-adrenal and parasympathetic structures in different degree of exogenous hypoxia. Patol Fiziol Eksp Ter. 2, 21-25. (in Russian)

Khmel'nitskii, O.K., Khmel'nitskaia, N.M., Tararak, T., et al. 2006. Impact of intermittent hypobaric hypoxia on the thyroid morphofunctional state in experimental hyperthyroidism. Arkh Patol. 68(6), 31-33. (in Russian)

Meerson, F.Z., Bogomolov, A.F. 1978. Mechanism of the decrease in basal metabolism during adaptation to hypoxia. Biull Eksp Biol Med. 86(9), 287-289.

Mirrakhimov, M.M., Rafibekova, Zh.S., Dzhumagulova, A.S., et al. 1985. Prevalence and clinical peculiarities of essential hypertension in a population living at high altitude. Cor Vasa. 27(1), 23-28. (in Russian)

Potievskaia, V.I., Chizhov, A. 1992. Effect of intermittent normobaric hypoxia on dynamics of disease state in patients with hypertension. Fiziol Zh. 38(5), 53-57. (in Russian)

Ruiz, L., Peñaloza, D. 1977. Altitude and hypertension. Mayo Clin Proc. 52(7), 442-445. (in Ukraine)

The Seventh Report of the Joint National Committee on Prevention, Detection, Evaluation, and Treatment of High Blood Pressure: the JNC 7 report. 2003. JAMA. 289(19), 2560-2572.

Vorob'ev, L.P., Chizhov, A., Potievskaia, V.I. 1993. The evaluation of the efficacy of adaptation to intermittent normobaric hypoxia as a method for treating hypertension. Vopr Kurortol Fizioter Lech Fiz Kult. 5, 9-12. (in Russian)

Young, J.H., Parler, P., Bristol, B., Klag, M.J. 2005. The coming epidemic: hypertension in rural Kyrgyzstan, Central Asia. J. Hum Hypertens. 19(2), 145-148.

Table 1: Changes in hemodynamic parameters (M ± m) under highland climatotherapy in patients with essential hypertension

Parameters	Before treatment	After treatment	P
SAD, mm Hg	163.7±3.8	132.6±3.1	<0.01
DAD, mm Hg	99.2±2.0	84.5±1.4	<0.01
HR, per min	68.6±1.5	59.7±2.0	<0.05
CI, $l/min/m^2$	3.37±0.18	3.12±0.13	NS
TPVR, din*s*m^5	1673±40.1	1435±38.7	<0.01

SAD - systolic blood pressure, DAD - diastolic blood pressure, HR - heart rate, CI - cardiac index, TPVR - total peripheral vascular resistance, NS – non significant

□□□

37

Genetics of High Altitude Pulmonary Hypertension

ALMAZ ALDASHEV
Institute of Molecular Biology and Medicine
Togolok Moldo street, 720040 Bishkek, Kyrgyzstan

INTRODUCTION

The normal adult pulmonary circulation is a low-pressure, high-flow vascular bed with a great capacity for vascular recruitment. A unique feature of the pulmonary, as opposed to the systemic, circulation is vasoconstriction observed at moderately decreased levels of alveolar oxygen. High-altitude pulmonary hypertension (HAPH) is a public health problem in the mountain areas of the world. Around 10 per cent of the world population is living in mountainous regions; millions of people are at risk to develop HAPH. According to data recently presented by Dr. Wu (Xining, China) only in China about 80 millions people live at high altitude and more than 16 millions living at Xinghai-Tibetan Plateau.

In Kyrgyzstan the Tien-Shan and Pamir mountains occupy about 90 per cent of the territory and about 5 per cent of the population lives at altitudes above 2500 m. The recent consensus statement on high altitude diseases defines HAPH as a clinical condition occurring in children and adults residing at altitudes above 2500 m are characterized by increased mean pulmonary arterial pressure (PAP) above 30 mm Hg at altitude of residency and the absence of excessive erythrocytosis. There has been some confusion in the literature and several terms for these diseases were historically used, including chronic mountain sickness, high altitude heart disease, hypoxic cor pulmonale, pediatric high altitude heart disease, subacute mountain sickness, high altitude pulmonary hypertension. HAPH has also characterized by vascular wall remodeling of pulmonary artery, comprised of mainly of hypertrophy of tunica media and adventitial thickening of pulmonary arteries. HAPH is complicated by right ventricular hypertrophy, which may progress to right ventricular failure. Genetic susceptibility to hypoxia-induced pulmonary hypertension is suggested by the large inter-individual variation in the pulmonary vascular response to hypoxia, which seems to have a strong hereditable component. For example, susceptibility to hypoxic pulmonary hypertension in cattle ("brisket disease") is an inherited trait. Among humans, large inter-individual differences in the magnitude of the pulmonary pressure response to the same levels of hypoxia and hypoxemia have been described and there appears to be an inverse relationship between the prevalence of HAPH within a highlander population and the length of time that ethnic group has spent at altitude. The incidence of HAPH is much lower in Tibetans than in the Han Chinese, a population who recently migrated to the Tibetan plateau. Interestingly, the Tibetans have lack of enhanced erythropoiesis inspite of low oxygen saturation and Andean and reduced pulmonary vascular musculature, both of which will minimize the rise in PAP during exposure to hypoxia.

The frequency of HAPH in indigenous Tibetans is also much lower than in indigenous Andeans, an observation that has been attributed to better adaptation of Tibetans to high altitude. Most researchers assumed that Tibetans have been at altitude for many more generations than Andeans and therefore have had more time to adapt; however, analysis of Y-chromosome haplotypes indicates that the antecedents of the extant Tibetan populations migrated from the lower Yellow River Basin about 6000 years ago, which is several thousand years later than conservative estimates place the occupation of the Andes. Also it was speculated that Tibetan population had a genetic selection for hypoxic environment for several thousands years without mixture of their genes pool. But according to ancient Chinese chronics Tibetan people originated in the 1st millennium BC from a mix of different ethnically and racially tribes like Khyan, Zhung, both of Asian origin and Dardes and Mones of Caucasean background who were living at different altitudes all over Xinghai-Tibetan Plateau. In the 5th century AD some Turkic Hunnu tribe occupied Tibet mixed with Tibetans. Thus the better ability of Tibetans for adaptation to hypoxia could be due to high penetrance and dominance of unique genetic traits inherited from their forefathers.

This hypothesis paves the way for focusing the study of gene polymorphisms responsible for these genetic traits beneficial for adaptive developmental responses to chronic hypoxia but also for searching of polymorphisms attributed to altitude-related illness. Polymorphisms are variant sites within genes that originated as a mutation in an individual and have since spread in the population (and are present at a greater frequency than can be accounted for by mutation rates alone). Association studies compare the frequency of alleles at polymorphic loci between groups of people with (i.e., cases) and without (i.e., controls) the condition. An allele that is statistically over represented in the affected cohort is considered to be associated with the condition. Most association studies look at variations in candidate genes, which are genes that encode components of pathways that are known (or suspected) to be involved in the etiology of the disease. A number of candidate gene polymorphisms have been assessed in the various forms of altitude-related illness.

A unique feature of the pulmonary, as opposed to the systemic circulation, is vasoconstriction observed at decreased levels of alveolar oxygen in both animals and humans. It is believed that the occurrence of hypoxic pulmonary vasoconstriction in intrapulmonary small arteries reflects the localization of molecular oxygen sensor and effectors like O2-sensitive K+ channels in pulmonary artery smooth muscle cells to the resistance pulmonary arteries. But this hypoxic pulmonary vascular response is although modulated by endothelial cells and release of endocrine and paracrine vasoconstrictors and vasodilators also having pro- and antiproliferative effects. The genetic background provides the basis of hypoxic response but in chronic hypoxia this response may be modified by programming during prenatal and postnatal life. Modifications which lead to imbalances of the effects of endocrine and paracrine vasoconstrictors and vasodilators are believed to be the cause of HAPH. These vasoactive substances implicated in the pathogenesis of HAPH include endothelin-1, angiotensin converting enzyme (ACE), angiotensin II, serotonin, nitric oxide (NO), bradykinin, catheholamines, leukotriens and prostaglandins.

Genetics of High Altitude Pulmonary Hypertension in Indigenous Population of Tian-Shan and Pamir

Ethnic Kyrgyz subjects (n=689) aged 17 to 86 years (mean age 37.8±0.9) who were residents of altitudes of 2500–4200 m above sea level in Tian-Shan and Pamir have been studied. All subjects underwent health screening by history, physical examination, spirometry, blood pressure measurement and ECG. Right heart catheterization has been performed in 147 highlanders aged 18-72 years (mean age 34.0±13.1) with symptoms of dyspnea or exercise limitation at altitude, ECG and echocardiographic signs of HAPH.

Right heart catheterization was performed in Bishkek (760 m above sea level) with a Swan-Ganz thermodilution catheter introduced via jugular vein. After 30 minutes of stabilization, baseline measurements were made of systolic, diastolic and mean pulmonary arterial pressure and pulmonary capillary wedge pressure. To assess the magnitude of the pressure response to acute hypoxia, subjects were required to breathe a hypoxic gas mixture (11% oxygen, 89% nitrogen) for 30 minutes, at the end of which hemodynamic measurements were repeated. Arterial oxygen saturation and ECG were monitored throughout. There were no significant complications from right heart catheterization in any of the subjects studied.

High altitude pulmonary hypertension was diagnosed if resting mean PAP was > 25 mmHg at sea level conditions or greater than 30 mm Hg when measured at high altitude environment.

The following genetic polymorphisms were studied in highlanders with and without HAPH: I/D, G2215A, G2350A (ACE gene), Arg16Gly and Glu27Gln (β_2-adrenoreceptor gene), 4a/b and G894T (eNOS gene), S/L polymorphism of serotonin transporter (5-HTT) gene. Allele and genotype distributions in the pulmonary hypertensive and control groups were compared using contingency tables, with the Fisher exact test when appropriate. Single factor analysis of variance and the nonparametric Kruskal–Wallis test were employed to compare mean pulmonary artery pressure in patients with different ACE genotypes.

Angiotensin-converting enzyme (ACE) is a candidate gene for the development of HAPH. ACE is a metalloproteinase catalyzing cleavage of a dipeptide from angiotensin-1 to produce a very potent vasoconstrictor angiotensin-2. ACE is also known to be involved in degradation of the vasodilator bradykinin. Beyond being a vasoconstrictor, angiotensin-2 stimulates the proliferation of vascular endothelial and smooth muscle cells, and is therefore one of the key players of processes of vascular remodelling.

Serum ACE levels drop in response to hypoxia, and the magnitude of the drop is greater in individuals that acclimatise well to hypoxia. The most commonly studied ACE polymorphism is an insertion (ACE-I) / deletion (ACE-D) of an intronic *Alu* repeat unit. The ACE-I allele is associated with lower circulating ACE levels. Several studies found the alternate allele (ACE-D) to be more common in subjects susceptible to high-altitude pulmonary oedema (HAPE), although other studies have failed to replicate the correlation. Surprisingly, reported a higher climbing success rate associated with the ACE-I allele and it has been shown to be overrepresented in accomplished climbers.

Our own experience is that the ACE-I allele is more common in Kyrgyz people with established HAPH. At first sight this may seem contradictory, given the data associating ACE-I with better acclimatization to hypoxia and climbing success but it is rationale. It can be envisaged that evolutionary selection pressure would act to select out subjects susceptible to acute and fatal altitude-associated illnesses like HAPE and high-altitude cerebral oedema (HACE). If the ACE-D allele is more common in HAPE-susceptible subjects, this allele would be selected against and the ACE-I allele would predominate. The association with HAPH, a more slowly developing disease, could be an epi-phenomenon. On the other hand, there is evidence that angiotensin-2 is protective in chronic hypoxia. Rabinovitch et al have reported that injection of angiotensin-2 was protected against the development of hypoxic pulmonary hypertension, vascular remodelling and right ventricular hypertrophy in chronically hypoxic rats, an effect mediated through stimulation of prostanoid production. Thus, the ACE-1, associated with lower angiotensin-2 levels, could increase susceptibility to HAPH. In other words, the genotype which is beneficial for early acclimatization to a hypoxic environment could be associated with high-altitude related diseases (HAPH) in older ages.

In addition to I/D polymorphism in the ACE gene, there are at least 12 other known variants in the gene, including two (G2215A N G2350A) that have more effect on plasma ACE variation than the I/D locus. There are several combinations of these polymorphisms in different haplotypes of ACE gene.

ACE haplotype frequencies can vary between populations especially those recently found or isolated and our study was the first to demonstrate it in Kyrgyz population. We have demonstrated for Kyrgyz population a linkage disequilibrium of D allele with A2215 and G2350, whereas I allele was linked with G2215 and A2350 alleles giving two haplotypes DAG and IGA. These two haplotypes (DAG and IGA) are the major ACE gene haplotypes found in Kyrgyzs with the frequencies 0.615 for DAG and 0.314 for IGA. The frequency of these two haplotypes is 0.93 whereas 5 other haplotypes presented very rare, with the frequency less than 0.1.

In Caucasions IGA and DAG haplotypes have almost the same frequencies and altogether give frequency of 0.6. The ACE gene haplotypes distribution in Kyrgyzs is very different from that demonstrated for Caucasions with a much higher frequency of IGA and DAG than the other five combinations. This particularity of ACE gene haplotypes distribution we could explain by relatively high genetic homogeneity of Kyrgyz population with a prominent "founder effect".

When we studied the distribution of these haplotypes in HAPH patients vs healthy Kyrgyz highlanders we found the same pattern for as I and D alleles. In patients with HAPH the IGA haplotype was more frequent compared to controls. Again, DAG haplotype had a greater effect on plasma ACE activity and systemic blood pressure than the individual alleles and DD genotype (higher ACE activity) was negatively associated with right ventricular hypertrophy in males with COPD. Thus our data strongly reveal that DAG haplotype in Kyrgyz highlanders is negatively associated with development of HAPH.

ACE is known to be involved in degradation of the vasodilator bradykinin which is also linked to the pathogenesis of hypoxic pulmonary hypertension. was shown that bradykinin B2 agonist demonstrates the capacity to reduce severe pulmonary hypertension, right ventricular hypertrophy and induces apoptosis of hyperproliferative cells in pre-capillary pulmonary arterioles in rats. We have demonstrated in our study the lack of association of +9/-9 polymorphism of bradykinin receptor B1 with the development of HAPH in Kyrgyz highlanders. The frequency of mutant allele +9 was very low and moreover homozygous +9/+9 genotype was not found in studied population.

Another candidate gene is b2-adrenergic receptor (β_2-AR) gene, located on chromosome 5. The well known polymorphisms of this gene are Arg16Gly and Gln27Glu, with the Gly16 allele related to desensitization of β_2-AR and the Gln27Gln genotype associated with decreased vascular response to β-agonists stimulation. We have demonstrated that severe right ventricular hypertrophy (II and III grade) correlated with b2-adrenergic receptor desensitization. Also desenitisation of β_2-AR has been reported in COPD patients with secondary pulmonary hypertension. Therefore we were very keen to study the association of Arg16Gly and Gln27Glu polymorphisms of β_2-AR gene with the development of HAPH in Kyrgyz highlanders.

Kyrgyz highlanders with HAPH were found to be more commonly homozygous for Gln27Gln compared to controls. The genotype Glu27Glu was more frequent in healthy highlanders compared to patients with HAPH. There was a trend towards a higher frequency of Gln27 allele in the HAPH group compared to controls (0.78 vs 0.66). The frequencies of Arg16 and Gly16 alleles in HAPH and controls were not significantly different.

The higher prevalence of the Gln27Gln genotypes in HAPH is consistent with our finding that HAPH patients have decreased response of β_2-AR to agonists. Decreased activity of β_2-AR caused the decreased vasodilatation response of pulmonary arteries to catecholamines and excessive pulmonary vasoconstriction in chronic hypoxia in Gln27 carriers.

Several studies have reported that serotonin (5-hydroxytryptamine, 5-HT) and its transporter (5-HTT) play a critical role in the pulmonary vascular smooth muscle hyperplasia and vascular remodeling associated with experimental hypoxic pulmonary hypertension (PH) and human primary PH. High expression of 5-HTT in pulmonary vessels of patients with idiopathic pulmonary arterial hypertension was associated with the insertion/deletion polymorphism of the promoter region of the 5-HTT gene. The long (L) allele is known to increase the transcription rate of 5-HTT gene and short (S) allele is associated with decreased expression of the transporter. The LL-genotype is associated with high level of 5-HTT expression and increased proliferation of pulmonary artery smooth muscle cells (PASMC). A large cohort study including 166 cases of familial pulmonary arterial hypertension and 83 idiopathic pulmonary arterial hypertension subjects has demonstrated that the 5HTT gene polymorphism does not correlate with age and diagnosis or survival of patients with idiopathic pulmonary arterial hypertension and only in familial pulmonary arterial hypertension patients did the LL genotype correlate with age at diagnosis. Another large cohort study did not find any significant evidence of association between the 5HTT gene and pulmonary arterial hypertension.

We have found that in Kyrgyz population the LL -genotype is very rare. We did not find any association of the L allele with HAPH. Moreover, the expression of 5HTT protein in platelets of HAPH patients and healthy Kyrgyz highlanders was not different.

Nitric oxide (NO) is a small bioactive molecule that is involved in a variety of physiological functions, including maintaining pulmonary vascular tone, and may play a role in altitude adaptation. High altitude indigenous populations e.g. Tibetans appear to maintain higher levels of exhaled NO, whereas HAPE-susceptible individuals have lower levels of exhaled NO. NO is synthesized from the amino acid L-arginine by a family of enzymes called nitric oxide synthases (NOS), including NOS3 (or endothelial NOS, eNOS) in the walls of blood vessels. There are known polymorphisms in the NOS3 gene which are associated with modulation of NO levels and with cardiovascular pathology: 4a tandem repeats which has been associated with lower exhaled NO and the G894T mutation which alters the amino acid sequence Glu298Asp and has been associated with hypertension. Association studies of NOS3 allele's distribution in HAPE patients have reported that the 4a and T894 were more common in Asian subjects with HAPE. Conversely, Weiss et al. (2003) found no association in Caucasians. The reason for this disagreement is unclear because the 4b allele is also associated with increased NO levels in Caucasians as well. The Kyrgyz population, which is also of Asian ancestry, shows no association of 4a and T894 alleles with HAPH. Interestingly the frequency of mutant eNOS alleles is very low in Kyrgyz highlanders, perhaps due to selection pressure against genotypes associated with HAPE or HACE.

The key gene which controls the expression of all hypoxia-regulated genes is HIF1a. We have studied the association of C1772T polymorphism of HIF1a gene with HAPH and did not find any association. C1772T mutation leads to substitution of Pro residue which is substrate for prolyl hydroxylase tagging of HIF1a for protease cleavage. In population of Kyrgyz highlanders the frequency of mutant allele was 24-times lower and no any 1772T/T mutant genotype was found. We have speculated that mutations of the main intracellular hypoxia-operating gene should be eliminated off the genome of the population of indigenous high-altitude dwellers.

While epidemiological data support the hypothesis that some humans are particularly susceptible to hypoxia, there are few genetic data to identify a root genetic cause. Few reliable data describe familial patterns of susceptibility and, although a number of genetic variants have been associated with high-altitude illnesses, the results are inconsistent between studies. The problem with HAPH is that it is a chronic not an acute illness, with a slow onset and that presents later in life. By the time a patient is diagnosed, the parents are usually deceased and the children not old enough to express the disease. In generally, family data for environmentally related illnesses can be difficult to obtain and genetic analysis of HAPH is in its infancy. The cohorts of well-phenotyped patients available to unpick the genetics are small and this limits the power of candidate gene studies.

But new developments genome studies, such as SNP databases and genome-wide haplotype mapping, and efficient new noninvasive methods of assessing pulmonary artery pressure, alongside increased understanding of the molecular mechanisms of oxygen sensing and hypoxic pulmonary vasoconstriction and vascular remodeling. These will provide opportunities to further our understanding of pathogenesis of HAPH, early diagnostics and developing of new remedies for its therapy.

References

Ahsan, A., Charu, R., Pasha, M.A., Norboo, T., Charu, R., Afrin, F., Ahsan, A., and Baig, M.A. 2004. eNOS allelic variants at the same locus associate with HAPE and adaptation. Thorax. 59, 1000–1002.

Aldashev, A.A. 1993. High altitude pulmonary hypertension and signal transduction in the cardiovascular system. In: Brody, J.S., Center, D.M., Tkachuk, V.A., eds. Signal transduction in Lung Cells, Vol. 65. New-York: Lung biology in health and disease, pp 459-482.

Aldashev, A.A. 2007. Gene polymorphism and high altitude pulmonary hypertension. Problems of High Altitude Medicine and biology. Springer. UK, pp. 151-168.

Aldashev, A.A., Borbugulov, U., Davletov, B., Mirrakhimov, M.M. 1989. Human adrenoceptor system response to the development of high altitude pulmonary arterial hypertension. J. Mol. Cell Cardiol. 21(1), 175-179.

Aldashev, A.A., Kojonazarov, B.K., Amatov, T.A., Sooronbaev, T.M., Mirrakhimov, M.M., Morrell, N.W., Wharton, J., Wilkins, M.R. 2005. Phosphodiesterase type 5 and high altitude pulmonary hypertension. Thorax. 60 (8), 683-687.

Aldashev, A.A., Sarybaev, A.Sh., Sydykov A.S., Kalmyrzaev, B.B., Kim, E.V., Mamanova, L.B., Maripov, R., Kojonazarov, B.K., Mirrakhimov, M.M., Wilkins, M.R, and Morrell, N.W. 2002. Characterization of High-Altitude Pulmonary Hypertension in the Kyrgyz: Association with Angiotensin-Converting Enzyme Genotype. Am. J. Respir. Crit. Care Med. 166, 1396-1402.

Beall C.M. 2000. Tibetan and Andean patterns of adaptation to high-altitude hypoxia. Hum. Biol. 72 (1), 201–228.

Beall, C.M., Andean. 2006. Tibet, and Ethiopian Patterns of Adaptation to High-Altitude Hypoxia. Journal of Integrative and Comparative Biology, Advance Access, 6, 1-7.

Busch, T., Bartsch, P., Pappert, D., Grunig, E., Hildebrandt, W., Elser, H., Falke, K.J., and Swenson, E.R. 2001. Hypoxia decreases exhaled nitric oxide in mountaineers susceptible to high-altitude pulmonary edema. Am. J. Respir. Crit. Care Med. 163, 368–373.

Charu, R., Stobdan, T., Ram, R.B., Khan, A.P., Qadar Pasha M.A., Norboo, T., Afrin, F. 2006. Susceptibility to high altitude pulmonary oedema: role of ACE and ET-1 polymorphisms.Thorax. 61(11), 1011-1012.

Cockcroft, J.R., Gazis, A.G., Cross, D.J., Wheatley, A., Dewar, J., Hall, I.P., Noon, J.P. (2000). Beta(2)-adrenoceptor polymorphism determines vascular reactivity in humans. Hypertension 36(3), 371-375

Dehnert,C., Weymann, J., Montgomery, H.E., Woods, D., Maggiorini, M., Scherrer, U., Gibbs, J.S and Bartsch, P. 2002. No association between high-altitude tolerance and the ACE I/D gene polymorphism. Med. Sci. Sports Exerc. 34, 1928–1933.

Droma, Y., Hanaoka, M., Hotta, J., Katsuyama, Y., Ota, M., Kobayashi, T. and Kubo, K. 2003. The r506 Q mutation of coagulation factor V gene in high altitude pulmonary-edema-susceptible subjects. High Alt. Med. Biol. 4(4), 497-498.

Eddahibi, S., Hanoun, N., Lanfumey, L., et al. 2000. Attenuated hypoxic pulmonary hypertension in mice lacking the 5-hydroxytryptamine transporter gene. J. Clin. Invest. 105, 1555–1562.

Eddahibi, S., Humbert, M., Fadel, E., et al. 2001. Serotonin transporter over expression is responsible for pulmonary artery smooth muscle hyperplasia in primary pulmonary hypertension. J. Clin. Invest. 108, 1141–1150.

Garruto, R.M., Chin, C.T., Weitz, C.A., Liu, J.C., Liu, R.L. and He, X. 2003. Hematological differences during growth among Tibetans and Han Chinese born and raised at high altitude in Qinghai, China. Am. J. Phys. Anthropol. 122, 171–183.

Ge, R.L. and Helun, H. 2001. Current concept of chronic mountain sickness: pulmonary hypertension-related high-altitude heart disease. Wilderness Environ. Med. 12, 190–194.

Gupta, M., Rao, K., Anand, I., Banerjee, A. and Boparai, M. 1992. Lack of smooth muscle in the small pulmonary arteries of the native Ladakhi. Is the Himalayan highlander adapted? Am. Rev. Respir. Dis. 145, 1201–1214.

Hingorani, A.D. 2003. Endothelial nitric oxide synthase polymorphisms and hypertension. Curr. Hypertens. Rep., 519–525.

Hoffmann, I.S., Tavares-Mordwinkin, R., Castejon, A.M., Alfieri, A.B. and Cubeddu, L.X. 2005. Endothelial nitric oxide synthase polymorphism, nitric oxide production, salt sensitivity and cardiovascular risk factors in Hispanics. J. Hum. Hypertens. 19, 233–240.

Hotta, J., Hanaoka, M., Droma, Y., Katsuyama, Y., Ota, M. and Kobayashi, T. 2004. Polymorphisms of renin–angiotensin system genes with high-altitude pulmonary edema in Japanese subjects. Chest. 126, 825–830.

Kojonazarov, B.K., Imanov, B.Z., Amatov, T.A., Mirrakhimov, M.M., Naeije, R., Wilkins, M.R., Aldashev, A.A. 2007. Noninvasive and invasive evaluation of pulmonary arterial pressure in highlanders. Eur. Respir. J. 29(2), 352-356.

Kumar, R., Pasha, Q., Khan, A.P. and Gupta, V. 2004. Renin angiotensin aldosterone system and ACE I/D gene polymorphism in high-altitude pulmonary edema. Aviat. Space Environ. Med. 75, 981–983.

León-Velarde, F., Maggiorini, M., Reeves, J.T., Aldashev, A., Asmus, I., Bernardi, L., Ge, R.L., Hackett, P., Kobayashi, T., Moore, L.G., Penaloza, D., Richalet, J.P., Roach, R., Wu, T., Vargas, E., Zubieta-Castillo, G., Zubieta-Calleja, G. 2005. Consensus statement on chronic and subacute high altitude diseases. High Alt. Med. Biol. 6(2), 147-57.

Lesch, K., Bengel, D., Heils, A., et al. 1996. Association of anxiety-related traits with a polymorphism in the serotonin transporter gene regulatory region. Science. 274, 1527–1531.

Machado, R.D., Koehler, R., Glissmeyer, E., Veal, C., Suntharalingam, J., Kim, M., Carlquist, J., Town, M., Elliott, C.G., Hoeper, M., Fijalkowska, A., Kurzyna, M., Thomson, J.R., Gibbs, S.R., Wilkins, M.R., Seeger, W., Morrell, N.W., Gruenig, E., Trembath, R.C., Janssen, B. 2006. Genetic association of the serotonin transporter in pulmonary arterial hypertension. Am. J. Respir. Crit. Care Med. 173(7), 793-797.

Milledge, J.S. and Catley, D.M. 1984. Angiotensin converting enzyme response to hypoxia in man: its role in altitude acclimatization. Clin. Sci. 67, 453–456.

Mirrakhimov, M., Winslow, R. 1996. The cardiovascular system at high altitude. Environmental physiology. Edited by Fregley, M., Blatteal, C. Oxford: Oxford University Press, pp. 1241-1258.

Montgomery, H.E., Marshall, R., Hemingway, H., Myerson, S., Clarkson, P., Dollery, C., Hayward, M., Holliman, D.E., Jubb, M., World, M., et al. 1997. Human gene for physical performance. Nature. 393, 221–222.

Moore, L.G. 2001. Human genetic adaptation to high altitude. High Alt. Med. Biol. 2, 257–279.

Morrell, N.W., Sarybaev, A.S., Alikhan, A., Mirrakhimov, M.M. and Aldashev, A.A. 1999. ACE genotype and risk of high altitude pulmonary hypertension in Kyrghyz highlanders. Lancet, pp. 353: 814.

Moudgil,R., Michelakis, E.D. and Archer, S.L. 2005. Hypoxic pulmonary vasoconstriction. J. Appl. Physiol. 98, 390–403.

Percy, M.J., Mooney, S.M., McMullin, M.F., Flores, A., Lappin, T.R., Lee, F.S. 2003. A common polymorphism in the oxygen-dependent degradation (ODD) domain of hypoxia inducible factor-1alpha (HIF-1alpha) does not impair Pro-564 hydroxylation Mol. Cancer. 2(1), 31.

Rabinovitch, M., Mullen, M., Rosenberg, H.C., Maruyama, K., O'Brodovich, H., Olley, P.M. 1988. Angiotensin II prevents hypoxic pulmonary hypertension and vascular changes in rat. Am. J. Physiol. 254(3), H500-508.

Rupert, J.L., Koehle, M.S. 2006. Evidence for genetic basis for altitude-related illness.High Alt. med. Biol. 7(2), 150-167.

Sagnella, G.A., Rotwell, M.J., Opinipla, A.K., Wick, P.D., Gook, D.G., Capuccio, F.P. 1999. A population study of ethnic variations in the angiotensin-converting enzyme I/D polymorphism: relationships with gender, hypertension and impaired qlucose metabolism. J. Hypertens. 17(5), 657-664.

Sayed-Tabatabaei, F.A., Schut, A.F., Hofman, A., Bertoli-Avella, A.M., Vergeer, J., Witteman, J.C. and van Duijn C.M. 2004. A study of gene–environment interaction on the gene for angiotensin converting enzyme: a combined functional and population based approach. J. Med. Genetics. 41, 99-103.

Su, B., Xiao, C., Deka, R., Seielstad, M.T., Kangwanpong, D., Xiao, J., Lu, D., Underhill, P., Cavalli-Sforza, L., Chakraborty, R. and Jin, L. 2000. Y chromosome haplotypes reveal prehistorical migrations to the Himalayas. Hum. Genetics. 107(6), 582–590.

Taraseviciene-Stewart, L., Scerbavicius, R., Stewart, J.M., Gera, L., Demura, Y., Cool, C., Kasper, M., Voelkel, N.F. 2005. Treatment of severe pulmonary hypertension: a bradykinin receptor 2 agonist B9972 causes reduction of pulmonary artery pressure and right ventricular hypertrophy. Peptides. 1292-3000.

Thomas, S., Bruce, C., Birkhead, A. and Wang, L. 2002. Effect of eNOS polymorphisms and coronary artery disease upon exhaled nitric oxide. J. Mol. Med. 80, 181–186.

Tsianos, G., Eleftheriou, K.I., Hawe, E., Woolrich, L., Watt, M., Watt, I., Peacock, A., Montgomery, H. and Grant, S. 2005. Performance at altitude and angiotensin I-converting enzyme genotype. Eur. J. Appl. Physiol. 93, 630–633.

van Suylen, R.J., Wouters, E.F., Pennings, H.J., Cheriex, E.C., van Pol, P.E., Ambergen, A.W., Vermelis, A.M., Daemen, M.J. 1999 .The DD genotype of the angiotensin converting enzyme gene is negatively associated with right ventricular hypertrophy in male patients with chronic obstructive pulmonary disease. Am. J. Respir. Crit. Care Med. 159(6), 1791-1795.

von Euler, U.S., Liljestrand, G. 1946. Observations on the pulmonary aterial blood pressure in the cat. Acta Physiol. Scand. 12, 301–320.

Wang, G.L., Semenza, G.L. 1993. General involvement of hypoxia-inducible factor 1 in transcriptional response to hypoxia. Proc. Natl. Acad. Sci. USA, 90(9), 4304-4308.

Weir, E.K., Tucker, A., Reeves, J.T., Will, D.H., Grover, R.F. 1974. The genetic factor influencing pulmonary hypertension in cattle at high altitude. Cardiovasc. Res. 8, 745 - 749.

Weiss, J., Haefeli, W.E., Gasse, C., Hoffmann, M.M., Weyman, J., Gibbs, S., Mansmann, U. and Bartsch, P. 2003. Lack of evidence for association of high altitude pulmonary edema and polymorphisms of the NO pathway. High Alt. Med. Biol. 4, 355–366.

Weitzenblum, E., Mammosser, M., Ehrhart, M. 1986. Evolution and prognosis of pulmonary hypertension in chronic obstructive pulmonary disease. Herz. 11, 147–154.

Wilkins, M.R., Aldashev, A.A., Morrell, N.W. 2002. Nitric oxide, phosphodiesterase inhibition, and adaptation to hypoxic conditions. Lancet. 359(9317), 1539-1540.

Willers, E.D., Newman, J.H., Loyd, J.E., Robbins, I.M., Wheeler, L.A., Prince, M.A., Stanton, K.C., Cogan, J.A., Runo, J.R., Byrne, D., Humbert, M., Simonneau, G., Sztrymf, B., Morse, J.A., Knowles, J.A., Roberts, K.E., McElroy, J.J., Barst, R.J., Phillips, J.A. 2005. Serotonin transporter polymorphisms in familial and idiopathic pulmonary arterial hypertension.Am. J. Respir. Crit. Care Med. 173(7), 798-802.

Wu, T., Wang, X., Wei, C., Cheng, H., Wang, X., Li, Y., Ge, D., Zhao, H., Young, P., Li, G. and Wang, Z. 2005. Hemoglobin levels in Qinghai–Tibet: different effects of gender for Tibetans vs. Han. J. Appl. Physiol. 98(2), 598–604.

Zhu, X., Bouzekri, N., Southam, L., Cooper, R., Adeyemo, A., McKenzie, C., Luke, A., Chen, G., Elston, R., Ward, R. 2001. Linkage and association analyses of angiotensin I-converting enzyme (ACE) gene polymorphism with ACE concentration and blood pressure. Am. J. Hum. Genet. 68, 1139-1148.

□□□

38

Results of Multicentric Trials of Novel High Altitude Products

R.P. TRIPATHI, ASEEM BHATNAGAR
Institute of Nuclear Medicine & Allied Sciences, DRDO, New Delhi

ABSTRACT

High altitude, particularly extreme altitude, presents a number of medical challenges to nullify the mortality, counter morbidity and to maintain optimal troop efficiency. These challenges persist despite medical advancements made in the area in the last few decades, calling for low cost, practical innovative out-of-box solutions. INMAS, DRDO, while focusing on deliverable-oriented research, has developed several concept-based biomedical products (n > 15, presently) applicable for high altitude medicine. Their efficacy is now supported by data from clinical trials, including user trials. Apart from novel products, few technologies and medical protocols are being developed. The focus of our research is on preventive aspects of medicine and increasing the comfort level and efficiency of the soldiers rather than on therapeutics aspects alone.

This presentation describes only those research products developed at INMAS that have reached a certain level of development with demonstrable proof of human safety & efficacy. In several cases, regulatory clearance and capacity to deliver sufficient number of prototype samples to the prospective users has been developed.

These novel products have been categorized into the following groups: a) Inhalation Therapy for Alveolar Deposition (ITAD) products, b) cold injury ointments, c) local acting gastric prokinator drugs, d) other products, e) new and important technologies. All have relevance and applicability at high altitude under various stressors. Many of these also have other spin-off advantages.

Inhalation Therapy for Alveolar Deposition (ITAD)

Most of the mountain ills can be traced back to pulmonary hypertension and reduced oxygen transfer across the lung membrane. Systemic therapy has already been reported unsuccessful in practice. When drug aerosols are inhaled using conventional systems, not more than 5-10% reaches the alveolar region, rendering this an inappropriate route for pharmaceutics delivery. For this reason, the use of conventional delivery systems like nebulisation chamber and Dry Powder Inhalation (DPI) is limited for treating tracheo-bronchiolar area. We have used nanotechnology principles and inexpensive pharmaceutical engineering methods with the delivery system (we call these Anukool I & II) as well as the respired formulation to produce 5 products which reduce pulmonary hypertension, and increase 'blood oxygen without giving oxygen'. These drugs, essentially based on vasodilator principles, have been found effective in management of high altitude pulmonary hypertension, non-acclimatized status, acute mountain sickness, and HAPE, apart from various other medical indications. Clinical studies, backed by appropriate imaging methods, have shown a response rate of 70-100% in different clinical situations. A number of clinical partners, including ITBP, 14 Corps, HAMRC, Patel Chest Institute, Directorate of Health, Himachal Govt., and PGI Chandigarh are participating in clinical trials which have been approved by Drug Controller General of India. In total, more than 800 case studies have been done till now. Results are definitive that their use as an emergency or maintenance therapy at HAA troops will be quite useful.

Abridged results of a few studies conducted successfully till now include:

a) REPORT 1 : Treatment of AMS by ITAD therapy

Benign acute mountain sickness is very frequent at high altitude. Presently acetazolamide based therapy is prescribed which is not very effective. We recently observed that adding ethanol to salbutamol

Table 1: Comparison of quantitative parameters of Group A(test formulation) and Group B (conventional treatment)

Parameter	Grp A (pre-treatment)	(48h)	Grp B (pretreatment)	(48 h)
LL Score Response	5.01	0.9*1	3.81	1.87*2
Rate(LL < 1) (36 h)Response		79.5%		10 %**1
Rate (48 h)		96%		18%**2
Mean Hospital stay		2 days		3 days**3
SPaO2	88.3%	93.64%*3	91.46%	90.12%*4
HR/min	85.15	79.5%*5	83.65	80.59*6
Systolic BP	127	125.7	124.18	128
Diastolic BP	98.68	84.95*7	83.06	84.31
Walk test Oxygen dip		- 1.75%		- 1.87%**4
Walk test HR increase		+ 4.75		+ 7.35%**5

Intragroup comparison
*1 P < .01
*2 P < .05
*3 P < .01
*4 P > .05 (insignificant)
*5 P < .05
*6 P > .05 (insignificant)
*7 P < .01

Intergroup comparison
**1 P < .001
**2 P < .001
**3 P < .01
**4 P > .05
**5 P < .05

sulphate respiratory fluid and using a large spacer produces submicron sized drug aerosols with enhanced lung deposition. This may results in reduced pulmonary artery pressure and increased blood oxygen. *Objectives:* to introduce 0.25% salbutamol sulphate Inhalation Therapy for Alveolar Deposition (ITAD) formulation for treating benign AMS, and to conduct phase-2 clinical trial with the formulation as a stand-alone therapy in comparison with acetazolamide based conventional treatment. *Methods*: After developing and characterizing the new salbutamol-sulphate formulation that has the property to produce submicron-sized or nano-sized inhalable particles upon nebulization, the test formulation was given as a stand-alone therapy to four times a day for 2-4 days to 44 subjects with benign mountain sickness. Control group of 32 AMS subjects was given acetazolamide-based oral treatment for the same duration. Efficacy of the two treatments was compared using Lake Louise score, pulse-oxymetry, hospital stay,1-min walk test, and response rate at 36 and 48 h. Response rate was defined as complete disappearance of symptoms. *Results*: In 48 h, Lake Louise score reduced very significantly in test group compared to conventional treatment (0.9 after 48 h from 5.1 compared to 1.9 from 3.8 using acetazalamide)(p <0.001). Complete response with the test formulation was significantly faster than control treatment at 36 & 48 h (p <0.001) resulting in shorter hospital stay (p < 0.01). Blood oxygen improved significantly with test formulation though the effect on heart rate and exercise capacity was marginal. Significantly, diastolic pressure also reduced within 2 days. The study suggests that stand-alone treatment with nebulized submicron-sized salbutamol sulphate is significantly superior to acetazolamide-based oral treatment for managing benign acute mountain sickness in medical units at high altitude.

b) REPORT 2: Use of salbutamol ITAD in patients of HAPE

Inhalation Therapy for Alveolar Deposition (ITAD) has been developed by INMAS and this device was given to HAMRC for field applications. From the studies conducted by INMAS it was concluded that ITAD of Salbutamol therapy increases blood oxygen by 7%, reduces heart rate by 8% and reduces pulmonary hypertension by 20-25%.

Aim

To evaluate the ITAD system in patients of HAPE

To modify the current treatment protocol of HAPE

MATERIALS AND METHODS

The ITAD system was evaluated for assessing the amelioration of symptoms and clinical improvement in patients of HAPE.

This study was conducted in 32 patients of HAPE admitted to Field Hospital at Leh for a period of two months. ITAD was given every 4 h along with the standard treatment protocol i.e. Disprin, Diamox and Oxygen. No other vasodilators or bronchodilators were given.

All the individuals were studied after taking written consent. Anthropometric measurements like height (Ht), weight (Wt) and body composition (BC) by impedance method was taken. Heart rate (HR), SpO_2 (Percentage saturation of Hb), Systolic blood pressure (SBP), Diastolic Blood pressure (DBP) and End tidal CO_2 ($EtCO_2$) were measured with a physiological Multiparameter monitor (MPM 5533, BPL, India). Pulmonary Artery Pressure (PAP) was non-invasively estimated by Doppler echocardiographic methods by Mylab CV 30 (Esaote, Italy). This group was compared to another control group of 32 age matched patients of HAPE who were on conventional treatment only. The ITAD group was studied on admission, after first ITAD (ITAD1), after ITAD which was administered 12 hrs after admission (ITAD3), on Day2 and 7 of admission. The control group was studied on Day of admission, Day2 and Day7.

The data was tabulated and analyzed student's unpaired t test for inter group variation and paired t test for intra group temporal variation of the variables being studied. $P<0.05$ was considered significant.

Table 1. Hape Patients on ITAD Salbutamol

	HR	SPO2	SBP	DBP	PAP	RR
On Admission	101.78	67.89	128.94	81.94	37.92	31.54
	±16.89	±9.1	±14.72	±10.17	±19.43	±11.32
After ITAD1	87.82	90.92	125.83	80.32	32.56	20.87
	±14.29	±2.1	±12.87	±6.32	±4.6	±8.3
After ITAD3	89.42	89.52	122.90	78.48	30.41	23.65
	±12.98	±3.6	±10.28	±10.38	±4.1	±3.89
Day 2	86.92	90.00	125.98	85.67	28.74	22.48
	±10.11	±2.81	±11.93	±5.92	±3.8	±6.23
Day 7	90.12	88.92	130.12	88.87	29.52	18.03
	±10.36	±4.26	±4.31	±5.61	±4.1	±4.81

Table 2. Hape Patients on Conventional Treatment

	HR	SPO_2	SBP	DBP	PAP	RR
On Admission	103.26	75.38	124.90	80.71	35.83	28.58
	±21.13	±13.52	±17.97	±11.84	±21.04	±8.31
Day 2	92.54	79.43	123.89	83.58	30.20	21.69
	±7.98	±11.84	±15.27	±9.54	±3.7	±6.93
Day 7	92.34	90.68	127.23	87.32	29.12	18.92
	±6.82	±3.1	±11.86	±7.38	±5.3	±5.38

Cold Injury Ointments

For want of better medications which have undergone rigorous medical scrutiny, troops have been using things like thrombophobe and a plethora of non-conforming medications for frostbite and other significant cold-injuries.

Instead of treating frostbite, a problem which is diminishing to insignificance in the Indian troops, we have focused on preventing / treating *mild* cold injuries. Virtually every body who is suffering from pain-numbness in the hand & feet at sub-zero temperature (which means a huge number of troops) will benefit from the medication, feel more comfortable and perform the tasks better. Thus, the focus has shifted from treating a severe condition involving a miniscule number of troops to enhancing the comfort level and efficiency in a huge number of subjects.

INMAS, DRDO has developed three ointment creams with different clinical indications which have been found useful in temperatures ranging from 0 to -40 C. Clinical response rates have ranged from 80-100% in the continuing MCT. Non-invasive imaging has been used to confirm deep penetration and protection of muscle tissue from the cold induced ischemia. The ointments are based on the concept of local action and contain several active molecules working towards negating the pathological changes expected in frostbite and other severe cold injuries. More than 400 subjects have already benefited in the clinical trials conducted by ITBP & HP Govt, which has been approved by the DCGI and its expert body. There is evidence to suggest that sensory-motor function of hand and feet improves dramatically after a few applications of the medications. We have reasons to believe that the ointments will be particularly beneficial for heli-pilots, motor vehicle drivers, sentries, LRPs, SRP troops in cold weather and in all troops in extreme cold weather at high altitude. Recently, pharma major Cipla manufactured and supplied 18000 ointment tubes for use in 14 Corps troops under a MoU with INMAS and helped establish the technology for continued defence use.

c) REPORT 3: Treatment of mild cold injury by Prefros ointment (0.5% salbutamol sulphate ointment)

The aim of the study was to evaluate One-week long treatment with Prefros ointment developed by INMAS, DRDO for treating mild to moderate cold injuries. Subjects were also drawn from sector Hq, Leh and two long range petrols (LRPs) controlled from the base center but ascending to 14000-16000 feet for 10 days. The study was conducted from December 2007 to March 2008. The ambient temperature ranged from -15 C to -26 C at Leh and -20 to -35 C in LRPs excluding chill factor. The 70 subjects at Leh (64 males, 6 females) completed the study while all 25 subjects in two LRPs (10 and 15 subjects, all males) completed the study. The mean age was 31 years +/-8.3. They had passed medical fitness test prior to the ascent. The four participating females were house-wives. All subjects had either mild (grade-1) or moderate (grade-2) cold injury. Grade-1 cold injury was defined as a) painful extremities attributable to the cold weather, b) numbness in the fingers & toes attributable to the cold, c) symptoms that get worse on nicotine / exposure and get better after receiving physical protection/external warmth. Grade-2 cold injury was defines as a) edema/blanching/cyanosis in fingers and toes attributable to the cold, and b) symptoms that got worse on nicotine / exposure and got better after receiving physical protection/external warmth. The subjects were instructed to apply the ointment twice-a-day on the active hand (and on one foot optionally). After 3-5 days, the subject had an option of using the ointment on all four extremities in addition to the nose and earlobes for another 5-7 days.

Response Assessment Methodology

Treatment response was assessed mid-course and at the end of study period semi-quantitatively by response in terms of a) individual sign and symptoms (nil, partial, complete, as per grading), b) skin temperature rise noted by thermometer or handshake test, c) HECT quantitative assessment, d) over all response rate, e) subjective feeling of the subjects, f) evidence of increase in comfort level and efficiency of performing work, and, f) adverse effects. Individual sign and symptoms consisted of a) pain/sensation in the ungloved and gloved hands, and on dipping hand in water (on scale 1-5 in control & test hand), b) Numbness, itching, edema, cyanosis on scale of 0-2 scale. Following the end of initial study period (3-5 days), assessment of improved signs or symptoms continued on individual basis in subjects where it was possible to continue the study.

Result Summary

- 4 of 95 persons had no response though their HECT test was positive suggesting improved hand-eye coordination / neuro-muscular function.
- 91 / 95 had total or definitive pain relief in 3 days.
- Response rate of itching was > 90% (3-5 days)
- *Numbness* : 85% (5-7 days). Almost all had improved hand-shake or HECT.
- *Cyanosis*: 3/4 in 5 days (complete response). Edema: 20/28 (complete response in 5-7 days). Others partial response in 1 week, who continued further treatment.

Instant Locally Acting Gastric Prokinetic Agents

Chronic gastric symptoms like dyspepsia, gastritis, nausea, gastric bloating, loss of appetite and consequent weight loss are very frequent feature in troops at high altitude, and represent a 'permanent' feature of life at significant altitudes. Loss of the moral and poor general health is a spin-off feeding on the original malaise of significant reduction in gastric emptying. Systematic and centrally acting drugs like domperidon and metochlopramide are not a viable option due to many reasons.

INMAS recently introduced two locally acting formulations that increase the gastric emptying time instantly and for short duration only to the extent of 200-500% for solids, and even more for liquid diet in a safe and effective physiological way. The two formulations act at tandem and are indicated as a combination in more severe cases. In mild cases, only one may suffice. Recently, 100% antinausea and 'anti-dyspeptic' effect was observed with a single dose in 50 cases with suspected acute mountain sickness at the altitude of 14500 feet in ITBP jawans, and it appeared to improve appetite the next morning. Though it appears that the formulations will be useful as a short term therapy in non-acclimatized status & acute mountain sickness, and as a long term maintenance therapy in troops residing at extreme altitudes / suffering from weight loss, definitive answers are expected only after one year of phase-2 trials presently under way at multiple centers.

d) REPORT 4: Use of gastric prokinetic agent (low dose locally acting nifedipine 500 mcg) in acute mountain sickness

Concept: A nifedipine based formulation has been invented that relaxes gastric pylorus and duodenal smooth muscles by virtue or local action. Increase in gastric empting upto 500% times has been observed on scintigraphy. At high altitude, one of the main causes of gastric problems associated with non-acclimatized status is increase in gastric retension, a phenomenon related to hypoxic conditions.

Aim: To evaluate the advantages of a single dose of the formulation in reducing AMS-related gastric symptomatology at a Himalayan plateau at 14500 feet. *Materials and methods:* Inductees arriving by road need to stay at this location for the night, starting from 6000 feet in the morning. More than 85% persons suffer from AMS at night, before proceeding to the destination next morning. Apart from substantial lake Louise score, these subjects have gastritis, dyspepsia, bloating of stomach, nausea-vomiting and other GIT related problems. To 50 of these inductees, one tablet of the novel prokinetic agent was given at night after noting the score on a questionnaire focusing on gastric symptoms only. To another 50 persons, no treatment was offered except SoS analgesics for treating headache. The score was compared the next morning w.r.t. the gastric symptoms the next morning, apart from general predisposition on a score of 1-10.

Results

- 48 / 50 subjects given the normal treatment had subjective relief next morning compared to 32/50 subjects receiving no treatment.
- Mean predisposition score in the treatment group was 7.5 compared to 5.3 for the control group. ($p < 0.01$)

- Gastric fullness and bloating reduced in almost all respondents receiving test medicine compared to 50% persistence in control group. Nausea reduction and apatite improvement were also significantly high (90% Vs 46%, 82% Vs 56%) with the test drug. Other factors were equally improved.

CONCLUSIONS

One or two tablets of the novel formulation just on arrival of the non-acclimatized persons at high altitude is helpful in reducing the gastric symptoms by the next morning.

Other Products & Processes

These include the following:

Light weight foldable stretcher: Technology to make LWFS has been matured and transferred to the industry. It is made of composite light weight material good enough to lift 120 kg. A conventional stretcher costs Rs 12000/- and weighs 10 kg or more. INMAS stretcher costs Rs 7000/- and weighs 6.6 kg. It can be folded into a back-pack and is ideal for carrying over mountains. Also, its material is non-conducting and does not feel extreme cold to touch like metals. INMAS has transferred the technology to two firms that can supply the product on demand.

Herbal UV protective cream: UV rays at high altitude are materially different than that present at plains. Consequently, commercial UV protective creams may not be useful at high altitude. Since UV skin exposure and consequent health problems in soldiers is very common, INMAS has developed a combination herbal cream. Its safety-efficacy on animals has been proved and clinical trials are planned with approval of army authorities. It will be an important advancement for this problem which persists despite all measures at high altitude.

High altitude boiling device: Food is improperly cooked at extreme altitude and precious fuel is lost due to low atmospheric pressure and extreme cold in the mountains. Internal temperature in pressure utensils is 85C or so. This contributes to poor digestion at posts located at extreme altitude. INMAS, with inputs from other labs, is developing a special high pressure utensil that is able to produce 115C temperature inside the utensil, making the procedure as easy as in plains. INMAS shall be shortly supplying 25-50 such utensils to the forces for user trials.

Carbon monoxide toxicity at high altitude: CO toxicity in confined spaces in cold weather, such as in rooms with bukharis and stoves, is still an unresolved area, causing serious acute or subacute sequele. Unfortunately limited research has been done in this important area. INMAS has initiated a multipronged research strategy: a) survey to detect the extent and seriousness of the problem, b) methods to reduce kerosene fume and CO concentration in the ambient atmosphere, c) methods / kit to quantify carboxy-Hb concentration, and d) methods to reduce carboxy-Hb burden in the body.

Other products and technologies : A number of other initiatives are being taken. To name a few: a) high altitude medical advisory, b)symposia on high altitude medical problems where direct communication is done with the jawans, c) methods to reduce the incidence of coaggulopathies.

Extracts from the Advisory

Approved treatment regimes for high altitude maladies

1. Tab acetazolamide, 250 mg t.d.s. for 2-3 days
2. Antacids like Gelucil & Digene, for 2-3 days.
3. H-2 blockers like ranitidine and omaprazole
4. Prefer paracetamol over other pain killers.
5. Prefer anti-nausea-analgesic combinations over plain pain killers
6. Avoid sedatives
7. Antistress / antioxidant medications (vit E, vit C, flavonoids) may be useful.
8. Micronutrients like Zinc, Selenium, Cobalt, may be useful
9. Chlorine tablets / boiled water should be insisted upon in small places and camps using raw water.
10. Avoid antihypertensives/ heart medicine just after landing

New treatment regimes (suggested by INMAS – presently in design / trial stage)

1. Methergin t.d.s. for 2-3 days, *for preventing / treating AMS*
2. Prokinetic agents like Domperidon *for treating nausea*
3. Probiotic formulations *for medium- and long-term gastric problems*
4. Oral Glycerol / intravenous mannitol *for treating altitude headache*

5. Anaerobe chemotherapy such as metronidazole for treating gastrointestinal problems as pulse therapy
6. Melatonin for sleep disturbance
7. ITAD therapy for treating AMS, HAPO, HACE
8. Gastric prokinetic agents for treating high altitude gastric problems
9. Cold injury ointments - for prevention and treatment.

Some important Do's and Don'ts mentioned in the Advisory include:

Travel

1. Break journey based on the final destination altitude
2. Reduce water intake to actual need
3. Don't overeat
4. Keep feet up periodically if possible, otherwise fold them
5. Travel light
6. Cover head with blanket / quilt during travel frequently
7. Hyperventilation and breathing exercises may be performed during travel
8. Eat honey before crossing high passes
9. Breath breaking (apnea) at full inspiration rather than at full expiration is useful at high passes and as a frequent respiratory exercise.

Food

1. Small multiple meals instead of three conventional meals when non-acclimatized
2. Early dinner / liquids only at night during acclimatization, extreme altitude
3. Breakfast heaviest, lunch moderate, dinner measly
4. Less water intake / mild dehydration better in non-acclimatized phase

5. Low protein (reverse of Atkin's diet), low fat diet
6. Saturated fat rather than unsaturated one for a few days
7. Graded increase in diet during acclimatization– relate it to post-meal tachycardia, breathlessness
8. Food should be neither too hot, nor too cold
9. Sip rather than gulp liquids.
10. Avoid too salty / too dry or concentrated diet
11. Reduce salt intake at extreme altitude
12. Don't lie down immediately after food
13. Food with particles less than 2 mm are preferable (Upma, porridge etc)
14. High roughage maintenance diet preferable
15. Curds, Srikhand, and yogurt- based food should be a regular feature
16. Abstain in non-acclimatized phase and during late evenings-night thereafter.
17. Don't mix food with chat/walk/shouting at extreme altitude
18. Use black salt instead of table salt
19. Eat honey during acclimatization period
20. Have food supplements
    a. *Vitamin tablets (B & C)*
    b. *Calcium &/or vit D*
    c. *Isabgol & allied products*
    d. *Sprouts*
    e. *Anti-oxidants, micronutrients, herbals*

Rest & Sleep

1. Prefer feet up, head down position
2. Right sided stance useful when full stomach
3. Cover face with shawl/ wool or any loose course matrix cloth
4. 1-3 hour gap between eating and sleeping is better
5. Humidify environment in the living quarters
6. Mouth breathing is advantageous in some situations

7. Prolonged breathlessness or chest heaviness after minimal exertion or at rest is abnormal and calls for medical intervention.
8. Avoid sleep apnea : *Measures to reduce sleep apnea*
   - *Sleep on side*
   - *Frequent gargling*
   - *Humidifying the ambient air*

Physical Work / Exercise / Strain

1. Graded exercise/ strain to be planned as per altitude
2. Antigravity exercises preferable
3. Step-and-shoot physical work rather than in one go should be the motto (during acclimatized period)
4. When feeling light-headed or heavy in the head, do not indulge in frivolous activities
5. All anti-hypoxemia measures to be taken if exertion is inevitable

Respiratory Exercises

1. Frequent mouth breathing when feeling breathlessness, or having head heaviness
2. Do Spirometry / conch blowing
3. Practice mouth organ playing / pinched lip breathing
4. Practice forced expiration followed by full inspiration periodically
5. Breath breaking at full inspiration rather than at full expiration is useful
6. Forced expiratory efforts help improve blood oxygen
7. Keep trousers belt slightly loose
8. Keep mouth wet by lozenges, sipping fluids, frequent gargles, and water washing.

Diagnostics & Therapeutics

1. Walk test helps identify persons with predisposition to AMS/ HAPO/non-acclimatized status (1/3/6 min walk test)
2. Be aware of inverse correlation between general cleanliness of a place and the probability of contamination in potable water.

❑❑❑

39

Adaptation and Mal-Adaptation to Ambient Hypoxia: Its Glocal Implication: A Ladakh Experience

TSERING NORBOO, TSERING STOBDAN
Ladakh Institute of Prevention, Leh-Ladakh, Jammu and Kashmir, India

ABSTRACT

The study of human physiology and evolutionary characteristics will be best carried out if it moves out of controlled laboratory conditions to matching field environment. The environment and populations living at high altitude is one such field setup. We studied Ladakhi, natives of Korzok village (4550 m) living in stressful hypoxic environment. Average hemoglobin of 17.46 (±2.18) g/dl & SpO_2 84 (±9.5) % in Ladakhi men at 4550m is a well known response to low oxygen saturation. However, a substantially higher value of FEV1/FVC and MMEF in native highlander has not been described anywhere at such elevation. Does such an improved lung mechanic helps in reducing work of breathing and facilitates adaptation? Scores of cognitive and ADL both neuro-cardiovascular functions were found to be much lower among Ladakh aged >70 yrs than comparable age group from Japan (lowlanders). As highlanders are continuously exposed to a high level of reactive oxygen species (ROS) this may be associated with neurodegenerative diseases, heart disease and cancer. Native Ladakhis have high prevalence of mild cognitive impairment, cancers and congenital and acquired heart disease. While studying Chronic Mountain Sickness using an internationally accepted symptom/sign scoring system, significant differences were noted between Andeans and Ladakhi population. The differences were also significant at gene expression levels. This suggests that a better scoring system specifically developed for the Himalayas might uncover comparable prevalence of CMS to Andean and also the specific environmental factors accounting for difference in signs and symptoms.

Key Words: Altitude; Adaptation: Chronic Mountain Sickness, Ladakh.

INTRODUCTION

High altitude (HA) environment is mainly characterized by hypobaric-hypoxia. The two terms i.e. HA acclimatization and adaptation are although looks synonymous to each other, differs significantly. While altitude acclimatization refers to the physiological responses that take place on ascending to HA (Tansey, 2008), adaptation is both genetic and nongenetic, consisting of physiological, biochemical, or behavioral adjustments (Hochachka *et al.*, 1999; Stobdan *et al.*, 2008). More than 50% out of approximately 140 million people worldwide living at altitude >2500m above sea level, are from Asia (Penaloza and rias-Stella, 2007). In the contemporary flat world situation, increasing interest in commerce, science, and adventure tourism in high mountains of the world is paralleled with new advances in the area of HA medicine and physiology and a burgeoning number of publications. It is thus appropriate that the subject of adaptation and maladaptation to ambient hypoxia is considered important in an international conference on novel approaches for food and health security at HA.

Human response to hypoxic environment is reported divergent. The three major HA regions with long-term resident populations are the Himalayas including the Tibetan Plateau (Tibetans, Ladakhi, and Sherpas), the Andes (Quechua and Ayamara), and Northern Africa (Ethiopians). The recent advances suggest a different and distinctive patterns of adaptation to the HA environment (Beall *et al.*, 2002; Beall, 2007) or variability due to genography (Allocco *et al.*, 2007). Perhaps attributed to different time intervals spent at altitude the Tibetans are considered adapted while Andean as acclimatize as they contrast in altitude response (Beall, 2000). While considerable studies on physiological and genetic adaptation to HA are based on resident populations of Andes, Tibet (Beall, 2007), Nepal (Santolaya *et al.*, 1989), Ethiopia (Beall *et al.*, 2002) and Europe etc, very little is known about

HA natives in India. Preliminary studies carried on the Ladakhi populations (Indian HA natives) reveals important features characteristic to HA adaptation. For example, similar to Andean highlanders, Ladakhi exhibit greater chest circumference than the lowland norms (Malik and Singh, 1978) and respiratory function (Malik and Singh, 1979; Wood *et al.*, 2003; Kapoor and Kapoor, 2005). In one of our previous finding it was reported that the progressive increase in carotid-body-weight with age among Ladakhi highlanders (Khan et al., 1988) is similar to other chronic hypoxic conditions (rias-Stella and Valcarcel, 1973;Heath et al., 1985). Contrary the difference was significant in its histological findings with prominence of dark cells in Ladakhi as compared to light chief cells in Andeans (Khan *et al.*, 1988).

As there are divergent views and contrasting findings on different HA populations around the world, we made an attempt to look for a possible conclusive answer in the natives of Ladakh from both the physiological and genetic perspectives. Our hypothesis of a perfect adaptation to HA were; 1) Normal hemoglobin, 2) Normal oxygen saturation, 3) Normal autonomic nervous system function, 4) Normal gene expression, 5) Normal cerebral vasomotor responses. Deviation from what is the normal for that altitude in absence of congenital or acquired cardio-pulmonary disease or thoracic cage deformity was our next proposed study on maladaptation to HA, viz Chronic Mountain Sickness (CMS), or Monge disease (Monge, 1942).

Adaptation: What is the Normal for that Altitude

Study subjects and sites: The initial study on cardio–pulmonary function was carried out in year 2003 on subjects aged above 17 and residing at altitude between 3300m and 4500m. The location and number of subject studied are depicted in Figure 1. This included the natives of Ladakh, Tibetan settled near Leh, and the nomads of Ladakh. This was a cross-sectional study, carried out in indoors or in shaded locations at rural clinic settings with service provision to facilitate participation. Informed consent was taken from each participant before the study. Ethical clearance was obtained from the Institutional Review Board of Ladakh Institute of Prevention.

Anthropometric measurements: Hemoglobin (Hb) concentration was determined on venipuncture blood, using a portable photometer (Hemocue AB, Sweden) calibrated to an accuracy of ±0.3 g/dl and linear over the range 0 - 23.5 g/dl. The apparatus had been validated in Ladakhi subjects in other field studies and the same was used in 2003, 2006 and onwards studies. End-tidal PCO_2 ($etPCO_2$), SpO_2, RR and pulse rate were measured using a powerful combination of capnography and pulse oximetry, NPB-75 (Nellcor Puritan Bennett Inc, USA). Accuracy of the measurement was ±2mmHg for $etPCO_2$ and ±2% for SpO_2 (over the range of 70 - 100%; unspecified accuracy below 70%). Blood pressure was measured using a mercury sphygmomanometer. BMI was calculated as weight in Kg/height m^2. Lactate concentration (mmol/L) in arterial blood sample was analyzed with i-STAT portable clinical analyzer with cartridge $^{CG}4^{+}$ (Abbott Point of Care Ltd. Canada). Pulmonary function tests (PFT) was performed in standing position using spirometer with accuracy ±3% from MicroMedical MicroLoop, UK. Prior studies have shown that the spirometer results are not affected by either climate or altitude (Pedersen *et al.*, 1994;Pollard *et al.*, 1996). Instructions for performing forced vital capacity (FVC) were explained to the subjects in their native language and subjects were asked to repeat the test until results were consistent with minimum three repetitions. In addition to measured values of lung volumes and flows, the spirometer also provided inbuilt data for comparison on predicted values based on age, height, gender as well as ethnicity (caucasian or non-caucasian), taking reference to normal values of the European coal & steel community (Quanjer *et al.*, 1993). We also compared our data to a more recent study that provides spirometric reference values of Caucasians, African-Americans and Mexican-Americans between the age group of 8 - 80 years old (Hankinson *et al.*, 1999).

Hemoglobin was highly variable in our subjects but generally elevated as seen in other HA natives. We found that 21/141 men studied (15%) had excessive polycythemia (Hb>20 g/dl). Assuming a mean cell Hemoglobin of 33%, hemoglobin of 20g/dl is associated with a hematocrit of 60%. Gender and age had predictable effects on spirometry, with slopes of parameter as a function of age similar to those found in Caucasians. Age did not differ significantly among altitude groups. Hb was higher in men for both groups with no difference between Tibetans and Ladakhi Table 1 &.2. PFT was remarkable for all parameters except peak expiratory flow (PEF) which was not different from predicted values except for Tibetan women at 4500m, where mean value was significantly less than predicted. At 4600m Tibetans had significantly higher peak flow than Ladakhis. Finding PEF values not higher than predicted was surprising given that reduced air density at high altitude should result in higher than predicted peak flows. In contrast the forced expiratory volume in one second (FEV1) and the mid-maximal expiratory flow (MMEF) were significantly elevated in both Ladakhis and Tibetans compared to % predicted for any race or to sojourners. The six sojourners studied at 4500m did not show elevated MMEF values. Both FEV1 and FVC were significantly higher than predicted in both Tibetans and Ladakhis at each altitude and there was no gender difference. The ratio FEV1/ FVC was also significantly higher than predicted for both Ladakhis and Tibetans, independent of gender at each altitude. Blood pressure was significantly higher than normal in both Tibetans and Ladakhis at the higher altitude of 4200m and 4500m with mean arterial pressure of approximately 100. Ladakhi men had higher diastolic BP than women (91vs 81) with no difference in systolic BP. There was no gender difference in BP for Tibetans.

Classically, during exercise at altitude, the limits of aerobic metabolism are exhausted before anaerobic pathways are recruited (Matheson *et al.*, 1991). This change from aerobic to anaerobic metabolism usually occurs at the anaerobic threshold, just before maximal O_2 consumption (VO_2) is reached. The subjects were instructed to perform metronome-guided (50/min) knee extension exercise (mild exercise). As a result of the exercise, VO_2 increased to only approximately 30% of the maximal VO_2 expected in an age-matched lowlander at sea level. The serum lactate increased significantly during exercise (from 0.81±o.11 to 1.06±0.24 mmol/L, $p<0.01$) (Ito et al., 2007), suggesting that Himalayan highlanders can recruit anaerobic metabolism at earlier stages of exercise. The increase in lactate was larger in the subjects with higher SpO_2 compared with those with lower SpO_2 indicating anaerobic parallels with aerobic metabolism. In contrast to lowlanders at sea level, mild exercise in hypoxic conditions increases serum lactate levels in Himalayan highlanders.

Neuro-cardiovascular function: The neuro-cardiovascular function was accessed in Ladakhi (n=40) with age older than 70 years (19 men and 21 women mean age 74.7 ±3.3 years and altitude 3500m) and compared with Japanese, living in Tosa town area (97 men and 227 women, mean age 80.7±4.7years and altitude 250m). Cardio-Ankle vascular index (CAVI), a new index of pulse wave velocity (PWV) as a gauge of arterial stiffness was measured between the right arm and ankle in a supine position, using a vasera CAVI instrument (Fukuda

Denshi co., Ltd., Tokyo Japan). A complete procedural details of the approach is described elsewhere (Otsuka *et al.*, 2005a; Otsuka *et al.*, 2005b). As mentioned in Otsuka *et al*, 2005, Kohs block design test and time estimation, up and go etc were carried to test different neuro-cardiopulmonary function. Functional reach test used to evaluate balance and manual dexterity was assessed using a panel with combination of 10 hooks, 10 big buttons and 5 small buttons.

Comparison of age matched CAVI in male and female subjects of Ladakh (altitude 3500m) and Tosa town (Altitude 250m) of Japan showed higher values in the Ladakhi subjects. The Table 3 and 4 depicted in Otsuka et al, 2005 details scores of cognitive and neurobehavioral function tests and ADL comparing the two populations. Results from the Kohs design block test were lower in Ladakh than in Tosa (9.0±8.8 vs 16.4 ±10.0, $p<0.0001$). Time estimation of 10 sec, or even more so of 60 sec was lower in Ladakh than in Tosa (TE10: 8.5 ±2.7sec vs 11.5 ±5.8sec; TE60: 42.7 ±16.8sec vs 57.1 ±23.0sec). ADL scores of the functional reach test were also significantly lower in Ladakhi than in subjects of Tosa town (Otsuka *et al.*, 2005a).

Intima-media thickness: Cardiovascular disease has a multifactorial aetiology, as is illustrated by the existence of numerous risk factors. Studies reports that chronic intermittent hypoxia (CIH) found to be associated with atherosclerosis, leading to a high cardiovascular risk (Savransky *et al.*, 2007). High animal fat diets and hypertension are also risk factors for arterial disease (Hornstra, 2001). In yet another study we measured intima-media thickness (IMT) in the human carotid artery using noninvasive ultrasound, a surrogate for subclinical atherosclerosis (Watson R *et al.*, 2007). IMT of the common carotid artery was measured in 96 subjects (20 female) from Korzok. The questionnaire revealed their high animal fat consumption and a high prevalence of hypertension. They also had normal BMIs and high daily workloads as describe earlier.

This field study suggests that compared to industrialized, normobaric populations, IMT is attenuated in Ladakh highlanders. The linear regression equation relating age to IMT was IMT = 0.006(Age) + 0.24 ($r=0.74$, $p<0.05$). While an age matched mean IMT values of controls from industrialized countries in normobaria ranged from 0.60–1.10 mm with slopes of 0.009–0.016 and intercepts from 0.12–0.19. This reduction occurs in spite of risk factors which increase IMT at sea level, and may be related to their aerobic load and genetics.

Modified Reed Rebreathing test: Cerebral blood flow is known to increase in response to hypoxia and to decrease with hypocapnia. Change in brain blood flow in response to such stimuli is termed cerebrovascular reactivity (CVR) usually defined as a change in cerebral blood flow (CBF) for a given unit change in $PaCO_2$ (Or its surrogate, $PetCO_2$). A modified Reed rebreathing test was performed to study CVR and autonomic nervous system function at high altitude (Slessarev M et al., 2007b). The subject breathed through a mouth piece connected to a 3-way valve that was initially open to ambient air. Subjects hyperventilate at ambient air for 5 minutes. At the end of 5 minutes, subjects are asked to exhale completely (to their FRC). The three way valve was then switched so that they breathe in and out of the rebreathing bag. Subjects were asked to take three deep breaths (to equilibrate CO_2 in the bag with that in the lung) and then they were asked to breathe normally. O_2 level in the bag was kept constant by a feedback system that monitors $PetO_2$, $PetCO_2$, $PetO_2$, SaO_2, HR, VE and CBF was constantly monitored. The test was terminated on the occurrence of any one of the following: 1) when the subject felt too uncomfortable to continue, 2) when $PetCO_2$ reaches 60 mmHg (or equivalent at altitude) or 3) when ventilation reaches 100 L/min. At the time of each test, 3ml of venous blood and 3ml of arterial blood from a radial artery was drawn. The blood was immediately analyzed using ah i-STAT portable clinical analyzer (Abbott Point of Care Ltd. Canada). Two rebreathing tests (one in background of hypoxia, and one in the background of hyperoxia) were completed in 40 minutes for each subject (including measuring of CVR).

There was no difference in the sensitiveness of the CBF response (% change from resting) to CO_2 (mm Hg change) between normal highlanders (mean ± SEM; hypoxic: 4.9 ±1.0%/mm Hg; hyperoxic: 3.7 ±0.4%/mm Hg) and highlanders with CMS (hypoxic: 4.7 ±0.7%/ mmhg: hyperoxic: 4.2 ±0.2 %/mmHg; P=0.835) with a linear relationship between CBF and CO2 (Slessarev M et al., 2007b). In addition we observed that sensitiveness of the CBF response to CO_2 were unaffected by PO_2 tensions in both groups (p=0.262). The linear relationship between CBF and CO_2 appears to be slightly greater than reported previously in sea level population (range 1.8-3.4%/mmHg) (Vovk et al., 2002). This relationship is unaffected by the presence of CMS or different tensions of PO_2; the latter suggest a relative CBF insensitivity to hypoxia in Himalayan highlanders.

Using modified rebreathing test, we compared respiratory chemoreflexes of 15 native Himalayan living at 4500m (Korzok Ladakh) with 24 lowlanders at hypoxic (50 mmHg) and hyperoxic (150 mmHg) isoxic PO_2 tensions (Slessarev M et al., 2007a). Highlanders had lower ventilatory sensitivities to CO_2 than lowlanders at both isoxic tensions (Hyoeroxic; 2.5 +/-0.4 vs. 4.2+/- 0.3 L/min /mm Hg; P=0.011; Hypoxic; 2.8+/-0.3 vs. 7.1+/-0.5 L/min /mm Hg; (P < 0.001) and the usual increase in ventilatory sensitivity induced by hypoxia in lowlanders was absent in highlanders (P=0.526). Both groups had reduced ventilatory recruitment thresholds with hypoxia and there were no differences in the sub-threshold ventilation between lowlanders and highlanders at both isoxic tensions. Thus non-chemoreflex drives to breathe did not differ between groups nor were they affected by PO_2. The highlanders were older (P< 0.061) shorter (P< 0.061) and lighter (P < 0.061) than lowlanders. Lower ventilatory recruitment threshold and sensitiveness to CO_2 seen in highlanders are an index of their altitude adaptation (Slessarev M et al., 2007a). The lack of an effect of hypoxia on CO_2 sensitivity in highlanders may explain the blunted hypoxic ventilatory responses. Our results therefore suggest that physiological mechanisms other than a hypoxia induced increase in ventilation may be involved in determining the adaptation of highlander to life at high altitude.

As altered baroreceptor function at altitude in sojourners, Andean (Moore et al., 2006) and Ethiopian natives (Gulli et al., 2007), has been documented, we also tested this measure in our Korzok population using same methods. A 60 degree head-up tilt (HUT) test is carried out for baroreceptor function mainly to study autonomic nervous system function at altitude. During HUT, cardiac output temporarily falls because of pooling of blood below the heart. To maintain adequate blood flow in the coronaries, there is an immediate vasodilatation initiated by endothelial mechanism. Additionally, a variety of baroreceptor induced vasoconstrictor mechanisms comes into play in different vascular beds, to maintain blood pressure and support blood flow to vital tissues such as the heart and the brain.

Mal-Adaptation: Chronic Mountain Sickness

A typical case: CMS a maladaptation syndrome occurring in persons long residing at high altitude, is characterized by excessive erythrocytosis and hypoxemia (Monge, 1942; Monge and Whittembury, 1976) and by reversibility on descend. It may be classified as primary (without identified cause) or secondary (due to underlying condition) (Leon-Velarde *et al.*, 2003). We for the first time reported a typical case description to highlight clinical features of CMS in Indian HA natives. A 53 year old man born and lived all his life at Korzok is a non-smoker, non-obese (BMI=23) with no thoracic cage deformity and no history of chronic obstructive disease or congenital heart disease. He was amongst the fittest highlander managing cattle, horses, livestock and accompanying trekkers to much higher altitude until the age of 48 years. In his 48th year, he developed dizziness, generalized physical weakness, blurring of vision, breathlessness on mild exertion, occasional headache and parasthesia in hand and feet. On examination in August 2006, he was found to have congested conjunctivae, cyanosis of lips and fingers, red wine mucosa, dilatation of veins in feet and hands and has had a symptom score of 26 which included Hb of 24.6 gm/dl and SpO_2 of 73%. His blood pressure was 140/100 mmHg, hematocrit 75% PCV, pH 7.39, pCO2 32.8mmHg, pO2 35 mm Hg, blood lactate concentration -1.02 mmol/L, HCO_3 20 mmol/L, BE (-5) mmol/L, Na +144 mmol/L, K +3.8mmol/L and blood glucose (random) 87 mg/dl. Echocardiography showed mild aortic regurgitation, trivial tricuspid regurgitation, and concentric right and left ventricular hypertrophy, findings consistent with hypertension and hypoxic lung disease. Despite these findings he showed remarkable orthostatic tolerance on 60 degree head-up tilt table (HUT) for 20 minutes. These intriguing findings in a patient with severe CMS demands physiological explanation. Has the highlander with high hematocrit value, not only has high packed cell volume but also has high blood volume, which enables him to withstand orthostatic stress.

Cross-sectional study: As Korzok village (240 km from Leh) was the highest located, altitude 4550m, with permanent human settlement amongst our study site, it was most suited for studying human adaptation and maladaptation to ambient hypoxia. We studied 91 male subjects since men were reported susceptible four times of women (Wu, 2005) and also females were further protected from CMS due to periodic menstrual cycle (Leon-Velarde *et al.*, 1997). The detail demographic features are as under Table 3. The criteria for inclusion were all men aged 18 years or older who were born at that altitude and who had not lived at low altitude for more than three months during the previous year. Detailed examination was done by physician, with particular intent to rule out conditions leading to secondary CMS. Questionnaire of signs and symptoms as designed by CMS scoring of international consensus group (Leon-Velarde *et al.*, 2003) was administered to check for severity. The average Hb of ladakhi men at 4550 m to be 17.46 (2.18) g/dl (±2SD) and SpO2 of 79.5 (±2SD) as cut off point for CMS scoring. Using international consensus group CMS scoring (Leon-Velarde *et al.*, 2003) 2% of our subjects had severe; 7 % moderate and 16% with mild CMS scoring (Figure 2). Blurring of vision and occasional diplopia, symptoms not included in the international consensus scoring system were an unusual complaint reported by 10% of our subjects. Although some findings on CMS among Tibetans suggest revival of criteria for defining this CMS (Ge and Helun, 2001), our findings of blurring of vision and occasional diplopia among our subjects further need specific study and explanation.

Finally we selected subjects for a case-control study consisting of individuals suffering from CMS and an equal age matched controls from complete 91 individuals. The study was part of an ambitious project to compare gene expression from all three HA populations of the world i.e. Andeans, Ethiopians and Himalayans (Xing *et al.*, 2008). Blood samples were obtained from each volunteer first in the ambient air of 4450m altitude. This was immediately followed by placing a highly efficient sequential gas delivery mask (Hi-OX80 Viasys Health care,yorba Linda CA USA) on their face supplied with O2 (flow adjusted such that end tidal PO2 was between 100 and 110 mmHg) from a K-size Oxygen tank. This was carried out to see the changes under sea level condition, as it was impossible to rapidly transport people off the Korzok highland plateau to low altitude. After one hour of hyperoxia a second set of blood sample was drawn from each subject. The total RNA of human white blood cells was extracted using the PAXgene Blood RNA kit (Qiagen, Germany) following the manufacturer's protocol. The detail method of molecular signatures (distinct pattern of gene expression) of hypoxia related genes, in white blood cells (WBC) is described elsewhere (Xing *et al.*, 2008).

The results were then compared with subjects from other HA populations (table 2; replicated from Xing *et al*, 2008). Xing *et al*, carried out a complete comparative study on a set of gene (Xing *et al.*, 2008). Table 4 and 5 of this article depicted a novel approach to assess the impact of specific genes on the CMS scores, using a sliding scale of CMS-scores derived from both controls and CMS patients. It is labeled as "impact table" and constructed while at subject breathing ambient air and also after 1 hour of hyperoxia. The PDP2 gene encoding a protein phosphatase that de-phosphorylates the E1-alpha subunit of pyruvate dehydrogenase and thus promotes pyruvate entry into the Krebs cycle was found to be the best predictor of CMS in a linear regression model among the Himalayans. However after one hour of hyperoxia, at the same altitude, the best predictor of CMS was PDK4 (pyruvate dehydrogenase kinase, isozyme 4) a gene whose product inhibits the pyruvate dehydrogenase complex by phosphorylating one of its subunits, thereby contributing to the regulation of glucose metabolism. All Himalayan (CMS patients and controls) had significantly higher expression levels of the genes analysed (except for NPHI) than Andean living at similar altitude. The quantitative comparisons of different gene expression in Andean and Himalayan altitude dwellers as depicted in figure 3 of Xing et al, 2008; shows that in ambient hypoxia all Himalayans (controls and CMS) had significantly higher gene expression when compared to Andeans except for HPH1. A comparison of symptomatology between CMS patients in the Andes and Himalayas is shown in figure 4 of Xing *et al,* 2008. Both populations were scored using an internationally accepted scoring system devised for the high Andes.

DISCUSSIONS

How humans adapted to ambient hypoxia may be better understood by studying their cardio-pulmonary function, physiology, and the hypoxia related genes in the matching ambient environment in which human had adapted. Hemoglobin was highly variable in our subjects but generally elevated as seen in other high altitude natives. Polycyathemia is a well recognized adaptation to high altitude that sometimes becomes excessive leading to CMS as illustrated in our typical case description. Winslow *et al,* suggest a pivotal role of hypoventilation in augmenting hypoxia and subsequent erythropoisis (Winslow *et al.,* 1981). Subsequent study provide a meta-analysis of data on Hemoglobin, arterial saturation, and ventilation to examine the hypothesis put forth by Winslow and Monge. (Reeves and Weil, 2001) They analysed hemoglobin concentration (response) as a function of arterial saturation (stimulus). They found that over the range of SaO_2 of 50-100% there was a significant inverse relationship between hemoglobin concentration and SaO_2 for both men (r=0.72) and women (r= 0.68). Our data cover a narrow range of SaO_2 (70-100 %) but show the same pattern.

The spirometry results in our study shows that Tibetans and Ladakhis have substantially higher values for FEV1/FVC and MMEF than predicted for Caucasians. As it is measured when the lung is about half empty, MMEF describes the function of mid size airways and, because it is presumed to be independent of effort, a reliable mechanics per se. Such improved lung mechanics could be an important adaptation to the sustained high levels of ventilation of these populations at HA by reducing the work of breathing. It would be of great interest to do a similar study in populations of Tibetans or Ladakhi living at or near sea level to assess whether this phenomenon is genotypic or phenotypic. Neuro-cardio-vascular function assessment in our study showed that the cognitive function (estimated by kohs block design test) and ADL function (Scored by the functional reach and button test) were worse at HA. A correlation of PWV (CAVI) with age showed that several subjects have very high CAVI values. There were several cases showing an extreme increase in CAVI.

In a preliminary report we found that Ladakhi highlanders agreed to increased carotid body-weight-with age (Heath *et al.*, 1985) features common to HA populations. The mechanism of CMS with normal lungs is believed to be blunted hypoxic ventilatory response (HVR) (Severinghaus *et al.*, 1966). The HVR decreases with age (Kronenberg and Drage, 1973) and with duration of stay (Weil *et al.*, 1971). In our Himalayan population the cerebral blood flow responses were similar to those in Ethiopians (Xing et al., 2008). They too have higher responses to CO_2 than those reported from sea level subjects and their cerebral circulation is insensitive to hypoxia17 Moreover, we found also a blunted hypoxic ventilatory response and suggestions that other mechanism in addition to hypoxia induced increase in ventilation, determine adaptation to "high life" in the Himalaya. We found that PDP2, a phosphates that de-phosphorylates the EI alpha subunit of pyruvate dehydrogenase and promotes pyruvate entry into Krebs cycle, is the best predictor of CMS. The molecular signature of CMS in the Himalayas is surprisingly found to be similar to that found in Andes; here too, PDP2 expression is the best predictor of CMS in ambient hypoxia.

The confluence of results from molecular studies, physiologic investigations and clinical experience in the Himalayan population of Ladakh suggests that environmental challenges including hypoxia, which disturb homeostasis, can lead to disease onset or accelerated aging. Further studies of these experiments of nature might find clinical utility at sea level besides helping planning health and environment policy at local level. Our endeavor in looking for a possible model adapted population is as yet unresolved but few of our recent findings need attention. While our focus in the present paper is on the salient features of few of our recent efforts in understanding human adaptation and mal-adaptation to ambient hypoxia, studied on Ladakhi population, we do have faith in its glocal implications.

ACKNOWLEDGMENTS

We are highly grateful to all the collaborators from USA, Japan, UK and Canada. The studies would not have been possible without the whole hearted involvements of Dr. N. Angchuk and Dr. D. Dolma from SNM hospital, Leh, India. Acknowledgements are due for all the volunteers for their patience and perseverance while facing long experiments.

DECLARATIONS

The article is a review on the combine efforts and global analysis of our past experiments. Most of the studies are either published in international journals or were presented as posters with abstract published in the proceedings of conferences.

References

Allocco, D.J., Song, Q., Gibbons, G.H., Ramoni, M.F. and Kohane, I.S. 2007. Geography and genography: prediction of continental origin using randomly selected single nucleotide polymorphisms. BMC.Genomics. 8, 68.

Beall, C.M. 2000. Tibetan and Andean contrasts in adaptation to high-altitude hypoxia. Adv.Exp.Med.Biol. 475, 63-74.

Beall, C.M. 2007. Two routes to functional adaptation: Tibetan and Andean high-altitude natives. Proc. Natl. Acad. Sci. U.S.A. 104(1), 8655-8660.

Beall, C.M., Decker, M.J., Brittenham, G.M., Kushner, I., Gebremedhin, A. and Strohl, K.P. 2002. An Ethiopian pattern of human adaptation to high-altitude hypoxia. Proc. Natl. Acad. Sci. U.S.A. 99(26), 17215-17218.

Ge, R.L. and Helun, G. 2001. Current concept of chronic mountain sickness: pulmonary hypertension-related high-altitude heart disease. Wilderness. Environ. Med. 12(3), 190-194.

Gulli, G., Claydon, V.E., Slessarev, M., Zenebe, G., Gebremedhin, A., Rivera-Ch, M., Appenzeller, O. and Hainsworth, R. 2007. Autonomic regulation during orthostatic stress in highlanders: comparison with sea-level residents. Exp. Physiol. 92(2), 427-435.

Hankinson, J.L., Odencrantz, J.R. and Fedan, K.B. 1999, Spirometric reference values from a sample of the general U.S. population. Am. J. Respir. Crit. Care Med. 159(1), 179-187.

Heath, D., Smith, P., Fitch, R. and Harris, P. 1985. Comparative pathology of the enlarged carotid body. J. Comp. Pathol. 95(2), 259-271.

Hochachka, P.W., Rupert, J.L. and Monge, C. 1999. Adaptation and conservation of physiological systems in the evolution of human hypoxia tolerance. Comp. Biochem. Physiol. A Mol. Integr. Physiol. 124(1), 1-17.

Hornstra, G. 2001. Influence of dietary fat type on arterial thrombosis tendency. J. Nutr. Health Aging. 5(3), 160-166.

Ito, S., et al. 2007. Serum Lactate Increased With Mild Exercise In Native Himalayan Highlanders, in Robert, C. Roach., Peter, D. Wagner. and Peter, H. Hackett. Eds., pp. 328.

Kapoor, S. and Kapoor, A.K. 2005. Body structure and respiratory efficiency among high altitude Himalayan populations. Coll. Antropol. 29(1), 37-43.

Khan, Q., Heath, D., Smith, P. and Norboo, T. 1988. The histology of the carotid bodies in highlanders from Ladakh. Int. J. Biometeorol. 32(4), 254-259.

Kronenberg, R.S. and Drage, C.W. 1973. Attenuation of the ventilatory and heart rate responses to hypoxia and hypercapnia with aging in normal men. J. Clin. Invest. 52(8), 1812-1819.

Leon-Velarde, F., McCullough, R.G., McCullough, R.E. and Reeves, J.T.. 2003. Proposal for scoring severity in chronic mountain sickness (CMS). Background and conclusions of the CMS Working Group. Adv. Exp. Med. Biol. 543, 339-354.

Leon-Velarde, F., Ramos, M.A., Hernandez, J.A., De, I.D., Munoz, L.S., Gaffo, A., Cordova, S., Durand, D. and Monge, C. 1997. The role of menopause in the development of chronic mountain sickness. Am. J. Physiol. 272(1-2), R90-R94.

Malik, S.L. and Singh, I.P. 1978. Growth trends among male Bods of Ladakh – a high altitude population. Am. J. Phys. Anthropol. 48(2), 171-175.

Malik, S.L. and Singh, I.P. 1979. Lung function in highlander Bods of Ladakh. Am. J. Phys. Anthropol. 51(3), 383-388.

Matheson, G.O., Allen, P.S., Ellinger, D.C., Hanstock, C.C., Gheorghiu, D., McKenzie, D.C., Stanley, C., Parkhouse, W.S. and Hochachka, P.W.. 1991. Skeletal muscle metabolism and work capacity: a 31P-NMR study of Andean natives and lowlanders. J. Appl. Physiol. 70(5), 1963-1976.

Monge, C. 1942. Life in the Andes and chronic mountain sickness. Science. 95(2456), 79-84.

Monge, C.C. and Whittembury, J. 1976. Chronic mountain sickness: Johns.Hopkins. Med. J. 139, 87-89.

Moore, J.P., Claydon, V.E., Norcliffe, L.J., Rivera-Ch, M.C., Leon-Velarde, F., Appenzeller, O. and Hainsworth, R. 2006. Carotid baroreflex regulation of vascular resistance in high-altitude Andean natives with and without chronic mountain sickness. Exp. Physiol. 91(5), 907-913.

Otsuka, K., et al. 2005a. Chronoecological health watch of arterial stiffness and neuro-cardio-pulmonary function in elderly community at high altitude (3524 m), compared with Japanese town. Biomed. Pharmacother. 59(1), S58-S67.

Otsuka, K., et al. 2005b. Effect of aging on blood pressure in Leh-Ladakh, a high-altitude (3524 m) community, by comparison with a Japanese town. Biomed. Pharmacother. 59(1), S54-S57.

Pedersen, O.F., Miller, M.R., Sigsgaard, T., Tidley, M. and Harding, R.M. 1994. Portable peak flow meters: physical characteristics, influence of temperature, altitude, and humidity. Eur. Respir. J. 7(5), 991-997.

Penaloza, D. and rias-Stella, J. 2007. The heart and pulmonary circulation at high altitudes: healthy highlanders and chronic mountain sickness. Circulation. 115(9), 1132-1146.

Pollard, A.J., Mason, N.P., Barry, P.W., Pollard, R.C., Collier, D.J., Fraser, R.S., Miller, M.R. and Milledge, J.S. 1996. Effect of altitude on spirometric parameters and the performance of peak flow meters. Thorax. 51(2), 175-178.

Quanjer, P.H., et al. 1993. Symbols, abbreviations and units. Working Party Standardization of Lung Function Tests, European Community for Steel and Coal. Eur. Respir. J.16, 85-100.

Reeves, J.T. and Weil, J.V. 2001. Chronic mountain sickness. A view from the crow's nest. Adv. Exp. Med. Biol. 502, 419-437.

Rias-Stella, J. and Valcarcel, J. 1973. The human carotid body at high altitudes. Pathol. Microbiol. Basel. 39(3), 292-297.

Santolaya, R.B., Lahiri, S., Alfaro, R.T. and Schoene, R.B. 1989. Respiratory adaptation in the highest inhabitants and highest Sherpa mountaineers. Respir. Physiol. 77(2), 253-262.

Savransky, V., Nanayakkara, A., Li, J., Bevans, S., Smith, P.L., Rodriguez, A. and Polotsky, V.Y. 2007. Chronic intermittent hypoxia induces atherosclerosis. Am. J. Respir. Crit. Care Med. 175(12), 1290-1297.

Severinghaus, J.W., Bainton, C.R. and Carcelen, A. 1966. Respiratory insensitivity to hypoxia in chronically hypoxic man. Respir. Physiol. 1(3), 308-334.

Slessarev, M., et al. 2007a. Chemo-reflex control of breathing in Himalayan and sea level residence, in Robert, C. Roach., Peter, D. Wagner and Peter, H. Hackett Eds., pp. 328.

Slessarev, M., et al. 2007b. The cerebral blood flow response to CO_2 in Himalayan Highlanders with and without chrotnic mountain sickness, in Robert, C. Roach, PDW and Peter, H. Hackett. Eds., pp. 326.

Stobdan, T., Karar, J. and Pasha, M.A. 2008. High altitude adaptation: genetic perspectives. High Alt. Med. Biol. 9(2), 140-147.

Tansey, E.A. 2008. Teaching the physiology of adaptation to hypoxic stress with the aid of a classic paper on high altitude by Houston and Riley. Adv. Physiol. Educ. 32(1), 11-17.

Vovk, A., Cunningham, D.A., Kowalchuk, J.M., Paterson, D.H. and Duffin, J. 2002. Cerebral blood flow responses to changes in oxygen and carbon dioxide in humans. Can. J. Physiol. Pharmacol. 80(8), 819-827.

Watson, R., Greene, R., Prisman, E., Slessarev, M., Ito, S., Norboo, T., Stobdan, T., Dolma, D., Norboo, A., Kunzang, M., Fisher, J., Appenzeller, O. 2007. Intima-media thickness in Ladakh highlanders. FASEB Journal. 21, 770-771.

Weil, J.V., Byrne-Quinn, E., Sodal, I.E., Filley, G.F. and Grover, R.F. 1971. Acquired attenuation of chemoreceptor function in chronically hypoxic man at high altitude. J. Clin. Invest. 50(1), 186-195.

Winslow, R.M., Monge, C.C., Statham, N.J., Gibson, C.G., Charache, S., Whittembury, J., Moran, O. and Berger, R.L. 1981. Variability of oxygen affinity of blood: human subjects native to high altitude. J. Appl. Physiol. 51(6), 1411-1416.

Wood, S., Norboo, T., Lilly, M., Yoneda, K. and Eldridge, M. 2003, Cardiopulmonary function in high altitude residents of Ladakh. High Alt. Med. Biol. 4(4), 445-454.

Wu, T.Y. 2005. Chronic mountain sickness on the Qinghai-Tibetan plateau. Chin Med. J. (Engl.). 118(2), 161-168.

Xing, G., et al. 2008. Adaptation and mal-adaptation to ambient hypoxia. Andean, Ethiopian and Himalayan patterns. PLoS. ONE. 3(6), 2342.

□□□

Authors Index

□□□

ABOUT THE EDITORS

Shashi Bala Singh is the Director, Defence Institute of High Altitude Research, DRDO, Leh-Ladakh, India. She obtained her Ph.D. in Physiology from All India Institute of Medical Sciences (AIIMS), Delhi in 1986. She joined DRDO at Defence Institute of Physiology & Allied Sciences, Delhi in 1990 and has vast experience in the field of high altitude human physiology. She has made excellent contributions in the field of Neurosciences. She took up the challenge to find out the solution to the problem of hypophagia and cognitive functions impairment at high altitude. Her work has led to the elucidation of some basic mechanisms involved in hypophagia at high altitude in terms of taste receptor sensitivity changes. The impairment in cognitive functions is mainly due to the oxidative stress leading to neuro-degeneration in brain structure involved in learning and memory. She is the life member of International Brain Research Organization (IBRO), Association of Physiologist and Pharmacologists of India (APPI), Physiological Society of India (PSI), and Indian Academic of Neuroscience. She is the fellow of Indian Association of Biomedical Scientist (FIABMS) and Indian Academy of Neuroscience (IAN). She has over 60 publications to her credit in reputed international and national journals. Shashi Bala Singh is the recipient of several awards and honours viz. Dr. J.N. Maitra Memorial Oration award in 2003, Bharat Nirman Talented Ladies award in 1995, DRDO Laboratory Scientist of the year award in 2006, Surg Rear Admiral M.S. Malhotra Prize in 2007 and Prof. Baldev Singh Oration award in 2007. Recently she has received DRDO Technology Spin Off Award -2007 from Prime Minister of India in May 2008.

Om Prakash Chaurasia obtained his Ph. D. degree in Botany from Magadh University Bodh Gaya, Bihar in 1992. He has joined DRDO in 1993 and posted in Defence Institute of High Altitude Research (DRDO), Leh (formerly Field Research Laboratory). He has also worked on flora of Andaman & Nicobar Islands. He has extensively surveyed trans Himalayan belts of Ladakh and Lahaul-Spiti and documented the fragile plant biodiversity and its ethnobotanical wealth. He has developed certain herbal products using high altitude plants. He has 05 patents, published 40 research papers in various national & international journals and authored books on Cold Desert Plants in 05 volumes and Ethnobotany & Plants of trans Himalaya. Presently he is working as Scientist 'E' in this institute and is actively engaged in R & D on various aspects on medicinal & aromatic plants of trans Himalayan region. He is life member of Journal of Plant Genetic Resources, Society of Ethnobotany, and Hill Horticulture and Development Society.

Ashish Yadav is a Scientist in Defence Institute of High Altitude Research, Defence Research & Development Organisation, Leh-Ladakh, India, since October 2007. He obtained his Ph.D. degree in Horticulture from G.B. Pant University of Agriculture & Technology, Pantnagar, India, in 2003 with specialization in Conservation and Evaluation of Temperate Fruit crops. He also earned Post Doctoral Research Fellowship from 2004-2007 at Fort Valley State University, Georgia, USA, in a United States Department of Agriculture funded project. His research dealt with the development of biotechnology protocols for *in vitro* plant regeneration amenable to genetic transformation in order to enhance cold hardiness and adaptation of exotic fruit & medicinal plants, and development of protocols for *in vitro* mass multiplication utilizing liquid lab rocker bioreactors. His current research is focused on sustainable production and quality improvement of fruit crops in high altitude cold desert conditions. Till date he has 12 peer-reviewed publications in reputed journals and 33 research presentations in various national and international conferences. He is also member of Hill Horticulture and Development Society (HHDS), American Consortium for Medicinal and Aromatic Plants (ACMAP) and People for Ethical Treatments of Animals (PETA).

Agnes M. Rimando is a Research Chemist with the United States Department of Agriculture, Agricultural Research Service, Natural Products Utilization Research Unit in Oxford, Mississippi, U.S.A. Born in the Philippines, she obtained her Ph.D. degree in Pharmacognosy (core natural products chemistry) from the University of Illinois at Chicago in 1993. To date, she has more than 100 publications, including 5 patents. She is editor or co-editor of 3 American Chemical Society (ACS) Symposium Series books and a guest editor of a special issue of the journal *Molecular Nutrition and Food Research*, volume 52(S1), 2008. She is in the editorial board of two scientific journals. Agnes Rimando has been active with the Agricultural and Food Chemistry Division (AGFD) of the ACS, and has served in various official positions, including Chair of AGFD in 2007. She is Chair of the American Chemical Society, Olemiss Local Section for 2008.

Thomas H. Terrill has been a Research Professional at Fort Valley State University (FVSU) since 1992. He is author and/or coauthor of over 35 refereed journal articles, numerous proceedings papers and abstracts and 1 book chapter. He is the coordinator for the Southern Consortium on Small Ruminant Parasite Control and a research collaborator with a number of national and international institutions/agencies in the southern USA, Puerto Rico, the US Virgin Islands, and South Africa. He is a member of several professional organizations and is currently serving as an Associate Editor for Agronomy Journal and as Editor of the American Forage and Grassland Council monthly publication Forage Progress. He is the FVSU technical representative for the USDA Western Regional Project W-102, "Integrated methods of parasite control for improved livestock nutrition". His teaching assignments include graduate courses in the Animal Science Master's program at FVSU. He is Principal Investigator for a number of grants supporting research in alternative forages for goats and epidemiology and control of small ruminant nematode parasites. His research is focused on development of year-round goat grazing systems, epidemiology and anthelmintic resistance of gastrointestinal nematodes, and use of novel, non-chemical parasitic nematode control methods, including nematode-trapping fungi, copper oxide wire particles, and condensed tannin (CT)-containing forages. His particular area of expertise is CT analysis methodology and evaluation of CT-containing forages in ruminant diets.